KAPLAN

RELATED KAPLAN BOOKS

College Admissions and Financial Aid

Access America's Guide to Studying in the U.S.A.
Newsweek College Catalog
Parent's Guide to College Admissions
Road to College
Scholarships 1998
What to Study: 101 Fields in a Flash
You Can Afford College
Yale Daily News Guide to Succeeding in College

Test Preparation

ACT 1998
SAT & PSAT 1998
SAT II: Chemistry
SAT II: Mathematics
SAT II: Writing
SAT Math Workbook
SAT or ACT? Test Your Best
SAT Verbal Workbook

KAPLAN®

SAT* II: BIOLOGY

1998–99

By Claire Aldridge, Ph.D.
and the
Staff of Kaplan Educational Centers

Simon & Schuster

Kaplan Books
Published by
Kaplan Educational Centers and Simon & Schuster
1230 Avenue of the Americas
New York, New York 10020

Special thanks to: Doreen Beauregard, Deborah Bond-Upson, Gerard Capistrano, Gordon Drummond, Bola Famuyiwa, Dr. James Major, Kiernan McGuire, and Sara Pearl

Executive Director, Pre-College Programs: Seppy Basili
Project Editor: Julie Schmidt
Cover Design: Cheung Tai
Interior Page Design: Jobim Rose
Production Editor: Maude Spekes
Managing Editor: Brent Gallenberger
Executive Editor: Del Franz

Manufactured in the United States of America.
Published simultaneously in Canada.

March 1998
10 9 8 7 6 5 4 3 2

ISBN 0-684-84850-3
ISSN 1096-4800

CONTENTS

About the Author

Claire Aldridge received her bachelor of science degree in biomedical science from Texas A&M University; she graduated magna cum laude. She earned her Ph.D. from the Department of Immunology and the University Program in Genetics at Duke University in 1996.

Claire was an invited speaker at the Ninth International Congress of Immunology in San Francisco, CA, in July 1995. She is an author of *Microbiology and Immunology*, a Kaplan publication, and has written numerous scientific articles.

She is currently serving as biology coordinator for Kaplan Educational Centers. She lives in Austin, Texas.

How to Use This Book

For more than 50 years, Kaplan has prepared students to take SATs. Our team of researchers and editors knows more about SAT preparation than anyone else, and you'll find their accumulated experience and knowledge in this book. As you work your way through the chapters, we'll show you precisely what knowledge and skills you'll need in order to do your very best on the SAT II: Biology Subject Test. You'll discover the most effective way to tackle each type of question, and you'll reinforce your studies with lots of practice questions. At the beginning of the biology review section, you'll find a lengthy diagnostic practice test, and each chapter in this section ends with a short quiz. At the back of the book you'll find two full-length, formatted tests with answer keys, scoring instructions, and detailed explanations. In addition, the Kaplan Advantage™ Stress Management System section contains helpful tips on beating test stress while you're preparing for the test and on pulling off a victory on Test Day.

The Best Prep

Kaplan's three full-length practice tests give you a great prep experience for the SAT II: Biology Subject Test.

Get Ready to Prep

If possible, work your way through this book bit by bit over the course of a few weeks. Cramming the week before the test is not a good idea. You probably won't absorb much information if you try to pack it in at the last minute.

Start your preparation for the SAT II: Biology Subject Test by reading the Kaplan Advantage Stress Management System. The stress-relief tips will help you stay calm and centered. Keep using the tips that work for you before, during, and after the test.

Learn the Basics

The first thing you need to do is find out what's on the SAT II: Biology Test. In the first section of this book, "The Basics," we'll provide you with background information about the SAT II: Subject Test and what it's used for. We'll also give you the lowdown on all the typical kinds of questions that are asked on the test and how best to tackle them.

Biology Review

Once you have the big picture, it's time to focus on the biology that's tested. The second section of this book, "Biology Review," gives you a succinct review of the biology you need to know to answer questions on Test Day. Each chapter in this section deals with a major subdivision of biology and focuses on concepts crucial to a full understanding of organisms and their interactions with the environment.

Note that throughout these biology review chapters, we use a select group of organisms to illustrate a variety of different concepts. We have chosen these organisms, the hydra, the earthworm, the grasshopper, and the human, because they exemplify different evolutionary stages particularly well. The hydra belongs to the Coelenterates, which is a phylum of simple organisms that developed relatively early in the evolution of organisms. Earthworms of the phylum Annelida are characteristic of an intermediate, more complex evolutionary stage, while grasshoppers (phylum Arthropoda) developed later than Annelida and are somewhat more complicated in structure. Finally, humans, the most highly evolved organisms of phylum Vertebrata, made their appearance fairly recently in evolutionary history. In general, humans exemplify the most complex adaptations that organisms have made to their environment.

The biology review section begins with a lengthy diagnostic test. If your time is limited, you can use the diagnostic test to bypass the material you already know well enough and to zero in on what you need to work on. Each chapter in this section also ends with a follow-up quiz, complete with answers and explanations. When you feel you have mastered the material in a chapter, take the follow-up quiz to make sure. Don't forget to use the glossary at the end of the book to brush up on vital definitions, and the index to look up lengthier explanations of troublesome concepts.

Finding Your Way

How you use this book depends on which test you're taking and how much time you have. Let's take a look at three typical students who are planning to take the SAT II: Biology Subject Test. Note that our hypothetical students use this book in three different ways. Which student's study plan best matches your situation?

"I'm taking the Biology Subject Test a month from today."

Angela has plenty of time to prep for the test. If you're like Angela, and you have at least two weeks to prepare, then we recommend that you do everything in this book that relates to the test you're taking.

"I'm taking the Biology Subject Test in a week."

If you're like Bill, you'll need a shortcut. If you have fewer than two weeks but more than two days to prepare, then we recommend that you use the diagnostic test to determine which chapters you can safely skim, or even skip.

"Help! It's two days before Test Day!"

Eric is in a panic. But you don't need to freak, even if you're in Eric's situation. Use our Panic Plan to get through this book. If you have only a day or two to prepare for the test, then you don't have time to prepare thoroughly. But that doesn't mean you should just give up and not prepare at all. There's still a lot you can do to improve your potential score. First and foremost, you should become familiar with the test. Read the introductory section in this book. And if you do nothing else, you should at least sit down and work through one of the full-length practice tests at the back of this book under reasonably testlike conditions.

When you finish the practice test, check your answers and look at the explanations to the questions you didn't get right. When you come across a topic that you only half remember, turn to the appropriate chapter in this book for a quick review. When you come across a topic you don't remember or understand at all, skip it. You don't have time to learn and assimilate completely new material. At least you'll know to skip any similar question you might encounter on the actual SAT II: Biology Subject Test.

The Icons

As you work your way through this book, you'll see the following helpful icons used repeatedly in the side bars in the margins of the text. Here's what they mean.

Basic Concepts. This icon highlights basic concepts that are explained in more detail in the accompanying text. Use these side bars to quickly determine what is being discussed in the text, and to get a wider perspective on the topic at hand.

College Board Publications

The College Board has released some test questions that you might want to look at. The free pamphlet *Taking the SAT II: Subject Tests* has a few sample questions. Even better, the College Board's book *The Official Guide to SAT II: Subject Tests* has full-length, in-format tests.

A Closer Look. Check out the sidebars that display this icon for additional facts and/or examples associated with a particular topic. Often, you'll be surprised by what you read here.

Study Tips. This icon appears next to information that can help you to grasp biological concepts more easily. Here you'll find "Don't Mix These Up on Test Day" sidebars, which point out important, easily confused concepts that are very likely to turn up on your SAT II: Biology test, as well as mnemonics that can help you to memorize important facts.

Test Strategies. This icon highlights Kaplan test-taking strategies that can help you boost your score.

Quiz. You'll find this icon with sidebars that feature "quick quizzes"—questions that will get your brain working on tough concepts even as you read about them.

Take the Practice Tests

At the back of the book are two full-length biology practice tests. The best way to use these tests is to take them under testlike conditions. Don't just drop in and do a random question here and there. Use these tests to gain experience with the complete testing experience, including pacing and endurance. You can do these tests at any time. You don't have to save them all until after you've read this whole book. Just be sure to save at least one test for your dress rehearsal some time in the last week before Test Day.

Take a Break Before Test Day

If possible, don't study the night before the test. Relax! Read a book or watch a movie. Get a good night's sleep. Eat a light breakfast the morning of the test, and quickly review a few questions if you feel like it (just enough to get your mind on the test). Walk into the test center with confidence—you're ready!

A Special Note for International Students

Approximately 500,000 international students pursued academic degrees at the undergraduate, graduate, or professional school level at U.S. universities during the 1995–1996 academic year, according to the Institute of International Education's *Open Doors* report. Almost 50 percent of these students were studying for a bachelor's or first university degree. This trend of pursuing higher education in the United States is expected to continue well into the next century. Business, management, engineering, and the physical and life sciences are particularly popular majors for students coming to the United States from other countries.

If you are not from the United States, but are considering attending a U.S. college or university, here's what you'll need to get started:

- If English is not your first language, start there. You'll probably need to take the Test of English as a Foreign Language (TOEFL) and the Test of Written English (TWE), or show some other evidence that you are fully proficient in English in order to complete an academic degree program. Colleges and universities in the United States will differ on what they consider to be an acceptable TOEFL score. A minimum TOEFL score of 550 or better is often expected by the more prestigious and competitive institutions. Because American undergraduate programs require all students to take a certain number of general education courses, all students, even math and computer science students, need to be able to communicate well in spoken and written English.

- You might also need to take the SAT or the American College Test (ACT). Many undergraduate institutions in the United States require both the SAT and TOEFL of international students.

- There are over 2,700 accredited colleges and universities in the United States, so selecting the correct undergraduate school can be a confusing task for anyone. You will need to get help from a good advisor or at least a good college guide that explains the different types of programs and gives you some information on how to choose wisely. Since admission to many undergraduate programs is quite competitive, you might also want to select three or four colleges and complete applications for each school.

- You should begin the application process at least a year in advance. An increasing number of schools accept applications year-round. In any case, find out the application deadlines, and plan accordingly. Although September (the fall semester) is the traditional time to begin university study in the United States, at most schools you can also enter in January (the spring semester).
- Finally, you will need to obtain an I-20 Certificate of Eligibility in order to obtain an F-1 Student Visa to study in the United States. This you will request from the university. The school will send you the I-20 document once you have been accepted.

For an overview of the undergraduate admissions process, see the appendix on college admissions in this book. For details about admissions requirements, curriculum, and other vital information on top colleges and universities, see Kaplan's *Road to College.*

Access America™

If you need more help with the complex process of undergraduate school admissions and information about the variety of programs available, you might be interested in Kaplan's Access America program.

Kaplan created Access America to assist students and professionals from outside the United States who want to enter the U.S. university system. The program was designed for students who have received the bulk of their primary and secondary education outside the United States in a language other than English. Access America also has programs for obtaining professional certification in the United States. Here's a brief description of some of the help available through Access America.

The TOEFL Plus Program

At the heart of the Access America program is the intensive TOEFL Plus Academic English program. This comprehensive English course prepares students to achieve a high level of proficiency in English in order to successfully complete an academic degree. The TOEFL Plus course combines personalized instruction with guided self-study to help students gain this proficiency in a short time. Certificates of Achievement in English are awarded to certify each student's level of proficiency.

Undergraduate School/SAT Preparation

If your goal is to complete a bachelor of arts (B.A.) or bachelor of science (B.S.) degree in the United States, Kaplan will help you prepare for the SAT or ACT, while helping you understand the American system of education.

Applying to Access America

To get more information, or to apply for admission to any of Kaplan's programs for international students or professionals, you can write to us at:

Kaplan Educational Centers
International Admissions Dept.
888 Seventh Avenue
New York, NY 10106

You can call us at (800) 522-7770 from within the United States, or at 01-212-262-4980 outside the United States. Our fax number is 01-212-957-654. Our E-mail address is world@kaplan.com. You can also get more information or even apply through the Internet at http://www.kaplan.com/intl.

THE BASICS

SAT II: Biology Dream Question

CHAPTER ONE

ABOUT THE SAT II: SUBJECT TESTS

You're serious about going to the college of your choice. You wouldn't have opened this book otherwise. You've made a wise choice, because this book can help you to achieve your goal. It'll show you how to score your best on the SAT II: Biology Subject Test. But before turning to the biology review, let's look at the SAT II as a whole.

Frequently Asked Questions

The following background information about the SAT II is important to keep in mind as you get ready to prep for the SAT II: Biology Subject Test.

What Is the SAT II?

Known until 1994 as the College Board Achievement Tests, the SAT II is actually a set of more than 20 different Subject Tests. These tests are designed to measure what you have learned in such subjects as Literature, American History and Social Studies, Biology, and Spanish. Each test lasts one hour and consists entirely of multiple-choice questions, except for the Writing Test, which has a 20-minute essay section and a 40-minute multiple-choice section. On any one test date, you can take up to three Subject Tests.

How Does the SAT II Differ from the SAT I?

SAT I is largely a test of verbal and math skills. True, you need to know some vocabulary and some formulas for the SAT I; but it's designed to measure how well you read and think rather than how much you remember. The SAT II tests are very different. They're designed to measure what you know about specific disciplines. Sure, critical reading and thinking skills play a part on these tests, but their main purpose is to determine exactly what you know about writing, math, history, chemistry, and so on.

"What Does That Spell?"

Originally, *SAT* stood for *Scholastic Aptitude Test.* When the test changed a few years ago, the official name was changed to *Scholastic Assessment Test.* In 1997, the test makers announced that *SAT* no longer stands for anything, officially.

Dual Role

Colleges use your SAT II scores in both admissions and placement decisions.

How Do Colleges Use the SAT II?

Many people will tell you that the SATs (I and II alike) measure only your ability to perform on standardized exams—that they measure neither your reading and thinking skills nor your level of knowledge. Maybe they're right. But these people don't work for colleges. Those schools that require SATs feel that they are an important indicator of your ability to succeed in college. Specifically, they use your scores in one or both of two ways: to help them make admissions and/or placement decisions.

Like the SAT I, the SAT II tests provide schools with a standard measure of academic performance, which they use to compare you with applicants from different high schools and different educational backgrounds. This information helps them to decide whether you're ready to handle their curriculum.

Call Your Colleges

Many colleges require you to take certain SAT II tests. Check with all of the schools you're interested in applying to before deciding which tests to take.

SAT II scores may also be used to decide what course of study is appropriate for you once you've been admitted. A low score on the Writing Test, for example, might mean that you have to take a remedial English course. Conversely, a high score on an SAT II: Mathematics Test might mean that you'll be exempted from an introductory math course.

Which SAT II Tests Should I Take?

The simple answer is: those that you'll do well on. High scores, after all, can only help your chances for admission. Unfortunately, many colleges demand that you take particular tests, usually the Writing Test and/or one of the Mathematics Tests. Some schools will give you a degree of choice in the matter, especially if they want you to take a total of three tests. Before you register to take any tests, therefore, check with the colleges you're interested in to find out exactly which tests they require. Don't rely on high school guidance counselors or admissions handbooks for this information. They might not give you accurate or current information.

When Are the SAT II Tests Administered?

Count to Three

You can take up to three SAT II tests in one day. The Writing Test must be taken first.

Most of the SAT II Tests are administered six times a year: in October, November, December, January, May, and June. A few of the tests are offered less frequently. Due to admissions deadlines, many colleges insist that you take the SAT II no later than December or January of your senior year in high school. You may even have to take it sooner if you're interested in applying for "early admission" to a school. Those schools that use scores for placement decisions only may allow you to take the SAT II as late as May or June of your senior year. You should check with colleges to find out which test dates are most appropriate for you.

How Do I Register for the SAT II?

The College Board administers the SAT II tests, so you must sign up for the tests with them. The easiest way to register is to obtain copies of the *SAT Registration Bulletin* and *Taking the SAT II: Subject Tests*. These publications contain all of the necessary information, including current test dates and fees. They can be obtained at any high school guidance office or directly from the College Board.

You can also register by telephone. If you choose this option, you should still read the College Board publications carefully before you make any decisions.

Do the Legwork

Want to register or get more info? You can get copies of the *SAT Registration Bulletin* and *Taking the SAT II: Subject Tests* from the College Board. If you have a credit card, you can also register for the SAT II by phone, or online at http://www.collegeboard.org.

College Board SAT Program
P.O. Box 6200
Princeton, NJ 08541-6200
(609) 771-7600

How Are the SAT II Tests Scored?

Like the SAT I, the SAT II tests are scored on a 200–800 scale.

What's a "Good" Score?

That's tricky. The obvious answer is: the score that the colleges of your choice demand. Keep in mind, though, that SAT II scores are just one piece of information that colleges will use to evaluate you. The decision to accept or reject you will be based on many criteria, including your high school transcript, your SAT I scores, your recommendations, your personal statement, your interview (where applicable), your extracurricular activities, and the like. So, failure to achieve the necessary score doesn't automatically mean that your chances of getting in have been damaged. For those who really want a numerical benchmark, a score of 600 is considered very solid.

A College Board service known as Score Choice offers you the chance to see your scores before anyone else. If you're unhappy with a score, you don't have to send it along to colleges. If you decide to take advantage of this service, you'll need to take your SAT II tests well in advance of college deadlines. At the very least, using Score Choice will slow down the reporting process. You may also want to retake one or more tests. Two more points to bear in mind:

- Once you've released a score, it can't be withheld in the future.
- If you use Score Choice, you lose the privilege of having some scores sent to schools for free.

For more information about Score Choice, contact the College Board.

Pack Your Bag

Gather your test materials the day before the test. You'll need:

- Your admission ticket
- A proper form of I.D.
- Some sharpened No. 2 pencils
- A good eraser

What Should I Bring to the SAT II?

It's a good idea to get your test materials together the day before the tests. You'll need an admission ticket; a form of identification (check the *Registration Bulletin* to find out what is permissible); a few sharpened No. 2 pencils; a good eraser; and a scientific calculator (for Math Level IC or IIC). If you'll be registering as a standby, collect the appropriate forms beforehand. Also, make sure that you know how to get to the test center.

SAT II Mastery

Now that you know a little about the SAT II tests, it's time to let you in on a few basic test taking skills and strategies that can improve your performance on them. You should practice these skills and strategies as you prepare for the SAT II.

Use the Structure of the Test to Your Advantage.

The SAT II tests are different from the tests that you're used to taking. On your high school tests, you probably go through the questions in order. You probably spend more time on hard questions than on easy ones, since hard questions are generally worth more points. And you often show your work, since your teachers tell you that how you approach questions is as important as getting the right answers.

None of this applies to the SAT II tests. You can benefit from moving around within the tests, hard questions are worth the same as easy ones, and it doesn't matter how you answer the questions—only what your answers are.

Don't Get Lost

Learn SAT II directions as you prepare for the tests. You'll have more time to spend answering the questions on Test Day.

The SAT II tests are highly predictable. Because the format and directions of the SAT II tests remain unchanged from test to test, you can learn the setup of each test in advance. On Test Day, the various question types on each test shouldn't be new to you.

One of the easiest things you can do to help your performance on the SAT II tests is to understand the directions before taking the test. Since the instructions are always the same, there's no reason to waste a lot of time on Test Day reading them. Learn them beforehand as you work through this book and the College Board publications.

Not all of the questions on the SAT II tests are equally difficult. The questions often get harder as you work through different parts of a test. This pattern can work to your benefit. Try to be aware of where you are in a test.

When working on more basic problems, you can generally trust your first impulse—the obvious answer is likely to be correct. As you get to the end of a test section, you need to be a bit more suspicious. Now the answers probably won't come as quickly and easily—if they do, look again because the obvious answers may be wrong. Watch out for answers that just "look right." They may be distractors—wrong answer choices deliberately meant to entice you.

There's no mandatory order to the questions on the SAT II. You're allowed to skip around on the SAT II tests. High scorers know this fact. They move through the tests efficiently. They don't dwell on any one question, even a hard one, until they've tried every question at least once.

When you run into questions that look tough, circle them in your test booklet and skip them for the time being. Go back and try again after you've answered the easier ones if you've got time. After a second look, troublesome questions can turn out to be remarkably simple.

If you've started to answer a question but get confused, quit and go on to the next question. Persistence might pay off in high school, but it usually hurts your SAT II scores. Don't spend so much time answering one hard question that you use up three or four questions' worth of time. That'll cost you points, especially if you don't even get the hard question right.

You can use the so-called guessing penalty to your advantage. You might have heard it said that the SAT II has a "guessing penalty." That's a misnomer. It's really a *wrong-answer penalty*. If you guess wrong, you get a small penalty. If you guess right, you get full credit.

The fact is, if you can eliminate one or more answer choices as definitely wrong, you'll turn the odds in your favor and actually come out ahead by guessing. The fractional points that you lose are meant to offset the points you might get "accidentally" by guessing the correct answer. With practice, however, you'll see that it's often easy to eliminate *several* answer choices on some of the questions.

The answer grid has no heart. It sounds simple, but it's extremely important: Don't make mistakes filling out your answer grid. When time is short, it's easy to get confused going back and forth between your test booklet and your grid. If you know the answers, but misgrid, you won't get the points. Here's how to avoid mistakes.

Always circle the questions you skip. Put a big circle in your test booklet around any question numbers that you skip. When you go back, these questions will be easy to relocate. Also, if you accidentally skip a box on the grid, you'll be able to check your grid against your booklet to see where you went wrong.

Always circle the answers you choose. Circling your answers in the test booklet makes it easier to check your grid against your booklet.

Grid five or more answers at once. Don't transfer your answers to the grid after every question. Transfer them after every five questions. That way, you won't keep breaking your concentration to mark the grid. You'll save time and gain accuracy.

Leap Ahead

You should do the questions in the order that's best for you. Don't pass up the opportunity to score easy points by wasting time on hard questions. Skip hard questions until you've gone through every question once. Come back to them later.

Guessing Rule

Don't guess, unless you can eliminate at least one answer choice. Don't leave a question blank unless you have absolutely no idea how to answer it.

Hit the Spot

A common cause of major SAT II disasters is filling in all of the questions with the right answers—in the wrong spots. Every time you skip a question, circle it in your test booklet and make doubly sure that you skip it on the answer grid as well.

Approaching SAT II Questions

Apart from knowing the setup of the SAT II tests that you'll be taking, you've got to have a system for attacking the questions. You wouldn't travel around an unfamiliar city without a map, and you shouldn't approach the SAT II without a plan. What follows is the best method for approaching SAT II questions systematically.

Think about the questions before you look at the answers. The test makers love to put distractors among the answer choices. Distractors are answers that look like they're correct, but aren't. If you jump right into the answer choices without thinking first about what you're looking for, you're much more likely to fall for one of these traps.

Guess—when you can eliminate at least one answer choice. You already know that the "guessing penalty" can work in your favor. Don't simply skip questions that you can't answer. Spend some time with them in order to see whether you can eliminate any of the answer choices. If you can, it pays for you to guess.

Pace yourself. The SAT II tests give you a lot of questions in a short period of time. To get through the tests, you can't spend too much time on any single question. Keep moving through the tests at a good speed. If you run into a hard question, circle it in your test booklet, skip it, and come back to it later if you have time.

You don't have to spend the same amount of time on every question. Ideally, you should be able to work through the easier questions at a brisk, steady clip, and use a little more time on the harder questions. One caution: Don't rush through basic questions just to save time for the harder ones. The basic questions are points in your pocket, and you're better off not getting to some harder questions if it means losing easy points because of careless mistakes. Remember, you don't earn any extra credit for answering hard questions.

Locate quick points if you're running out of time. Some questions can be done more quickly than others because they require less work or because choices can be eliminated more easily. If you start to run out of time, look for these quicker questions.

When you take the SAT II: Subject Tests, you have one clear objective in mind: to score as many points as you can. It's that simple. The rest of this book is dedicated to helping you to do that on the SAT II: Biology Subject Test.

Think First

Always try to think of the answer to a question before you shop among the answer choices. If you've got some idea of what you're looking for, you'll be less likely to be fooled by "trap" choices.

Speed Limit

Work quickly on easier questions to leave more time for harder questions. But not so quickly that you lose points by making careless errors. And it's okay to leave some questions blank if you have to—even if you leave a few blank, you can still get a high score.

CHAPTER TWO

GETTING READY FOR THE SAT II: BIOLOGY SUBJECT TEST

Now that you know the basics about the SAT II: Subject Tests, it's time to focus on the biology test. What's on it? How is it scored? After reading this chapter, you'll know just what to expect on Test Day.

Since November 1997, the SAT II: Biology Subject test has been offered in two different versions—the regular Biology Subject Test and the Biology E/M (Ecological/Molecular) Subject Test. Both versions of the test expect you to have a mastery of the concepts and principles covered in a one-year, college-prep biology class. We will describe each test in turn, starting with the more established version, the Biology Subject Test.

Content at a Glance

The SAT II: Biology Subject Test covers one year of college-prep biology. Both versions of the test cover five main topics: cellular and molecular biology, ecology, classical genetics, organismal biology, and evolution and diversity. The Biology E/M (Ecological/Molecular) test, however, places particular emphasis on ecology and molecular biology.

The Biology Subject Test

This test consists of 95 multiple-choice questions, which you will be given one hour to answer. Check out the table below for a list of the topics covered by these questions. The table will also tell you the percentage of the test devoted to each topic.

Topics	Percentage of the test
Cellular and Molecular Biology	30%
Cell Structure and Organization	
Photosynthesis	
Cellular Respiration	
Enzymes	
Molecular Genetics	
Biosynthesis	
Biological Chemistry	
Ecology	15%
Energy Flow	
Nutrient Cycles	
Populations	
Communities	
Ecosystems	
Biomes	

Topics	Percentage of the test
Classical Genetics	10%
Mendelian Genetics	
Meiosis	
Inheritance Patterns	
Organismal Biology	30%
Reproduction	
Physiology	
Animal Behavior	
Evolution and Diversity	15%
Origin of Life	
Evidence of Evolution	
Natural Selection and Speciation	
Patterns of Evolution	
Classification and Diversity	
Prokaryotes, Protists, and Fungi	
Plants	
Animals	

Basic Skills

Three basic skills are tested on the SAT II: Biology Subject Test:

- Recalling information
- Applying knowledge
- Synthesizing information

Three basic skills are tested on this exam. First of all, the *ability to recall knowledge* will form the basis for approximately 35 percent of the questions. This means that your ability to remember specific facts, your mastery of terminology, and your comfort with straightforward knowledge will be examined on around 33 of the questions.

The second skill tested is your ability to *apply your biology knowledge to unfamiliar situations*. Thirty-three (35 percent) of the questions on the exam are devoted to this area. These questions will test how well you understand concepts and your ability to reformulate information in a variety of ways. In other words, you will be required to express a given piece of information in different forms; you may be asked to compare graphical data to written data, for example. These questions will also test how well you can solve problems, particularly those dealing with mathematical relationships.

The third question type explores your proficiency at *synthesizing biological information*. These questions, which make up 30 percent of the exam (29 questions), will require you to make inferences and deductions from qualitative and quantitative data, such as data you might accumulate doing an experiment in the laboratory, and to then integrate that data to form conclusions. The data may be in paragraph form, like those word problems you hated in fourth grade, or it may be in graph or chart form. These questions will also examine your ability to recognize unstated assumptions—you will need to be prepared to think about what is implied in the setup of the experiment or the question stem.

As you can see from the table above, the SAT II: Biology Subject test covers a broad range of topics. It requires you to think about those topics in ways that you may not have done before. As a result, it is likely that some of the questions on your subject test will explore topics that you did not cover in your biology class. If this is the case, do not be alarmed; there is so much to biology that you cannot possibly cover everything in a year, although it may seem like you've learned the entire field at the time. If you encounter this problem while you are taking the diagnostic test, plan to spend a little extra time on the chapters that cover your area(s) of weakness, so that on test day you are completely familiar with these topics.

While preparing for this exam, you should also make sure that you understand common algebraic concepts such as ratios and proportions and, more importantly, that you are able to apply these concepts to the word problems and data interpretation questions you will surely see on test day. You will not be allowed to use a calculator on this exam, but don't worry; the math should be nothing more complicated than simple calculations involving multiplication or division.

Recent SAT II: Biology Scores

In the 1996-97 administrations of the SAT II: Biology Subject Test, students in the 99th percentile scored 790 or above. The mean score for all students taking the SAT II: Biology was 596.

Source: The College Board's *Handbook for the SAT® Program, 1996-97*

The Biology E/M Test

As we have seen, the Biology Subject Test consists of 95 questions that test your knowledge of a wide range of biology topics. The new Biology E/M Test has a total of 80 multiple-choice questions that you are given one hour to answer: 60 "common core" questions and 20 ecological or molecular questions. The purpose of this new exam is to evaluate students' mastery of basic biological principles, just as the original Biology Subject Test has always aimed to do; however, the E/M Test places particular emphasis on the fields of ecology and molecular biology.

The Biology E/M Test is not offered as frequently as the Biology Subject Test, so make sure to register in time if you are interested in trying your luck at this test. You will register for the "Biology E/M Test" with the College Board, although you will actually be answering *either* the "E" (ecological) *or* the "M" (molecular) section of the exam on Test Day. When this big day rolls around, you will be given the opportunity to indicate whether you want to take the Biology-E or the Biology-M option; you may not take both on the same day.

The "common core" questions of the E/M Test cover more or less the same content as the Biology Subject Test. The ecological and molecular questions, on the other hand, explore the fields of ecology and molecular biology respectively. Both of these options also assume a solid understanding of evolution and diversity. The ecological section may draw from the field of classical genetics, and you will probably find questions on molecular and bacterial genetics in the molecular section of the exam.

More Info on the Biology E/M Test

For more information on the new Biology E/M Test,check out page 35 of *Taking the SAT II: Subject Test*. Your high school guidance counsellor should have a copy of this booklet. You can also have one sent to you by calling the College Board at (609) 771-7600.

Here is a more detailed breakdown of the format of the Biology E/M Test:

Topics	Percentage of the test
Common Core Questions:	
Cellular and Molecular Biology	12%
Ecology	12%
Classical Genetics	10%
Organismal Biology	30%
Evolution and Diversity	11%
Ecology/Evolution Section (Biology-E Test)	25%
Molecular/Evolution Section (Biology-M Test)	25%

We saw earlier that three basic skills are tested on the Biology Subject Test—the ability to recall information, the ability to apply knowledge, and the ability to synthesize information. The Biology E/M Test centers around the same skills. However, the E/M Test places relatively little emphasis on the more simple types of recall questions, which basically require you to memorize facts. You will find that the new E/M Test is more concerned than the older test with getting you to apply your knowledge and to synthesize biological information. Hence your problem solving skills will probably be engaged to a greater extent on this new exam.

This book will help you to prepare for *both* the Biology Subject Test and the Biology E/M Test. We recommend that you carefully study the biology review section and answer all the quiz and practice test questions in order to maximize your score on both versions of the exam. However, because the Biology E/M Test specializes in the fields of ecology and molecular biology, and because it stresses problem solving skills, you may want to make certain adjustments to your study plan if you intend to select this new test option.

Test Tip

Our test strategies won't make up for a weakness in a given area, but they will help you to manage your time effectively and maximize points.

First of all, you are obviously going to want to make sure that you are very familiar with either the ecology or the molecular biology covered on the exam. There is little point in selecting the E/M option if you are not particularly strong in these fields. And whether you select the Biology-E or the Biology-M section on Test Day, don't neglect classical genetics and evolution and diversity in your studies. These topics are likely to crop up in both of these sections.

You can prep yourself for the problem solving emphasis, meanwhile, by honing your mathematical skills and by getting as much experience as you can in the laboratory. The E/M Test will ask you to use simple algebraic concepts, and it assumes that you are comfortable with the metric system. It will also present a great deal of experimental data for you to synthesize, so the more familiarity you have with experimental situations, the better.

Scoring Information

Both the Biology Subject Test and the Biology E/M Test are scored in a range from 200–800, just like a section of the SAT I exam. Your raw score is calculated by subtracting $\frac{1}{4}$ of the number of questions you got wrong from the number of questions you got right. For example, if you answered 70 questions correctly and 25 incorrectly, your raw score would be:

Number correct	70.00
$\frac{1}{4}$ × Number incorrect	<u>– 6.25</u>
Raw Score	63.75 (64)

This raw score is then compared to the scores of all the other test takers to calculate a scaled score. This scaling accounts for any slight variations in difficulty between test administrations. On a recent administration, you could miss two questions and still receive a scaled score of 800; a raw score of 65 translated into a 650. So, it is possible to miss a few questions and still receive a competitive score.

Question Types

Whether you choose the Biology Subject Test or the Biology E/M Test, you will encounter two main types of multiple choice questions: classification questions and five-choice completion questions. Make sure you feel comfortable with both types and their directions before test day. Don't waste time reading directions when you are being timed! We have also included a selection of sample ecology and molecular biology questions, to give you an idea of what to expect on the new E/M Test.

Classification Questions

Classification questions consist of five lettered choices that are used in all of the questions that follow. Typically, the five choices will test your knowledge of ideas, biological laws, organism names, graphs, or of some other type of data presentation. Following the five choices will be three to five statements that can be functions of the choices, definitions, descriptive characteristics, or conditions that would favor the data set in question. Each of the five choices may be used more than once, so do not eliminate an answer just because you have already used it.

To familiarize yourself with this question type, read through the directions, and attempt to answer questions 1–4 below. Check your answers against the in-depth explanations that follow the question set.

Test Strategy

On Test Day, do classification questions first; they require less reading and will give you the most points for your time invested.

Next, do the Type 1 and Type 2 Five-Choice Completion questions. Again, you will get a lot of points for the amount of time you invest.

Directions: Each set of lettered choices below refers to the numbered statements immediately following it. Select the one lettered choice that best fits each statement, and then fill in the corresponding oval on the answer sheet. A choice may be used once, more than once, or not at all in each set.

Questions 1–4:

A. Prophase
B. Metaphase
C. Anaphase
D. Interphase
E. Anaphase I

Test Strategy

Don't eliminate an answer choice just because you've used it. *Answer choices can be used more than once.*

1. the stage during which a cell's DNA is replicated
2. the stage during which homologous pairs of chromosomes are pulled to opposite poles of a cell
3. a stage in meiosis
4. the stage during which a cell's chromosomes condense

Explanations. The questions in this group deal with mitosis, the mechanism that regulates a cell's ability to replicate itself, and meiosis, the mechanism that regulates a cell's ability to produce gametes. A cell will spend roughly 90 percent of its time in interphase. This phase may be broken down into a number of different stages. During the G_1 stage, the cell doubles in size, and new organelles such as mitochondria, ribosomes, the endoplasmic reticulum, and centrioles are produced. In the next stage, the S stage, all of the DNA is replicated (see question 1) so that during division, a complete copy of the genome can be distributed to both daughter cells. Following S stage is the G_2 stage, during which the cell continues to grow in size.

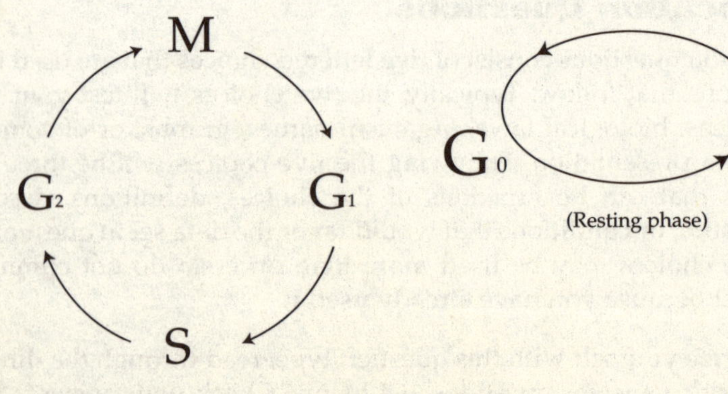

The Cell Cycle

During mitosis, the cell divides and distributes its DNA to its two daughter cells, such that each cell receives a complete copy of the original genome. Almost all cells, except for nerve cells and specialized muscle cells, can undergo mitosis. This process may be broken down into the following four stages:

- *Prophase.* The chromosomes condense (see question 4). The centriole pairs separate and move toward opposite poles of the cell, while the spindle apparatus forms between them. The nuclear membrane dissolves and the spindle fibers enter the nucleus.
- *Metaphase.* The chromosomes align at the metaphase plate and prepare to separate.
- *Anaphase.* The centromeres split so that each chromatid has its own centromere, and the sister chromatids are pulled toward the opposite poles of the cell (see question 2).
- *Telophase.* The spindle apparatus disappears, and new nuclear membranes are formed.

During meiosis, on the other hand, the gametocyte's chromosomes are replicated during the S phase of the cell cycle, just as in mitosis. In the first round of division, Meiosis I, also known as the reduction division, the cell produces two intermediate daughter cells that are haploid. In Meiosis II, the duplicated chromosomes split, resulting in four genetically distinct haploid gametes. In Anaphase I of Meiosis I, the homologous pairs of chromosomes are pulled to opposite poles of the cell. This process, called disjunction, accounts for the Mendelian law of independent assortment.

In light of all this, the correct answers for this question set are (D), (E), (E), and (A).

Five-Choice Completion Questions

These are common multiple-choice questions, and there are four types of them. The following directions apply to all four types.

Directions: Each of the questions or incomplete statements below is followed by five suggested answers or completions. Select the one that is best in each case and then fill in the corresponding oval on the answer sheet.

Type 1 Questions

These questions have a unique solution. The unique solution is often the only correct answer or the best answer. Sometimes, though, it will be the most inappropriate answer. These question types will have NOT, EXCEPT, or LEAST in capital letters somewhere in the stimulus. Here are two examples:

Four Types

Remember that there are four types of five-choice completion questions:

1. Unique solution
2. Roman numeral questions
3. Figure or diagram identification
4. Experimental data presentation

Questions 5–6:

5. If the victim of an automobile accident suffered isolated damage to the cerebellum, which of the following would most likely occur?

   A. loss of voluntary muscle contraction
   B. loss of sensation in the extremities
   C. loss of muscular coordination
   D. loss of speech
   E. loss of hearing

6. Which of the following is NOT a type of genetic mutation?

   A. point
   B. silent
   C. insertion
   D. frameshift
   E. malignant

Explanations. To answer question 5, you need to know that the cerebellum is located in the hindbrain, along with the pons and the medulla. All higher brain sensory neurons and motor neurons pass through the hindbrain. The main function of the cerebellum is to coordinate unconscious movement, such as hand-eye coordination, posture, and balance. Therefore, damage to the cerebellum would most likely affect (C) muscular coordination. As for the remaining choices, the loss of voluntary muscle contraction, sensation in the extremities, speech, or hearing may be caused by damage to specific areas of the cerebrum. The cerebrum is located in the forebrain and is divided into two hemispheres, the left and the right. This organ is responsible for the coordination of most voluntary activities, sensation, and "higher functions," including speech and cognition. Sensation of the extremities may also be controlled in part by the spinal cord.

Let's move on to question 6. Point mutations occur when a single nucleotide base is substituted for another nucleotide base. A silent mutation is a point mutation that occurs in a noncoding region, or when the mutation does not change the amino acid sequence due to the degeneracy of the genetic code. Meanwhile, a frame shift mutation is either an insertion or a deletion of a number of nucleotides. These mutations have serious effects on the protein coded for, since nucleotides are read as a series of triplets. The addition or subtraction of nucleotides (except in multiples of three) will change the reading frame of the mRNA. Finally, (E), malignant, is not a type of mutation. A mutation may lead to a cell becoming malignant (cancerous), but it does not necessarily do so.

Type 2 questions

Typically, these have three to five roman numerals following each question. One or more of these roman numerals may prove to be the correct answer(s). Following these roman numerals, you will encounter five lettered choices with various combinations of the roman numerals. You must select the combination that includes all of the correct answers and excludes all of the incorrect answers. Try your hand at the sample question below.

Test Strategy

Once you eliminate a roman numeral, make sure to eliminate all lettered answer choices with that roman numeral.

8. Which of the following are characteristic of animal cells, but NOT bacterial cells?

   I. They are eukaryotic.
   II. They possess ribosomes.
   III. They possess cell walls.
   IV. They reproduce asexually.

A. I only
B. I and II only
C. I, II, and III only
D. II, III, and IV only
E. none of the above

Test Strategy

Use the structure of a roman numeral question to your advantage. Eliminate choices as soon as you find them to be inconsistent with the truth or falsehood of a statement in the stimulus. Similarly, consider only those choices that include a statement that you've already determined to be true.

Explanation. This question requires you to know the differences between the basic animal cell, the fungal cell, and the bacterial cell. All animal (and plant) cells are eukaryotic, while all bacterial cells are prokaryotic. Bacteria also have cell walls made of peptidoglycans, and they reproduce asexually via binary fission. Both cell types produce proteins with ribosomes, although they differ in size and composition. Therefore, the correct answer is (A).

Type 3 questions

Organized in sets around a figure or a diagram, each question is nevertheless independent of the other questions in its set. These questions test your knowledge of morphology and function of a variety of biological structures. Typically, these questions are not as difficult as Type 4 questions, the experimental data questions. Here are three examples:

What's The Sinoatrial Node?

The sinoatrial node is another name for the pacemaker of the heart.

Questions 9–11 refer to the following diagram.

HUMAN HEART

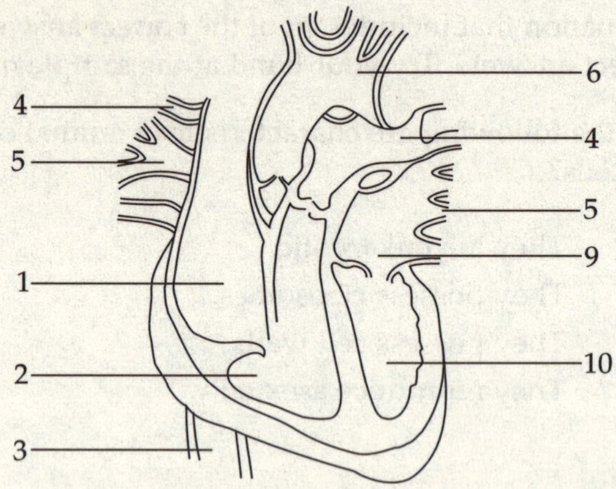

9. Which part of the heart carries the most oxygenated blood?

   A. 4
   B. 5
   C. 1 and 2
   D. 6
   E. 10

10. Where is the pacemaker of the heart?

   A. 1
   B. 2
   C. 9
   D. 10
   E. 8

11. Which sequence is the correct flow of blood through the heart?

   A. 3 —> 9 —> 10 —> 4 —> 5 —> 1 —> 2
   B. 1 —> 2 —> 5 —> 4 —> 9 —> 10 —> 6
   C. 1 —> 2 —> 4 —> 5 —> 9 —> 10 —> 6
   D. 1 —> 2 —> 3 —> 4 —> 5 —> 9 —> 10
   E. 3 —> 2 —> 1 —> 4 —> 5 —> 9 —> 10

Explanations. Let's begin with question 9. The heart is the driving force of the circulatory system. The right and left halves can be viewed as two separate pumps; the right side of the heart pumps deoxygenated blood into pulmonary circulation (toward the lungs), while the left side pumps oxygenated blood into systemic circulation (throughout the body). The two upper chambers are called atria, while the two lower chambers are called ventricles. The former are thin-walled, while the latter are extremely muscular. The left ventricle is more muscular than the right ventricle because it is responsible for generating the force that propels systemic circulation, and because it pumps against high resistance. The most oxygenated blood is in the pulmonary veins. These veins return to the heart from the lungs, where the blood was oxygenated. This is the only type of vein in the body that carries deoxygenated blood. In a similar fashion, the pulmonary artery carries deoxygenated blood to the lungs, and is the only artery in the body that carries deoxygenated blood. Therefore, the answer is (B).

As for question 10, an ordinary cardiac contraction originates in, and is regulated by, the sinoatrial node (SA node) which is also known as the pacemaker. This small mass of specialized tissue is located in the wall of the right atrium. It spreads impulses through both atria, stimulating them to contract simultaneously. (A) is the answer here.

Finally, in question 11, the correct sequence of blood flow begins in the right atrium and then travels into the right ventricle. From there, the blood flows into the pulmonary arteries, and proceeds to the lungs to be oxygenated. It returns via the pulmonary veins and flows into the left atrium. From there, it flows into the left ventricle and is pumped throughout the body, starting at the aorta. The correct choice is (C).

Type 4 Questions

These are also organized in sets, but center on an experiment, chart, graph, or other experimental data presentation. These questions aim to assess how well you apply your scientific skills to unfamiliar situations. As in Type 3 questions, each question is independent of the others in its set. These questions are typically found during the latter part of the test, and are probably the most difficult ones you will encounter. They test your ability to identify a problem, evaluate experimental situations, suggest hypotheses, interpret data, make inferences and draw conclusions, check the logical consistency of hypotheses based on your observations, and select the appropriate procedure for further study. Four sample questions are presented below.

Test Strategy

Type 4 questions will often concentrate on trends or outliers in charts and graphs. Make an effort to pay close attention to these factors when skimming through the questions on Test Day.

Pedigree Analysis

Note the following important facts about pedigrees:

1. Autosomal recessive traits skip a generation.
2. Autosomal dominant traits are found in every generation.
3. Sex-linked traits show gender skewing.

Questions 12–15 are based on the following pedigree:

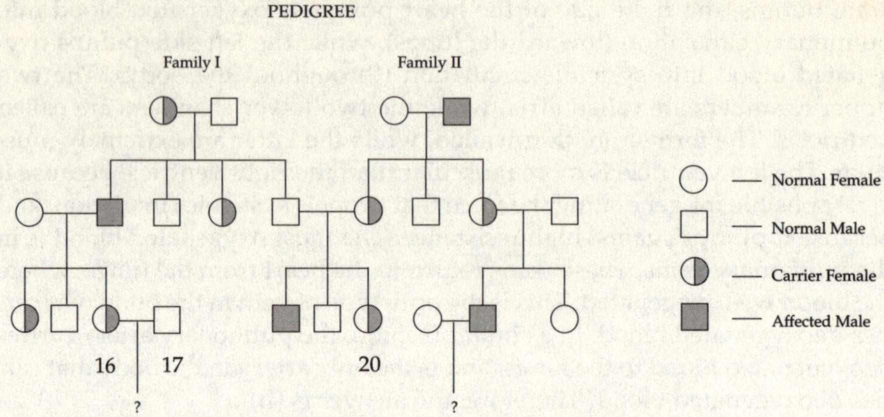

12. Based on the pedigree, what kind of trait is this?

    A. autosomal recessive
    B. sex-linked recessive
    C. autosomal dominant
    D. sex-linked dominant
    E. heterozygous

13. What is the probability that 16 and 17 will have an unaffected daughter?

    A. 50%
    B. 0%
    C. 100%
    D. 25%
    E. 75%

14. This trait is rare in females because

    A. male-specific hormones trigger the disease
    B. it is a dominant trait
    C. the gene is on the *Y* chromosome
    D. females must receive two faulty *X* chromosomes
    E. none of the above

15. What is the genotype of individual 20?

A. *XX*
B. *Aa*
C. X_aX
D. *AA*
E. *aa*

Explanations. In answer to question 12, we can establish that this is an sex-linked recessive disorder (B). Sex-linked recessives typically affect only males, while females serve as carriers. They are differentiated from autosomal dominant traits in that they show up in every generation, with no gender skewing. Autosomal recessives also show no gender skewing, but skip generations.

As for questions 13 and 14, the pattern of inheritance for a sex-linked recessive is somewhat complicated. Since the gene is carried on the *X* chromosome, and males pass the *X* chromosome only to their daughters, affected males cannot pass the trait to their male offspring. Affected males will pass the gene to all of their daughters. However, unless the daughter also receives the genes from her mother, she will be a phenotypically normally carrier of the trait. Since all of the daughter's male children will receive their only *X* chromosome from her, half of her sons will receive the recessive sex-linked allele. Thus, sex-linked recessives generally affect only males; they cannot be passed from father to son, but can be passed from father to grandson via a daughter who is a carrier, thereby skipping a generation. All of the daughters will receive a functional *X* chromosome from their father, so they will all be outwardly normal. However, 50 percent of the daughters will receive an affected *X* chromosome from their mother, so they will be carriers and run the risk of their sons having the disorder. (C) is the correct choice for question 13, and (D) is the answer to question 14.

Individual 20 is a carrier female. Since the trait in these questions is sex-linked and recessive, this female must be X_aX. If you chose (C) in question 15, you were on the right track.

Sample Questions for the Ecological/Molecular Sections

As previously mentioned, the new ecological and molecular sections of the E/M Biology Test feature the same question types that you will encounter on the original Biology Subject Test. The sample questions below offer insights into the biology content that is explored in these sections.

Test Strategy

Look for opposing answers in the answer selections. If two answers are close in wording or if they contain opposite ideas, there is a strong possibility that one of them is the correct answer.

By the same token, if two answers mean basically the same thing, then they cannot both be correct—you can eliminate both answer choices.

Ecological Section

The first two sets of sample questions will focus on ecology as well as evolution and diversity.

Questions 1–4:

Within a particular species, six populations are either separated by a variety of geographical barriers or are able to interbreed. The diagram below shows which populations are able to interbreed and which are isolated.

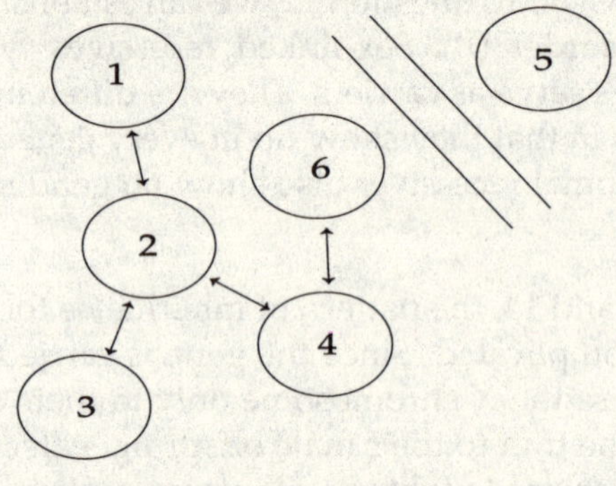

Use this information to answer questions 1–4.

1. Which of the following populations may be in Hardy-Weinberg equilibrium?

   A. 1
   B. 2
   C. 3
   D. 4
   E. 5

2. The gain or loss of alleles that will occur between populations 4 and 6 is known as

   A. gene flow
   B. genetic drift
   C. assortative mating
   D. natural selection
   E. mutation

3. Population 5 is wiped out by a flood. Only 10 percent of the population survives. This type of occurrence is known as

   A. gene flow
   B. founder effect
   C. bottleneck effect
   D. macroevolution
   E. geographical variation

4. New selective pressures brought about by the introduction of a lethal pathogen emerge for this species. Which population has the greatest chance of surviving these pressures?

   A. 1
   B. 2
   C. 3
   D. 4
   E. 5

Questions 5–8:

Although goldfish have an optimal water temperature, they can tolerate higher or lower water temperatures if the change in water temperature to which they are subjected is gradual rather than abrupt. Swimming speed is used in the graph below to determine the general health of the fish in a variety of different water temperatures.

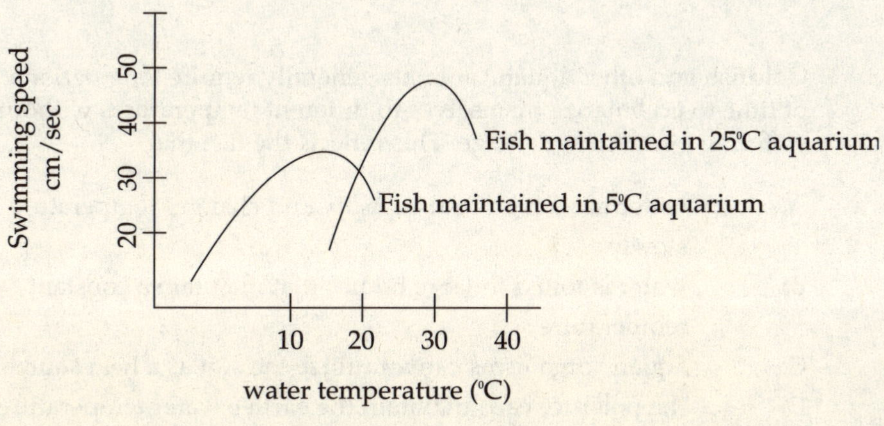

Test Strategy

Realize that both versions of the SAT II: Biology Test emphasize general trends and basic biology concepts. The test makers are probably not going to give you a question/graph that would take a rocket scientist 30 minutes to figure out. So, look for trends and outliers in graphs. If a value or a plot is vastly different from the others, it is likely that there will be a question about it.

5. What is the optimal temperature for maintenance of goldfish?

A. 5°C
B. 10°C
C. 15°C
D. 25°C
E. 35°C

6. A fish that is maintained in a 5°C aquarium is placed in a 30°C aquarium. It would

A. swim at 20 cm/sec
B. swim at 50 cm/sec
C. swim at 40 cm/sec
D. be unable to adjust and suffer harm
E. be unable to remove oxygen from the water

7. Terrestrial organisms are often subjected to abrupt changes in the external temperature of their environment. As a result,

A. they suffer from a decrease in motility in warm weather
B. they are selected to tolerate such changes
C. they are able to control their external temperature
D. they suffer serious harm when abrupt temperature changes occur
E. they alter the morphology of their bodies

8. Goldfish and other aquatic animals generally require long periods of time to acclimatize themselves to different temperatures without suffering permanent damage. This reflects the fact that

A. water has a high heat capacity and changes temperature slowly
B. water is found in large bodies that maintain a constant temperature
C. aquatic organisms cannot utilize the sun as a heat source
D. the polar ice caps maintain the earth's water temperature
E. salt water is generally much warmer than fresh water

Explanations. Questions 1–5 center around the topic of population genetics. In order for Hardy-Weinberg equilibrium to be maintained, the following requirements must be met. There must be a large population; there must be no net mutations; no random mating may occur; and no migration may take place. In question 1, population 5 is the only isolated population in which there is no possibility that net migration of genes will occur. Therefore, (E) is the answer to this question.

As for question 2, the gain or loss of alleles in a non-isolated population is known as gene flow (A). Gene flow occurs when fertile individuals migrate, or when gametes such as pollen grains are transferred. Genetic drift, meanwhile, refers to the changes in a gene pool that can be attributed to chance. Assortative mating, on the other hand, is the term assigned to the process by which individuals mate with partners who resemble themselves phenotypically, as when blister beetles select mates that are similar to them in size.

Examples of genetic drift include the bottleneck effect, the founder effect, gene flow, and mutations. In question 3, the botteneck effect (C) occurs when a disaster (such as an earthquake, a floor, or a fire) reduces the size of a population in a drastic and unselective manner. The small surviving population is unlikely to be representative of the original population in terms of its genetic makeup.

A population can overcome new selective pressures brought on by the introduction of a new pathogen through genetic variability. Since population 2 (B) is able to receive new alleles by breeding with three other populations, it has the highest genetic variability of the choices in question 4. It would therefore have the greatest chance of surviving new selective pressures.

Let's move on to the second question set. According to the chart, the goldfish at 25°C have the fastest swimming speed, an indication of general wellbeing. Of the fish described in this question set, these fish must therefore be the closest to their optimal temperature. The answer to question 5 is (D).

Fish are able to adjust to temperature changes without suffering serious harm only if these temperature changes occur gradually. For this reason, the fish in question 6 would most likely be unable to adjust to the new temperature, and would suffer harm (D).

As for question 7, terrestrial organisms live in environments that undergo abrupt changes in temperature. It is therefore logical to surmise that the environment selects for individuals who possess the capacity to tolerate rapid changes in temperature (B). Organisms that cannot tolerate such changes are unlikely to survive and reproduce on land. In aquatic environments, on the other hand, temperature changes occur slowly due to the high heat capacity of water. In contrast to terrestrial organisms, goldfish and other aquatic animals have not been selected to quickly acclimatize to temperature changes, and the answer to question 8 is (A).

The Power of Chance

Genetic drift is the changes in a gene pool that can be attributed to chance. The following are types of genetic drift:

- Bottleneck effect
- Founder effect

Molecular Section

Questions 9–12 refer to the diagram below.

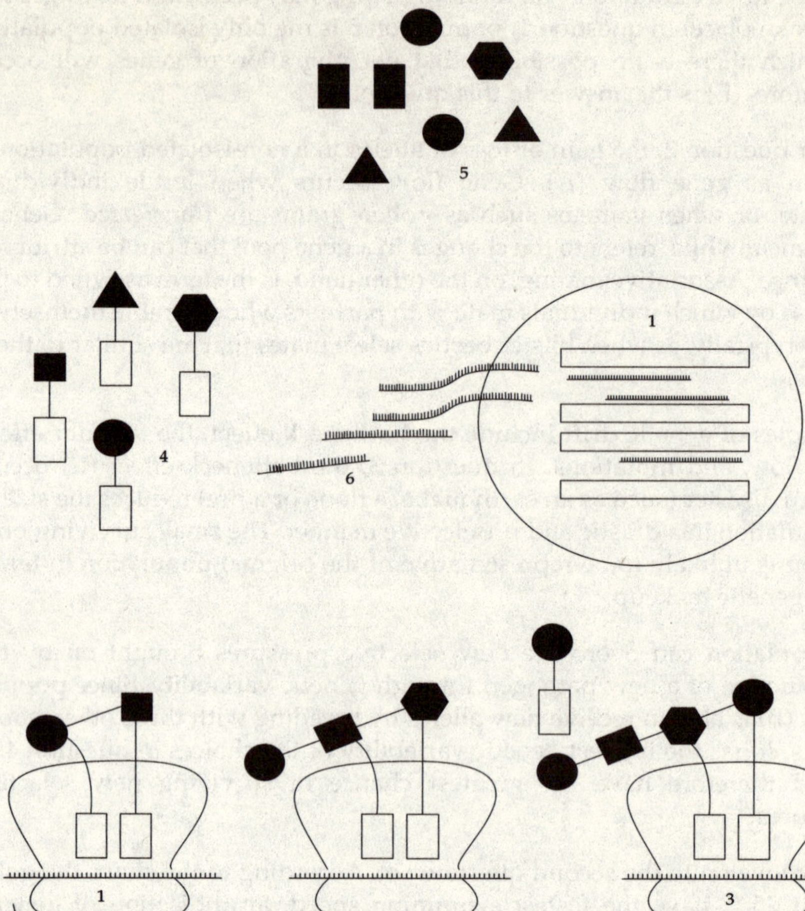

9. In the diagram above, what type of bond is being formed?

A. hydrogen
B. peptide
C. ionic
D. unbreakable
E. no bonds are being formed

10. What process is occurring at 1?

 A. elongation
 B. translocation
 C. translation
 D. transcription
 E. termination

11. Which of the following are made up of both protein and RNA?

 A. 3
 B. 4
 C. 5
 D. 3 and 6
 E. 4 and 6

12. A mutation in a gene that codes for a tRNA molecule may cause which of the following?

 A. autolysis of the cell
 B. premature ending of transcription
 C. a change in amino acid specificity for that tRNA
 D. mutated ribosomes
 E. a smaller number of free amino acids

Questions 13–16:

In your town, people who drank water from a particular well developed a serious bacterial infection. In order to isolate the causative agent, you perform the following experiment.

You take a water sample from the well and inoculate a nutrient agar plate (Plate I) and a nutrient agar plate containing tetracycline, an antibiotic (Plate II). You also inoculate a nutrient agar plate with a sample you isolated from one of your patients (Plate III). As a control, you inoculate a plate of nutrient agar with distilled water (Plate IV). The results are depicted as the figures below.

Test Strategy

Predict your answer before you go to the answer choices so that you don't get persuaded by the wrong answers you will find there. This helps to boost your confidence and protects you from persuasive or tricky incorrect choices. Most wrong answer choices are logical twists on the correct choice.

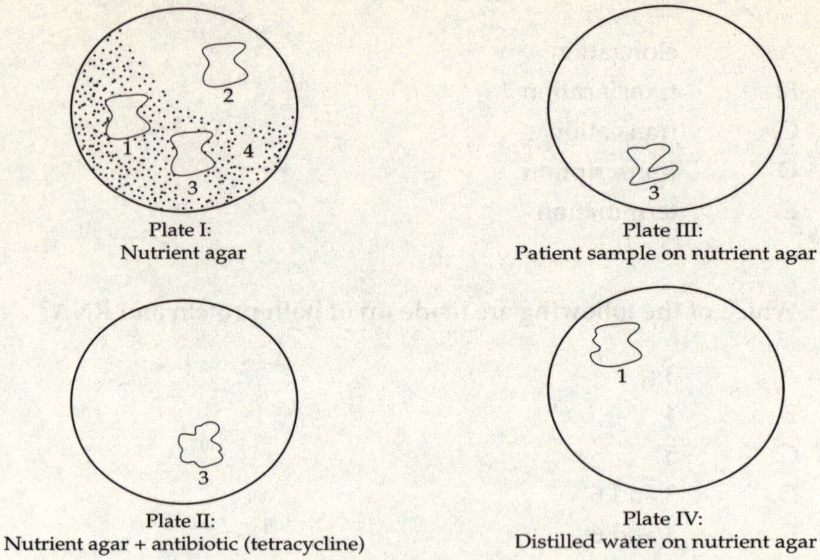

13. How do you explain the presence of Bacteria 1 on the plate inoculated with distilled water (Plate IV)?

    A. Bacteria I always grows in the absence of Bacteria 2.
    B. It is infected by a bacteriophage.
    C. The antibiotic killed all the other types of bacteria.
    D. Bacteria 1 is the only anaerobe.
    E. Contamination occurred during inoculation.

14. Which of the following explains the growth pattern of Plate II?

    A. Bacteria 2 produces an antibiotic to which Bacterias 1, 3, and 4 are susceptible.
    B. Bacteria 2 produces an antibiotic to which Bacteria 4 is susceptible.
    C. Bacteria 2 produces an antibiotic to which Bacterias 1 and 4 are susceptible.
    D. Bacteria 4 requires a protein produced by Bacteria 1.
    E. Bacteria 4 requires a protein produced by Bacteria 3.

15. You discover that Bacteria 3 from other sources is sometimes susceptible to tetracycline. However, when these bacteria are cultured with the Bacteria 3 strain isolated from the well, they become resistant to tetracycline. This is because

A. the well makes all bacteria resistant to tetracycline
B. Bacteria 3 mutates easily into a resistant form
C. the gene for resistance is located on a plasmid that can be passed to other members of the same strain
D. Bacteria 3 is easily infected by lytic bacteriophages
E. Bacteria 3 produces its own antibiotic

16. To develop a therapy for your patients, you

A. repeatedly inoculate Plate II with a variety of antibiotics until you find one that kills Bacteria 3
B. treat patients with the antibiotic produced by Bacteria 2
C. treat patients with tetracycline
D. treat patients with a bacteriophage
E. none of the above

Explanations. Questions 9–12 explore the process of protein synthesis. In question 9, the type of bond being formed is (B) a peptide bond. This covalent bond is formed when tRNA (transport RNA) brings two amino acids into close proximity with each other. The tRNA's anticodons bind to the appropriate codon, or sequence, on the mRNA.

Transcription (D) is the process occurring in question 10. In transcription, information coded in the base sequence of DNA is transcribed onto a strand of mRNA (messenger RNA). mRNA leaves the nucleus through nuclear pores. The remaining events of protein synthesis take place in the cytoplasm.

As for question 11, structure 3 (A) is the correct answer. This ribosome is composed of two subunits, which in their turn consist of proteins and RNA that bind during protein synthesis. The ribosome has three binding sites—one for mRNA and two for tRNA. mRNA (structure 6) and tRNA (structure 4) are made up of RNA exclusively.

A mutation in a tRNA molecule may affect the anticodon, which would affect the specificity of that tRNA for its corresponding mRNA. Such a mutation may also affect the tRNA's site of amino acid attachment; this would affect the specificity of this tRNA for its amino acid. Hence the answer to question 12 is (C). Since stop codons are recognized by the ribosome and not the tRNA, a mutation in the tRNA could not result in a pre-

Protein Synthesis

Protein synthesis is the process by which combinations of amino acids build up complex proteins. It may be broken down into the following stages:

- Transcription
- Post-transcriptional RNA processing
- Translation

mature termination of translation. A premature ending of transcription (B) would be the result of a mutation in DNA, not tRNA. These mutations, also known as nonsense mutations, change the DNA sequence from one that codes for an amino acid into one that signals the end of the mRNA.

In the experiment on which questions 13–16 are based, there is only one explanation for the presence of Bacteria 1 on Plate IV: Contamination has occurred during inoculation. (E) is the correct response to question 13.

In question 14, Bacteria 4 grows all over the nutrient plate, except in the regions surrounding Bacteria 2. The logical conclusion is that Bacteria 2 produces an antibiotic that kills Bacteria 4 (B). We don't know whether or not this antibiotic can also kill Bacterias 1 and 3, since they are not in close enough proximity to Bacteria 2 to be affected by it.

If the well strain of Bacteria 3 can transfer antibiotic resistance as it does in question 15, it must possess a plasmid, a circular piece of extrachromosomal DNA, that it can transfer to other members of its species (C). Plasmids often contain genes that code for antibiotic resistance, and they are used extensively in genetic engineering.

Finally, in question 16, it would be necessary to screen a number of different antibiotics to find one capable of killing Bacteria 3, the most likely causative agent of the illness, before embarking on a course of treatment (A). It has already been proven that Bacteria 3 is resistant to tetracycline (Plate II), and there is no evidence that the antibiotic produced by Bacteria 2 is capable of killing Bacteria 3 (Plate I).

Test Strategy

Eliminate answers and guess!

Now that you've worked you've worked your way through our sample SAT II: Biology questions, you should be ready to tackle our practice diagnostic test. This test will probe your knowledge of the various biology topics covered on both versions of the SAT II: Biology exam. Use it to identify areas that you have not completely mastered, and plan to review the chapters that deal with these topics particularly carefully. Good luck!

BIOLOGY REVIEW

ANSWER SHEET
FOR THE DIAGNOSTIC PRACTICE TEST

1 Ⓐ Ⓑ Ⓒ Ⓓ Ⓔ	26 Ⓐ Ⓑ Ⓒ Ⓓ Ⓔ	51 Ⓐ Ⓑ Ⓒ Ⓓ Ⓔ	76 Ⓐ Ⓑ Ⓒ Ⓓ Ⓔ
2 Ⓐ Ⓑ Ⓒ Ⓓ Ⓔ	27 Ⓐ Ⓑ Ⓒ Ⓓ Ⓔ	52 Ⓐ Ⓑ Ⓒ Ⓓ Ⓔ	77 Ⓐ Ⓑ Ⓒ Ⓓ Ⓔ
3 Ⓐ Ⓑ Ⓒ Ⓓ Ⓔ	28 Ⓐ Ⓑ Ⓒ Ⓓ Ⓔ	53 Ⓐ Ⓑ Ⓒ Ⓓ Ⓔ	78 Ⓐ Ⓑ Ⓒ Ⓓ Ⓔ
4 Ⓐ Ⓑ Ⓒ Ⓓ Ⓔ	29 Ⓐ Ⓑ Ⓒ Ⓓ Ⓔ	54 Ⓐ Ⓑ Ⓒ Ⓓ Ⓔ	79 Ⓐ Ⓑ Ⓒ Ⓓ Ⓔ
5 Ⓐ Ⓑ Ⓒ Ⓓ Ⓔ	30 Ⓐ Ⓑ Ⓒ Ⓓ Ⓔ	55 Ⓐ Ⓑ Ⓒ Ⓓ Ⓔ	80 Ⓐ Ⓑ Ⓒ Ⓓ Ⓔ
6 Ⓐ Ⓑ Ⓒ Ⓓ Ⓔ	31 Ⓐ Ⓑ Ⓒ Ⓓ Ⓔ	56 Ⓐ Ⓑ Ⓒ Ⓓ Ⓔ	81 Ⓐ Ⓑ Ⓒ Ⓓ Ⓔ
7 Ⓐ Ⓑ Ⓒ Ⓓ Ⓔ	32 Ⓐ Ⓑ Ⓒ Ⓓ Ⓔ	57 Ⓐ Ⓑ Ⓒ Ⓓ Ⓔ	82 Ⓐ Ⓑ Ⓒ Ⓓ Ⓔ
8 Ⓐ Ⓑ Ⓒ Ⓓ Ⓔ	33 Ⓐ Ⓑ Ⓒ Ⓓ Ⓔ	58 Ⓐ Ⓑ Ⓒ Ⓓ Ⓔ	83 Ⓐ Ⓑ Ⓒ Ⓓ Ⓔ
9 Ⓐ Ⓑ Ⓒ Ⓓ Ⓔ	34 Ⓐ Ⓑ Ⓒ Ⓓ Ⓔ	59 Ⓐ Ⓑ Ⓒ Ⓓ Ⓔ	84 Ⓐ Ⓑ Ⓒ Ⓓ Ⓔ
10 Ⓐ Ⓑ Ⓒ Ⓓ Ⓔ	35 Ⓐ Ⓑ Ⓒ Ⓓ Ⓔ	60 Ⓐ Ⓑ Ⓒ Ⓓ Ⓔ	85 Ⓐ Ⓑ Ⓒ Ⓓ Ⓔ
11 Ⓐ Ⓑ Ⓒ Ⓓ Ⓔ	36 Ⓐ Ⓑ Ⓒ Ⓓ Ⓔ	61 Ⓐ Ⓑ Ⓒ Ⓓ Ⓔ	86 Ⓐ Ⓑ Ⓒ Ⓓ Ⓔ
12 Ⓐ Ⓑ Ⓒ Ⓓ Ⓔ	37 Ⓐ Ⓑ Ⓒ Ⓓ Ⓔ	62 Ⓐ Ⓑ Ⓒ Ⓓ Ⓔ	87 Ⓐ Ⓑ Ⓒ Ⓓ Ⓔ
13 Ⓐ Ⓑ Ⓒ Ⓓ Ⓔ	38 Ⓐ Ⓑ Ⓒ Ⓓ Ⓔ	63 Ⓐ Ⓑ Ⓒ Ⓓ Ⓔ	88 Ⓐ Ⓑ Ⓒ Ⓓ Ⓔ
14 Ⓐ Ⓑ Ⓒ Ⓓ Ⓔ	39 Ⓐ Ⓑ Ⓒ Ⓓ Ⓔ	64 Ⓐ Ⓑ Ⓒ Ⓓ Ⓔ	89 Ⓐ Ⓑ Ⓒ Ⓓ Ⓔ
15 Ⓐ Ⓑ Ⓒ Ⓓ Ⓔ	40 Ⓐ Ⓑ Ⓒ Ⓓ Ⓔ	65 Ⓐ Ⓑ Ⓒ Ⓓ Ⓔ	90 Ⓐ Ⓑ Ⓒ Ⓓ Ⓔ
16 Ⓐ Ⓑ Ⓒ Ⓓ Ⓔ	41 Ⓐ Ⓑ Ⓒ Ⓓ Ⓔ	66 Ⓐ Ⓑ Ⓒ Ⓓ Ⓔ	91 Ⓐ Ⓑ Ⓒ Ⓓ Ⓔ
17 Ⓐ Ⓑ Ⓒ Ⓓ Ⓔ	42 Ⓐ Ⓑ Ⓒ Ⓓ Ⓔ	67 Ⓐ Ⓑ Ⓒ Ⓓ Ⓔ	92 Ⓐ Ⓑ Ⓒ Ⓓ Ⓔ
18 Ⓐ Ⓑ Ⓒ Ⓓ Ⓔ	43 Ⓐ Ⓑ Ⓒ Ⓓ Ⓔ	68 Ⓐ Ⓑ Ⓒ Ⓓ Ⓔ	93 Ⓐ Ⓑ Ⓒ Ⓓ Ⓔ
19 Ⓐ Ⓑ Ⓒ Ⓓ Ⓔ	44 Ⓐ Ⓑ Ⓒ Ⓓ Ⓔ	69 Ⓐ Ⓑ Ⓒ Ⓓ Ⓔ	94 Ⓐ Ⓑ Ⓒ Ⓓ Ⓔ
20 Ⓐ Ⓑ Ⓒ Ⓓ Ⓔ	45 Ⓐ Ⓑ Ⓒ Ⓓ Ⓔ	70 Ⓐ Ⓑ Ⓒ Ⓓ Ⓔ	95 Ⓐ Ⓑ Ⓒ Ⓓ Ⓔ
21 Ⓐ Ⓑ Ⓒ Ⓓ Ⓔ	46 Ⓐ Ⓑ Ⓒ Ⓓ Ⓔ	71 Ⓐ Ⓑ Ⓒ Ⓓ Ⓔ	
22 Ⓐ Ⓑ Ⓒ Ⓓ Ⓔ	47 Ⓐ Ⓑ Ⓒ Ⓓ Ⓔ	72 Ⓐ Ⓑ Ⓒ Ⓓ Ⓔ	
23 Ⓐ Ⓑ Ⓒ Ⓓ Ⓔ	48 Ⓐ Ⓑ Ⓒ Ⓓ Ⓔ	73 Ⓐ Ⓑ Ⓒ Ⓓ Ⓔ	
24 Ⓐ Ⓑ Ⓒ Ⓓ Ⓔ	49 Ⓐ Ⓑ Ⓒ Ⓓ Ⓔ	74 Ⓐ Ⓑ Ⓒ Ⓓ Ⓔ	
25 Ⓐ Ⓑ Ⓒ Ⓓ Ⓔ	50 Ⓐ Ⓑ Ⓒ Ⓓ Ⓔ	75 Ⓐ Ⓑ Ⓒ Ⓓ Ⓔ	

right

wrong

Use the answer key following the test to count up the number of questions you got right and the number you got wrong. (Remember to not count omitted questions as wrong.) The "Compute Your Score" section following the Answer Key will show you how to find your score.

Remove this answer sheet and use it to complete the Practice Test.

DIAGNOSTIC PRACTICE TEST

Part A

Directions: Each question or incomplete statement below is followed by five possible answers or completions, lettered A–E. Choose the answer that is the best in each case. Fill in the corresponding oval on your answer sheet.

1. Which of the following is a means of internal communication for an organism and coordinates the activities of its organ systems?

   (A) endocrine system
   (B) digestive system
   (C) circulatory system
   (D) skeletal system
   (E) muscular system

2. Air entering the lungs of a tracheotomy patient through a tracheotomy (a tube inserted directly into the trachea) is colder and drier than normal, which often causes lung crusting and infection. This occurs primarily because the air

   (A) enters the respiratory system too rapidly to be filtered
   (B) is not properly humidified by the larynx
   (C) does not flow through the nasal passageways
   (D) does not flow past the mouth and tongue
   (E) none of the above

3. Smooth muscle develops from which of the following germ layers?

   (A) endoderm
   (B) mesoderm
   (C) epiderm
   (D) ectoderm
   (E) none of the above

4. What is the function of a lysosome's membrane?

   (A) It provides an acidic environment for the lysosome's hydrolytic enzymes within the neutral environment of the cell.
   (B) It is continuous with the nuclear membrane, thereby linking the lysosome with the endoplasmic reticulum.
   (C) It is used as an alternative site of protein synthesis.
   (D) The cytochrome carriers of the electron transport chain are embedded within it.
   (E) It separates the nucleus from the cytoplasm.

GO ON TO THE NEXT PAGE

5. Which of the following statements regarding evolution is true?

(A) Certain phenotypes are more fit in certain environments than others.
(B) Migration helps keep the gene pool constant.
(C) The gene pool has been stable for thousands of years.
(D) Mutations always affect the fitness of an organism.
(E) all of the above

6. For the following organisms, which follows the correct sequence of evolution?

(A) sponge, flatworm, chordate, mollusk
(B) flatworm, mollusk, sponge, chordate
(C) sponge, flatworm, mollusk, chordate
(D) mollusk, sponge, flatworm, chordate
(E) flatworm, sponge, chordate, mollusk

7. Arthropods that have a variable number of appendages include which of the following organisms?

I. Arachnids
II. Insects
III. Crustaceans

(A) I only
(B) II only
(C) III only
(D) I and III
(E) II and III

8. Oogenesis is the process by which

(A) primary oocytes produce sperm
(B) primary oocytes produce eggs
(C) the egg implants in the uterus
(D) the egg is released from the ovary
(E) starfish regenerate limbs

9. An individual that has only one *X* chromosome is genotypically *XO*. This person

(A) cannot survive
(B) will have immature, ambiguous (both male and female) reproductive systems
(C) will be phenotypically female
(D) does not produce steroid hormones
(E) none of the above

10. In the process of fat emulsification, bile salts make fats more susceptible to the action of lipases by

(A) transporting fat globules to the region of lipase activity
(B) increasing the surface area of the fat globules
(C) functioning as a catalyst for the lipases
(D) lowering the pH of the small intestine
(E) none of the above

GO ON TO THE NEXT PAGE

11. In the evolutionary scale, which organism falls between an amphibian and a bird?

(A) monotreme
(B) lamprey eel
(C) reptile
(D) marsupial
(E) bony fish

12. Brain cells of the housefly *Musca domestica* have 6 pairs of chromosomes. Therefore, it can be concluded that

(A) the fly's diploid number is 24
(B) the fly's haploid number is 12
(C) the fly's haploid number is 3
(D) the fly's haploid number is 6
(E) the fly's haploid number is 24

13. A snake eats frogs, which eat insects. The snake is an example of a

(A) primary consumer
(B) secondary consumer
(C) producer
(D) tertiary consumer
(E) decomposer

14. Which of the following is an example of a scavenger?

I. fungi
II. vulture
III. hyena

(A) I only
(B) II only
(C) I and II
(D) I and III
(E) II and III

15. A black male mouse (I) is crossed with a black female mouse, and they produce 15 black and 5 white offspring. A different black male mouse (II) is crossed with the same female, and the offspring from this mating are 30 black mice. Which of the following must be true?

(A) The female mouse is homozygous.
(B) Male mouse II is heterozygous.
(C) Two of the mice are heterozygous.
(D) All the progeny of mouse II are homozygous.
(E) All three mice are homozygous.

16. A difference between fats and carbohydrates is that

(A) carbohydrates are always steroids
(B) carbohydrates have a H:O ratio of 2:1
(C) fats are known as starch
(D) fats have a H:O ratio with much more oxygen than hydrogen
(E) fats always contain nitrogen

GO ON TO THE NEXT PAGE

17. Water travels into and out of cells via

(A) carrier proteins
(B) symport systems
(C) diffusion
(D) osmosis
(E) active transport

18. Which of the following statements illustrates the principle of induction during vertebrate development?

(A) The presence of a notochord beneath the ectoderm results in the formation of a neural tube.
(B) A neuron synapses with another neuron via a neurotransmitter.
(C) The neural tube developes into the brain, the spinal chord, and the rest of the nervous system.
(D) Secretion of TSH stimulates the secretion of the hormone thyroxine.
(E) none of the above

19. What is the correct sequence of events in the development of the embryo?

(A) morula —> cleavage —> blastula —> gastrula
(B) cleavage —> morula —> blastula —> gastrula
(C) cleavage —> gastrula —> blastula —> morula
(D) blastula —> cleavage —> gastrula —> morula
(E) morula —> blastula —> cleavage —> gastrula

20. Members of a class are more alike than members of

(A) an order
(B) a phylum
(C) a genus
(D) a species
(E) a family

21. Pancreatic lipase is involved in the digestion of

(A) starch
(B) protein
(C) fat
(D) cellulose
(E) nucleic acids

GO ON TO THE NEXT PAGE

22. All of the following thermoregulatory mechanisms are involved in heat conservation EXCEPT

(A) goose bumps
(B) perspiration
(C) shivering
(D) blood vessel constriction
(E) all of the above

23. The mouse is known as *Mus musculus*. The *Mus* is the

(A) phylum
(B) class
(C) order
(D) genus
(E) species

24. Albinos have a genotype of *aa*, while all other members of population are either *AA* or *Aa*. The offspring of a cross between a heterozygous male and an albino female would be

(A) 100% albino
(B) 100% normal
(C) 50% normal, 50% albino
(D) 25% normal, 75% albino
(E) 75% normal, 25% albino

25. Which part of cellular respiration directly produces the most ATP during the breakdown of glucose?

(A) glycolysis
(B) anaerobic respiration
(C) Krebs cycle
(D) electron transport chain
(E) none of the above

26. Which of the following increases the surface area of ingested lipids?

(A) maltase
(B) cholecystokinin
(C) bile
(D) lipase
(E) HCl

27. The first organisms on Earth were thought to be

(A) autotrophs
(B) chemosynthetic
(C) heterotrophs
(D) poisoned by oxygen
(E) photosynthetic

28. Maple trees, apple trees, and orchids and palms are examples of

(A) gymnosperms
(B) bryophytes
(C) angiosperms
(D) chlorophytes
(E) rhodophytes

GO ON TO THE NEXT PAGE

29. Molds and yeast are classified as

(A) rhodophytes
(B) bryophytes
(C) fungi
(D) ciliates
(E) flagellates

30. In humans, brown eyes are dominant over blue eyes. In the cross of *BB* x *bb*, what percentage of the offspring will have brown eyes?

(A) 75%
(B) 50%
(C) 0%
(D) 100%
(E) 25%

31. In humans, a normal egg must contain

I. an *X* chromosome
II. 23 chromosomes
III. a *Y* chromosome

(A) I only
(B) II only
(C) III only
(D) I and II
(E) II and III

32. Which of the following are bilaterally symmetrical?

(A) planaria
(B) roundworms
(C) humans
(D) arthropods
(E) all of the above

33. A heterozygous female (*Hh*) is mated with a recessive male (*hh*). What percentage of their offspring are homozygous?

(A) 0%
(B) 25%
(C) 50%
(D) 75%
(E) 100%

34. Examples of parasitism include all of the following EXCEPT

(A) tick bird and rhinoceros
(B) virus and host cell
(C) tapeworm and man
(D) tuberculosis bacteria and man
(E) flukes and fish

GO ON TO THE NEXT PAGE

35. Meiosis differs from mitosis in that

I. two cell divisions take place
II. DNA replicates during Interphase
III. haploid cells are produced from diploid cells

(A) I only
(B) II only
(C) III only
(D) I and III
(E) I, II, and III

36. Stomata in plant leaves close at night to prevent the loss of

(A) O_2
(B) H_2O
(C) CO_2
(D) energy
(E) chlorophyll

37. When calcium binds troponin in muscle cells, the binding site for which of the following is exposed?

(A) tropomyosin
(B) myosin
(C) ATP
(D) ADP
(E) Pi

38. All organisms utilize

(A) CO_2
(B) the triplet code to produce proteins
(C) oxygen
(D) ADP as cellular energy
(E) membrane-bound organelles

39. In fruit flies, the gene for wing type is located on an autosomal chromosome. The allele for wild-type wings is dominant over the allele for vestigial wings. If a homozygous dominant male fly is crossed with a female with vestigial wings, what percentage of their female progeny are expected to have wild-type wings?

(A) 0%
(B) 25%
(C) 50%
(D) 75%
(E) 100%

40. We studied a pond and found water skimmers and minnows. Fifteen years later, the pond had filled in, resulting in swampy land; frogs and snakes were prevalent. This is an result of

(A) predation
(B) succession
(C) speciation
(D) natural disasters
(E) global warming

GO ON TO THE NEXT PAGE

41. Damsel flies and dragonflies can live in the same ecosystem because

(A) they occupy different niches
(B) they are commensal
(C) they are both insects
(D) dragonflies have stronger wing muscles than damsel flies
(E) they mate at different times of the year

42. Legumes are good for the soil because

(A) deer won't eat them
(B) they are photosynthetic
(C) they have nitrogen-fixing bacteria on their roots
(D) animals convert them to energy
(E) none of the above

43. If the DNA sequence is TACAGA, then the mRNA sequence is

(A) UCUAUG
(B) TACAGA
(C) AUGUCU
(D) UACAGA
(E) ATGTCT

44. Sexually reproducing species have a selective advantage over asexually reproducing species because sexual reproduction

(A) is more energy efficient
(B) allows for genetic diversity
(C) decreases the likelihood of mutations
(D) always decreases an offspring's survival ability
(E) can occur in any climate

45. The proteins which result in the phenotypes that we see are under the ultimate control of

(A) other proteins
(B) rRNA
(C) tRNA
(D) mRNA
(E) DNA

GO ON TO THE NEXT PAGE

Questions 46–50 refer to the following figure:

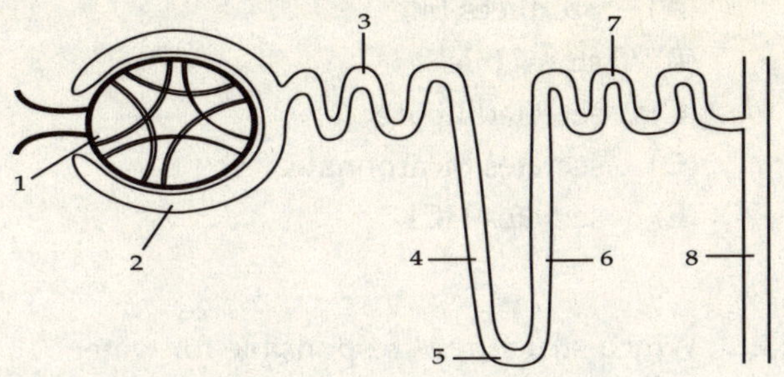

46. Structure 1 is known as the

(A) glomerulus
(B) Bowman's capsule
(C) proximal convoluted tubule
(D) loop of Henle
(E) distal convoluted tubule

47. Structure 3 is where

(A) urine is concentrated
(B) almost all glucose and amino acids are reabsorbed
(C) potassium is secreted
(D) blood is oxygenated
(E) the renal artery flows in

48. ADH acts on structure

(A) 4
(B) 5
(C) 6
(D) 7
(E) 8

49. The organ depicted is involved in

(A) digestion
(B) cellular respiration
(C) homeostasis
(D) digestion of lipids
(E) maintenance of the heartbeat

50. Which structure is part of the circulatory system?

(A) 1
(B) 2
(C) 3
(D) 4
(E) 5

GO ON TO THE NEXT PAGE

Questions 51–54 refer to the following figure:

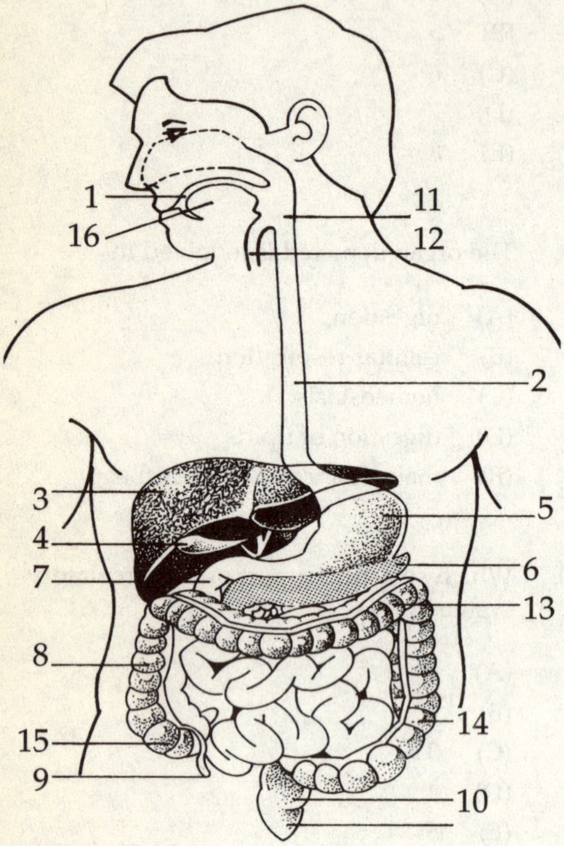

51. At which site does digestion of starches begin?

(A) 1
(B) 2
(C) 3
(D) 4
(E) 5

52. Structure 4

(A) produces bile
(B) stores bile
(C) secreted lipase
(D) secretes bicarbonate
(E) secretes HCl

53. Which structure is responsible for water absorption?

(A) 5
(B) 6
(C) 7
(D) 8
(E) 9

54. Which structure has both exocrine and endocrine function?

(A) 5
(B) 6
(C) 7
(D) 8
(E) 9

GO ON TO THE NEXT PAGE

Part B

Each set of choices A–E below should be compared to the numbered statements that follow it. Choose the lettered choice that best matches each numbered statement. Fill in the correct oval on your answer sheet. Remember that a choice may be used once, more than once, or not at all in each set.

Questions 55–58:

(A) growth hormone
(B) oxytocin
(C) progesterone
(D) aldosterone
(E) glucagon

55. increases uterine contractions during child birth
56. stimulates the release of glucose to the blood
57. induces water resorption in the kidneys
58. prepares the uterus for implantation of the fertilized egg

Questions 59–61:

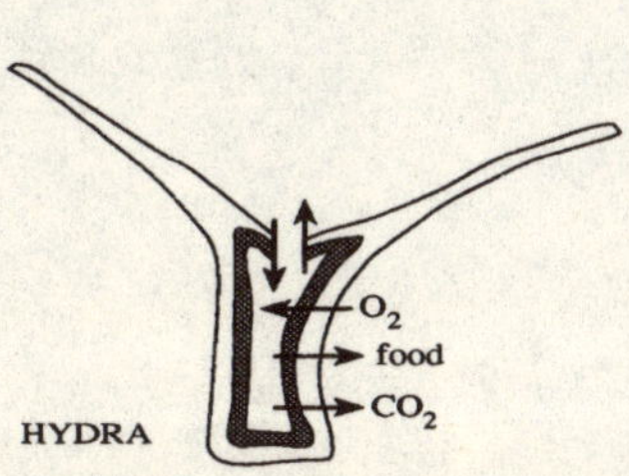

(A)

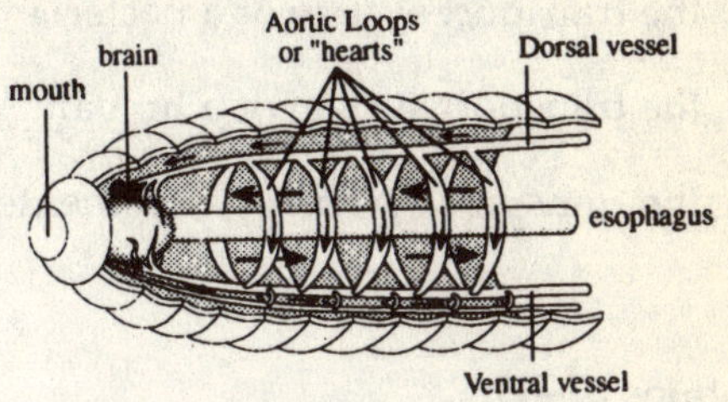

(B)

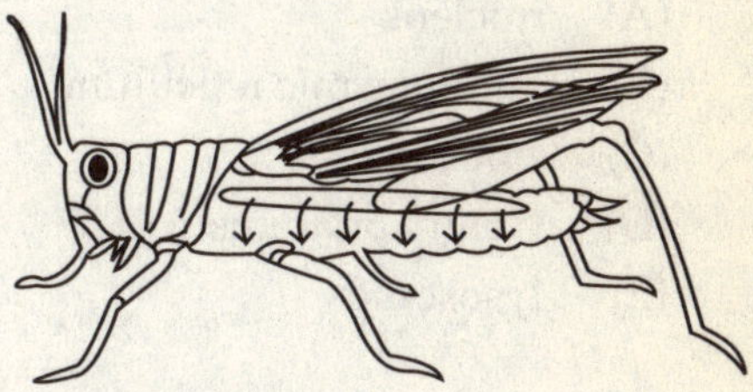

(C)

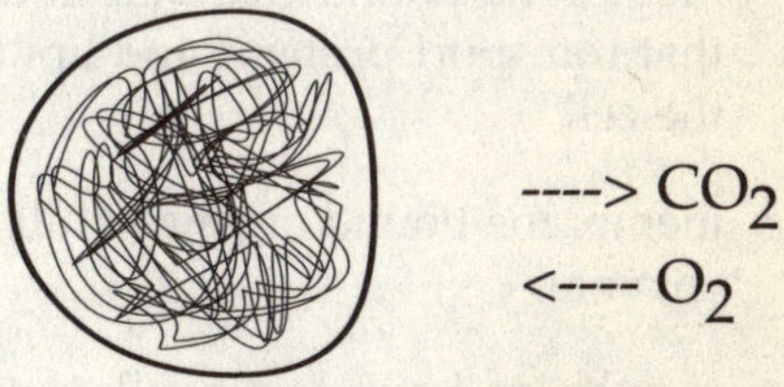

(D)

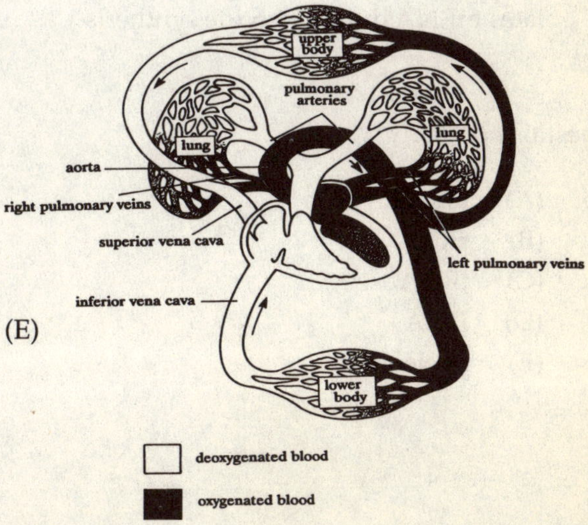

(E)

GO ON TO THE NEXT PAGE

59. the transport system of a bacteria

60. the transport system of a human

61. the transport system of a segmented worm

Questions 62–65:

(A) nucleus
(B) endoplasmic reticulum
(C) ribosomes
(D) Golgi apparatus
(E) lysosomes

62. membrane-bound channels in the cytoplasm that transport proteins and lipids throughout the cell

63. membrane-bound organelle full of hydrolytic enzymes

64. membrane-bound organelle that contains the chromosomes

65. this consists of RNA and proteins and translates mRNA during peptide synthesis

Questions 66–68:

(A) mRNA
(B) rRNA
(C) tRNA
(D) DNA
(E) nucleolus

66. site of rRNA synthesis

67. product of transcription, acts as protein template

68. binds specific amino acids and carries them to the ribosomes during protein synthesis

Questions 69–71:

(A) natural selection
(B) adaptive radiation
(C) vestigial structure
(D) migrations
(E) parallel evolution

69. variation within a species that gives it a competitive advantage

70. production of different species from a common ancestor

71. appears useless but had an ancestral function

GO ON TO THE NEXT PAGE

Questions 72–75:

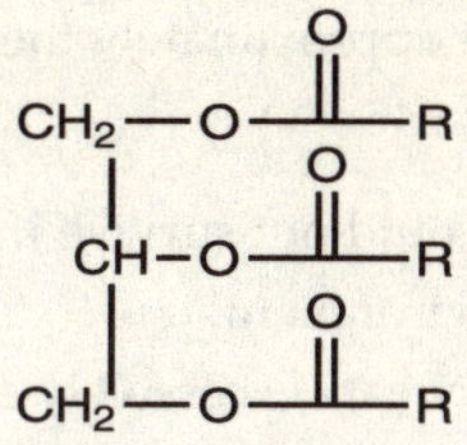

(A)

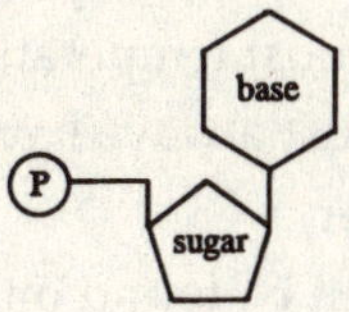

(B)

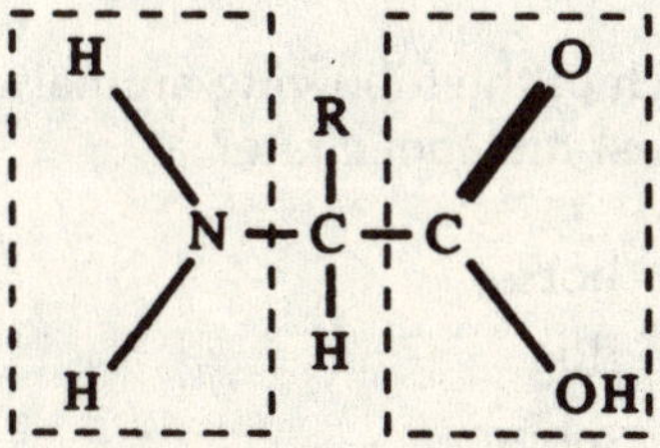

(C)

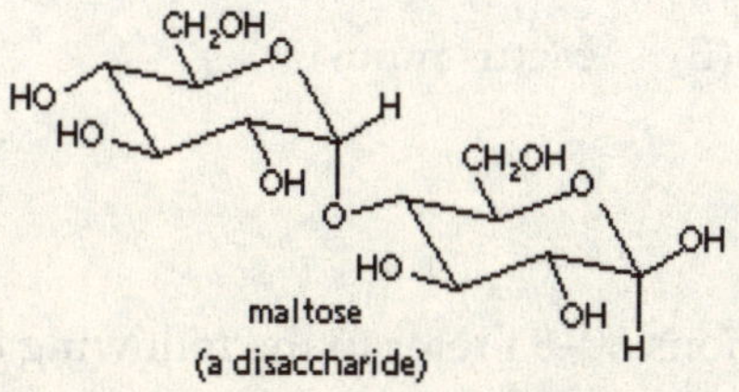

(D)

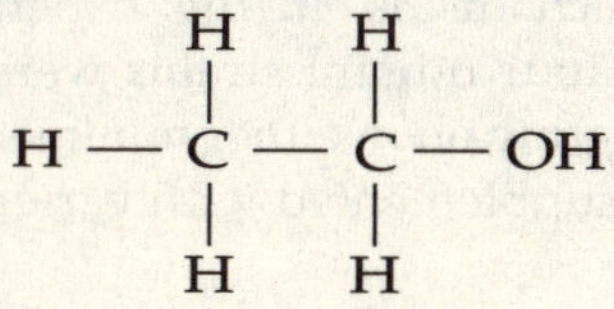

(E)

72. building block of DNA or RNA

73. building block of proteins

74. lipid

75. carbohydrate

Part C

Each of the following sets of questions is based on a laboratory or experimental situation. Begin by studying the description of each situation. Next, choose the best answer to each of the questions that follow it. Fill in the corresponding oval on your answer form.

Questions 76–79 refer to the following paragraph:

All birds and mammals are able to maintain relatively constant body temperatures, despite fluctuations in external temperature. These animals have evolved thermoregulatory mechanisms that help them to adapt to their environments. One such mechanism is the metabolic rate. A plot of the rate of oxygen consumption versus body weight for various mammals reveals that metabolic rate is inversely proportional to body weight (see figure). However, metabolic rate and the transfer of heat to the environment are directly proportional to the surface area-to-volume ratio of the animal. For instance, a shrew has a higher metabolic rate and a greater surface area-to-volume ratio than a horse, which means that the shrew generates more internal heat per gram of body weight and loses more heat to the environment. This makes it especially difficult for small animals to maintain a constant body temperature in cold weather.

GO ON TO THE NEXT PAGE

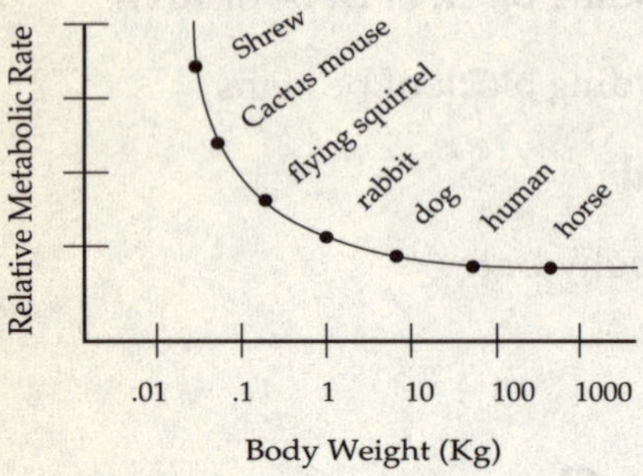

76. Small animals such as penguin chicks will huddle together when it is cold outside. The most likely explanation for this is that

(A) huddling increases the effective surface area-to-volume ratio and decreases the loss of body heat
(B) huddling decreases the effective surface area-to-volume ratio and increases the loss of body heat
(C) huddling is detrimental to penguin chicks
(D) huddling is an instinct in penguin chicks
(E) baby penguins huddle following imprinting

77. A mechanism mammals have developed to dissipate excess heat in hot weather is

(A) huddling
(B) sitting in the shade
(C) sweating
(D) burrowing
(E) none of the above

78. Which of the following is most likely true of an animal that must keep all of its vital organs at approximately the same temperature?

(A) It could not survive in a desert environment.
(B) It sweats excessively and becomes dehydrated in hot weather.
(C) Its body temperature is determined by the most temperature-sensitive organ.
(D) It must always have a large supply of water.
(E) It will be found only on the tundra.

79. Which of the following animals has the highest metabolic rate?

(A) horse
(B) dog
(C) rabbit
(D) flying squirrel
(E) cactus mouse

Questions 80–83 refer to the following experiment:

E. coli is a bacteria that can be used to study a variety of mechanisms. In this experiment, one wild type and four mutant strains were tested for their ability to grow on in either minimal media or media that was supplemented with various amino acids.

GO ON TO THE NEXT PAGE

Supplement	None	Arginine
Wild type	+	+
Strain 1	–	+
Strain 2	–	–
Strain 3	–	–
Strain 4	+	+

Supplement	Threonine	Histidine
Wild Type	+	+
Strain 1	–	–
Strain 2	+	–
Strain 3	–	+
Strain 4	+	+

A (+) sign indicates growth and a (–) sign indicates no growth.

80. Based on the table, strain 2

(A) cannot produce arginine
(B) cannot produce threonine
(C) cannot produce histidine
(D) cannot produce glycine
(E) can produce all amino acids

81. Which amino acid(s) does strain 3 NOT need to grow?

(A) arginine and histidine
(B) threonine and histidine
(C) histidine only
(D) arginine and threonine
(E) threonine only

82. The mutation in strain 1 that renders it incapable of growing without arginine occurs in

(A) DNA
(B) mRNA
(C) protein
(D) the anticodon region of the tRNA
(E) rRNA

83. The mutation in strain 4

(A) renders it dormant
(B) does not affect its ability to synthesize amino acids
(C) causes it to become haploid
(D) is identical to that in strain 1
(E) none of the above

GO ON TO THE NEXT PAGE

The following graphs depict growth in bacteria under a variety of different conditions. Use these graphs to answer the questions 84–86.

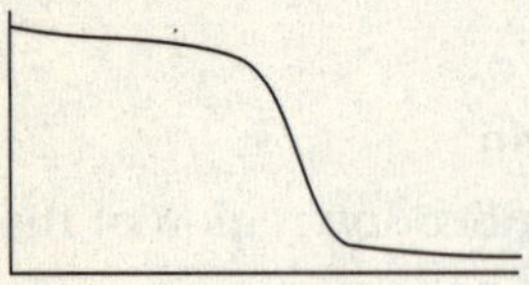
(1)

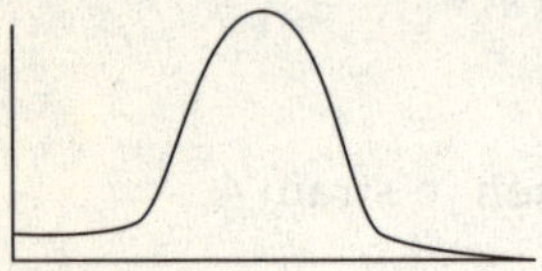
(2)

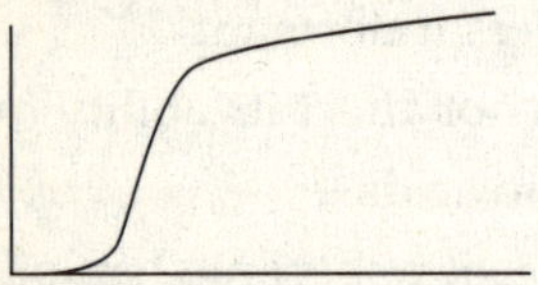
(3)

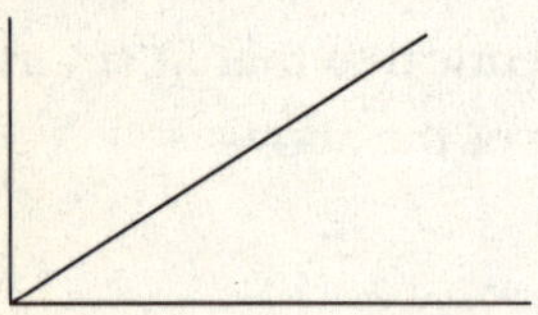
(4)

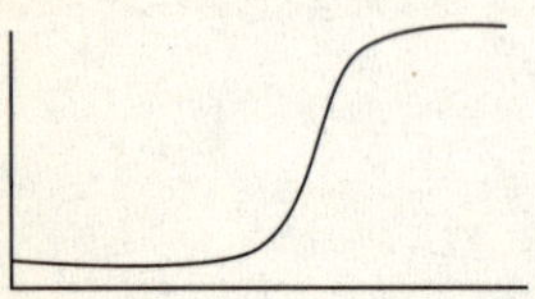
(5)

84. Which graph shows what happens when you introduce a bacteriophage that lyses the bacteria two weeks after the colonies are established?

(A) 1
(B) 2
(C) 3
(D) 4
(E) 5

85. Which graph describes what happens when you introduce very minimal nutrients for the first two weeks, then add surplus nutrients?

(A) 1
(B) 2
(C) 3
(D) 4
(E) 5

86. Which graph shows normal growth?

(A) 1
(B) 2
(C) 3
(D) 4
(E) 5

In her experiment to determine where the C, H, and O come from when plants produce carbohydrates, a scientist grew plants in the presence of a variety of radioactive compounds and used a Geiger counter to determine whether the starch produced by the plants was radioactive.

	Molecule	**Radioactive Starch**
Expt. 1	$^{14}CO_2$	+
Expt. 2	$H_2{}^{18}O$	–
Expt. 3	$C^{18}O_2$	+

Use this data to answer questions 87–92.

87. The source of oxygen in carbohydrates produced by plants is

(A) CO_2
(B) H_2O
(C) O_2
(D) none of the above
(E) all of the above

88. The oxygen released from the plant comes from

(A) CO_2
(B) H_2O
(C) O_2
(D) none of the above
(E) all of the above

89. If a plant utilized H_2SO_4 instead of water, the plant would release

(A) CO_2
(B) H_2O
(C) O_2
(D) SO_4
(E) none of the above

90. Which compound captures light energy in plants?

(A) O_2
(B) CO_2
(C) H_2O
(D) chlorophyll
(E) none of the above

91. The light reaction of photosynthesis occurs in the

(A) stoma
(B) thylakoid membranes
(C) mitochondria
(D) nucleus
(E) ribosomes

92. The Calvin cycle

(A) does not use light directly
(B) occurs in the cytoplasm
(C) releases CO_2
(D) produces ATP
(E) none of the above

A study of a meadow yielded the following data:

Organisms	**Number**
Hawk	3
Various birds	50
Spiders	500
Insects	3,000
Shrubs and other plants	>5,000

Use this data to answer questions 93–95.

93. Which organisms are the primary consumers?

(A) hawks
(B) birds
(C) spiders
(D) insects
(E) shrubs

GO ON TO THE NEXT PAGE

94. Which type of organism is not represented?

(A) primary producer
(B) secondary consumer
(C) tertiary consumer
(D) primary consumer
(E) decomposer

95. Which of the following is a pyramid of mass?

(A)

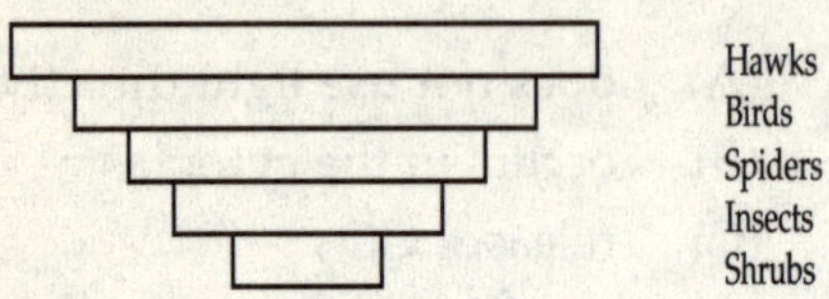

(B)

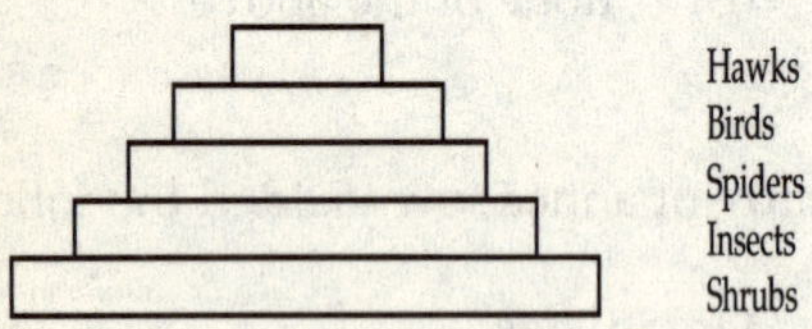

(C)

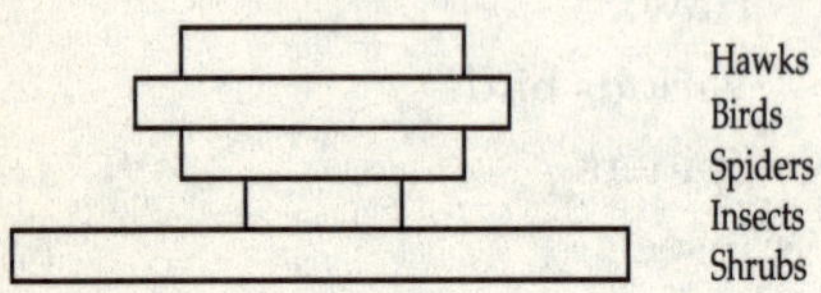

(D)

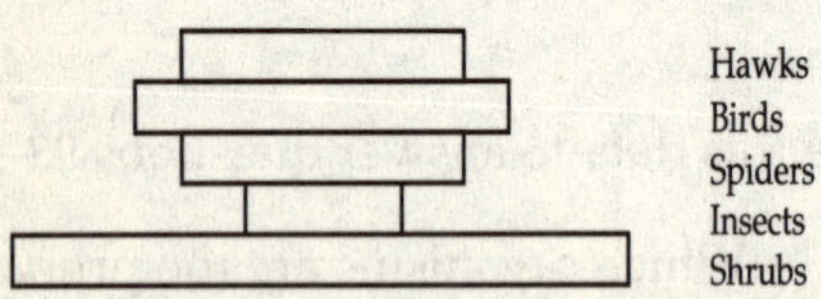

(E) none of the above

Turn the page for
answers and explanations to the
Diagnostic Practice Test.

Answer Key

1. A
2. C
3. B
4. A
5. A
6. C
7. C
8. B
9. C
10. B
11. C
12. D
13. D
14. E
15. C
16. B
17. D
18. A
19. B
20. B
21. C
22. B
23. D
24. C
25. D
26. C
27. C
28. C
29. C
30. D
31. D
32. E
33. C
34. A
35. D
36. B
37. B
38. B
39. E
40. B
41. A
42. C
43. C
44. B
45. E
46. A
47. B
48. E
49. C
50. A
51. A
52. B
53. D
54. B
55. B
56. E
57. D
58. C
59. D
60. E
61. B
62. B
63. E
64. A
65. C
66. E
67. A
68. C
69. A
70. B
71. C
72. B
73. C
74. A
75. D
76. A
77. C
78. C
79. E
80. B
81. D
82. A
83. B
84. B
85. E
86. C
87. A
88. B
89. D
90. D
91. B
92. A
93. D
94. E
95. B

Compute Your Practice Test Score

Step 1: Figure out your raw score. Refer to your answer sheet for the number right and the number wrong on the practice test you're scoring. (If you haven't checked your answers, do that now, using the answer key that follows the test.) You can use the chart below to figure out your raw score. Multiply the number wrong by 0.25 and subtract the result from the number right. Round the result to the nearest whole number. This is your raw score.

SAT II: Biology Practice Test

Number right		Number wrong		Raw score
☐	–	(0.25 x ☐)	=	☐ (rounded)

Step 2: Find your practice test score. Find your raw score in the left column of the table below. The score in the right column is your Practice Test score.

Find Your Practice Test Score

Raw	Scaled	Raw	Scaled	Raw	Scaled	Raw	Scaled	Raw	Scaled	Raw	Scaled
95	800	78	690	61	590	43	490	26	390	9	290
94	800	77	680	59	580	42	480	25	380	8	290
93	790	76	670	58	570	41	470	24	380	7	280
92	780	75	670	57	570	40	470	23	370	6	280
91	770	74	660	56	560	39	460	22	370	5	270
90	760	73	660	55	550	38	460	21	360	4	260
89	750	72	650	54	550	37	450	20	360	3	260
88	740	71	640	53	540	36	450	19	350	2	250
87	740	70	640	52	540	35	440	18	350	1	240
86	730	69	630	51	530	34	430	17	340	0	240
85	730	68	630	50	530	33	430	16	330	–1	230
84	720	67	620	49	520	32	420	15	330	–2	230
83	720	66	620	48	510	31	420	14	320	–3	220
82	710	65	610	47	510	30	410	13	320	–4	220
81	710	64	610	46	500	29	410	12	310	–5	210
80	700	63	600	45	500	28	400	11	300	–6	210
79	690	62	590	44	490	27	400	10	300	–7– –9	200

Practice Test Score

☐

A note on your practice test scores: Don't take these scores too literally. Practice test conditions cannot precisely mirror real test conditions. Your actual SAT II: Biology Subject Test score will almost certainly vary from your practice test scores. Your scores on the practice tests will give you a rough idea of your range on the actual exam.

Explanations to Diagnostic Practice Test

1. (A) The endocrine system is a system of internal communication for the body, allowing it to coordinate the activities of the organ systems through messengers known as hormones. These hormones are secreted by the endocrine glands and carried throughout the body by the blood.

2. (C) When a patient breathes through a tracheotomy, the air entering the respiratory system bypasses a very important area—the nasal cavities. In a normally breathing individual, the extensive surfaces of the nasal passageways warm and almost completely humidify the air, and particles are filtered out by nasal air turbulence. Since the air reaching the lungs of a tracheotomy patient has not been warmed or humidified, lung crusting and infection often result.

3. (B) The endoderm develops into the lungs, the gastrointestinal tract, and the lining of the bladder. Ectoderm develops into the brain and nervous system, the lens of the eye, the inner ear, hair, nails, sweat glands, the lining of the mouth and nose, and the skin. The mesoderm becomes everything else, including the musculoskeletal system, the reproductive system, the circulatory system, and the kidneys.

4. (A) See explanation (3) above for descriptions of the functions of the endoderm, the ectoderm, and the mesoderm. The final choice, the epiderm, is not an embryonic germ layer.

5. (A) In Darwin's theory of natural selection, some organisms in a species have variations that give them an advantage over other members of the species. These adaptations enable these organisms to survive in greater numbers than organisms which lack them.

6. (C) On an evolutionary scale, sponges are the most primitive organisms, followed by flatworms, then mollusks, and finally chordates.

7. (C) A crustacean is characterized by having segmented bodies with a variable number of appendages and by breathing through gills. Other members of Arthropoda are spiders, which have four pairs of legs, and insects, which have only three pairs of legs.

8. (B) Oogenesis is the process whereby primary oocytes undergo meiosis to produce one egg (or ovum) and two or three polar bodies.

9. (C) A Turner's female has the genotype *XO*; she carries only one *X* chromosome and does not mature sexually, but she is clearly female. Turner's patients have what is known as "streak gonads," or undeveloped gonads. These individuals are often shorter than normal and may have varying degrees of mental development problems.

10. (B) Emulsification, which occurs via the detergent action of bile salts, increases the surface area of fat globules exposed to the lipases. This process involves the breaking down of fat globules into smaller particles so that the lipases can act on their surfaces. Since lipases can only attack these surfaces and not other parts of the fat globules, the action of bile salts greatly enhances their capacity for fat digestion.

11. (C) Evolutionarily, reptiles fall between amphibians and birds. The lamprey eel and fish are more primitive than amphibians, while the mammalian monotreme and marsupials are more advanced than birds.

12. (D) If there are six pairs of chromosomes in a diploid cell, there are a total of 12 chromosomes in a $2n$ cell, and the haploid cell would have half that number; $n = 6$.

13. (D) Insects are typically primary consumers. The frog that eats these herbivorous insects is the secondary consumer, while the snake that eats the frogs is a tertiary consumer.

14. (E) A scavenger eats animals that have already been killed and, typically, is incapable of hunting. Vultures are scavengers, while fungi and saprophytic, decomposing dead plant material. Hyenas also act as scavengers, though they also hunt upon occasion.

15. (C) In the cross of Mouse I with the female, the ratio of the offspring is 3:1, indicating a *Bb* x *Bb* cross. Therefore, mouse I is *Bb*, while the female is *Bb*. In the second cross of mouse II and the female mouse, 100% of the offspring are black. Hence Mouse II must be homozygously dominant.

16. (B) Carbohydrates, known as starches or cellulose, have a H:O ratio of 2:1. Fats, in contrast, have a H:O ratio with much more hydrogen than oxygen.

17. (D) Osmosis is defined as the diffusion of water through semipermeable membranes from an area of lesser solute concentration to an area of greater solute concentration.

18. (A) In terms of vertebrate development, induction is defined as the process by which a particular group of cells causes the differentiation of another group of cells. (A) is an example of induction: The group of cells that forms the notochord induces the formation of the neural tube. Other examples of induction in vertebrate development include the formation of the eyes, in which the optic vesicles induce the ectoderm to thicken and form the lens placode, which induces the optic vesicle to form the optic cup. This in turn induces the lens placode to form the cornea and the lens.

19. (B) The zygote continues to divide from the cleavage stage until it becomes the morula, a solid ball of cells. This inverts to become the blastula, a hollow ball of cells. The blastula in its turn invaginates to become the gastrula.

20. (B) Remember the order of classification: kingdom, phylum, class, order, family, genus, and species. Members of a class are more alike than members of a kingdom or a phylum, but less alike than members of orders, families, genera, and species.

21. (C) Lipases are enzymes involved in lipid digestion. Lipids are a category of nonpolar organic substances that includes fats, oils, waxes, and steroids. Therefore, pancreatic lipase is an enzyme involved in the digestion of fat.

22. (B) Perspiration, or sweat, is the secretion of water, salts, and urea from the sweat pores of the skin. As the sweat comes into contact with air, it evaporates, thereby cooling the skin. Thus perspiration is a thermoregulatory mechanism involved in cooling down the body or heat dissipation, NOT heat conservation. All of the other choices are ways to either conserve or produce more heat to keep the body warm.

23. (D) The first half of the scientific name is the genus of the organism; the second half is the species of that organism.

24. (C) In an *Aa* x *aa* cross, 50% of the offspring will be *Aa* and the other 50% will be *aa*.

25. (D) The electron transport chain directly produces the most ATP. This is because NADH and

$FADH_2$, produced in the Krebs cycle, donate high energy electrons to the electron transport chain.

26. (C) Bile acts as an emulsifying agent, increasing the surface area of lipids to allow lipase to break them down into monoglycerides and free fatty acids. However, bile is not an enzyme.

27. (C) According to the heterotroph hypothesis, the first forms of life lacked the ability to synthesize their own nutrients, requiring preformed molecules. Gradually, the molecules spontaneously formed by these life forms began to prove inadequate to meet their energy needs, and autotrophs developed in response.

28. (C) Angiosperms are flowering plants that produce seeds enclosed in an ovary, e.g. the maple tree. Gymnosperms, on the other hand, do not have enclosed seeds; one example is the conifer.

29. (C) Fungi are typically saprophytic organisms and are divided into true fungi—such as yeast, molds, and mushrooms—and slime molds.

30. (D) In a *BB* x *bb* cross, 100% of the offspring will be *Bb*. Genotypically, they will be heterozygous, and phenotypically, they will have brown eyes.

31. (D) A normal egg will have a haploid (23) number of chromosomes and an *X* chromosome. *Y* chromosomes come only from sperm.

32. (E) All of these organisms are bilaterally symmetrical, which implies that they have two identical mirror images. Radially symmetrical organisms, which are symmetrical from top to bottom but not from left to right, include hydra, starfish, and jellyfish. These organisms are often round.

33. (C) In an *Hh* x *hh* cross, 50% of the offspring will be *Hh* and 50% will be *hh*.

34. (A) The tick bird and rhinoceros engage in a symbiotic relationship known as mutualism. The rhinoceros provides the tick bird with insects that live on its hide, while the tick bird keeps the rhinoceros parasite-free.

35. (D) Meiosis has two divisions that take a cell from diploid to haploid. Both mitosis and meiosis replicate DNA during interphase.

36. (B) The stomata close at night, when there is no light energy to catalyze the light reaction, in order to prevent loss of water through transpiration in the stomata.

37. (B) When calcium binds troponin, it exposes a site that allows actin to bind to myosin, causing the shortening of the sarcomere and muscle contraction.

38. (B) All organisms produce proteins through the triplet code. However, some of them do not need carbon dioxide or oxygen (and, in fact, some, such as tetanus, are poisoned by oxygen). All organisms use ATP as cellular energy. Only eukaryotes, however, have membrane-bound organelles.

39. (E) This is a basic cross using Drosophila melanogaster. You're told that the gene for wing type is located on an autosomal chromosome, a non–sex chromosome, which means that it is NOT inherited as a sex-linked trait. You're also told that the dominant allele codes for wild-type wings; "wild type" simply means that this is the phenotype that predominates in nature. The recessive allele codes for the vestigial wing type, which is a stumpy wing. The gender of the flies is of no relevance here; gender only comes into play for sex-linked traits. So, our male fly is homozygous dominant for wing type and our female fly is homozygous recessive. A cross between a homozygous dominant and a homozygous recessive yields 100% heterozygous individuals, which in this case means that 100% of the progeny will have wild-type wings.

40. (B) In an ecological succession, each stage causes changes that allow for the next stage to develop. If you start with a pond, it will ultimately become filled in with earth. Populations change until the climax community, a woodland, is reached. Frogs and snakes would be found in an intermediate stage of this ecological succession, when the pond is transformed into moist land.

41. (A) In the same ecosystem, there can never be two species in the same niche; if two species are found together here, they MUST occupy different niches.

42. (C) The nitrogen-fixing bacteria form nodules on the roots of legumes so that elemental nitrogen can be converted to usable nitrates.

43. (C) Uracil is found in RNA instead of thymine. It bonds with adenine through two hydrogen bonds, while cytosine bonds with guanine through three hydrogen bonds.

44. (B) Asexual reproduction is more efficient than sexual reproduction in terms of the number of offspring produced per reproduction, the amount of energy invested in this process, and the amount of time involved in the development of the young, both before and after birth. Overall, sexual reproduction is a much more time-consuming, energy-consuming process than its asexual counterpart.

Asexual reproduction, however, must rely heavily on mutation to introduce phenotypic variability in future generations, since it almost exclusively produces genetic clones of the parent organism. Sexual reproduction, on the other hand, involves the process of meiosis. Because meoisis consists of two rounds of cell division, it is marked by an increased likelihood of cross-over events, chromosome inversions, and nondisjunction events during the two anaphases. This implies that there is a much greater risk of mutation occurring in sexual reproduction, especially at the level of the entire chromosome.

The reason that species that reproduce sexually have a selective advantage over other species is that genetic recombination is a given with every fertilization. Two genetically unique nuclei, the sperm nucleus and the egg nucleus, fuse to form an equally genetically unique zygote. Phenotypic variability is in this way introduced into individuals in a population, which could be either advantageous or disadvantageous to these individuals. If the genetic recombination proves to be advantageous, the individual is more likely to survive and pass his genes onto future generations.

45. (E) DNA ultimately controls all genetic expression, as DNA serves as the blue print of the cell.

46. (A) The glomerulus is a network of capillaries in the Bowman's capsule of the kidney.

47. (B) In the proximal convoluted tubule, almost all of the amino acids, glucose, and important salts are actively resorbed from the glomerular filtrate back into the bloodstream.

48. (E) ADH acts to make the collecting duct more permeable to water, thereby increasing water absorption.

49. (C) The nephron is involved in maintaining fluid balance in the body and is crucial for maintenance of homeostasis. It also acts to excrete nitrogenous wastes as part of the excretory system.

50. (A) As stated in the explanation to question 46 above, the glomerulus is a network of capillaries in the Bowman's capsule of the kidney. Capillaries are part of the circulatory system.

51. (A) Digestion of starches begins in the mouth as salivary amylase breaks them down into maltose. This is why if you chew bread for long enough, it begins to taste sweet.

52. (B) The gall bladder stores bile produced by the liver. Bile emulsifies (or increases the surface area of) fats.

53. (D) The large intestine is responsible for most of the water and Vitamin K absorption that our body needs.

54. (B) The pancreas both releases digestive enzymes and bicarbonate ions (in what is termed exocrine function) and secretes glucagon and insulin to control blood glucose levels (endocrine function).

55. (B) Oxytocin is released by the posterior pituitary. It increases uterine contractions during childbirth.

56. (E) Glucagon is released by the pancreas. It breaks down glycogen into glucose, increasing the glucose concentration of the blood.

57. (D) Aldosterone increases sodium resorption in the nephron. Water and chloride ions follow passively, and water resorption is increased.

58. (C) Progesterone is secreted by the corpus luteum. Its function is to thicken the uterine lining, preparing it for implantation of the fertilized egg.

59. (D) Bacteria do not have a developed vascular system; instead, they exchange fluids, nutrients, and wastes with the environment.

60. (E) Humans have a four-chambered heart that oxygenates blood at the lungs. It transports fluids, nutrients, and wastes throughout the body via a system of vessels known as arteries, capillaries, and veins.

61. (B) Segmented worms have a closed circulatory system, with dorsal and ventral vessels and five aortic arches (hearts) that force the blood through the body.

62. (B) The endoplasmic reticulum is a network of channels that transports proteins throughout the cell. Some protein synthesis occurs here, on ribosomes attached to the membrane of the ER.

63. (E) Lysosomes are membrane-bound sacs of digestive enzymes with very low pH. Degradation of proteins occurs here.

64. (A) The nucleus is the brain of the cell; here, DNA is housed and transcribed into mRNA, which determines which proteins are produced and, ultimately, what the cell's function will be.

65. (C) The ribosome is either attached to the ER or floats freely in the cytoplasm. The site of protein synthesis, it is composed of proteins and rRNA.

66. (E) The nucleolus is a region of the nucleus where rRNA is synthesized.

67. (A) mRNA is coded directly from the DNA of the cell and acts as the template for protein translation.

68. (C) tRNA carries its specific amino acid to the ribosome, where it attaches to the growing polypeptide chain coded for by mRNA.

69. (A) Natural selection involves survival of the fittest. Some organisms with certain adaptations are more fit than other members of the same species, and are able to pass their genetic traits to their offspring more efficiently.

70. (B) Adaptive radiation may be defined as the development of a number of different species from a common ancestor as a result of differing environmental pressures.

71. (C) Vestigial structures appear useless, but in fact had a necessary ancestral function. One example is the appendix of man, which helped early humans to store and digest food, but now has no apparent function.

72. (B) A nucleotide is made up of a sugar, a phosphate group, and a nitrogenous base. This nucleotide is the building block of nucleic acids such as DNA and RNA.

73. (C) An amino acid is made up of an amino group and a carboxylic acids group. Amino acids serve as the building blocks of proteins.

74. (A) A lipid consists of three fatty acids (carboxyl groups) bound to one glycerol molecule.

75. (D) Carbohydrates are organic molecules with a H:O ratio of 2:1. Examples include fructose, glucose, sucrose, maltose, and galactose.

76. (A) An animal's primary goal in cold weather is to keep warm. It wants to limit the amount of heat lost to the environment, because the more heat it loses, the colder the animal feels. The amount of heat transferred from a body to its external environment is directly proportional to the amount of surface area exposed to the external environment. If heat is lost through exposed surface area, then it follows that the greater the surface area that is exposed, the greater the amount of heat is that will be lost. Smaller animals have a tougher time than larger ones trying to maintain a constant internal body temperature, because smaller animals have a greater surface area-to-volume ratio and a higher metabolic rate than large ones. Therefore, smaller animals both generate more heat and lose more heat to the environment than larger ones. If a group of small animals can somehow decrease their collective surface area-to-volume ratio, then they will lose less heat as a unit. This is accomplished through huddling.

77. (C) Sweating dissipates heat produced by the animals. Sitting in the shade and burrowing, meanwhile, cool the external environment rather than dissipating heat.

78. (C) You're being asked to draw a conclusion about what it means for all organs, including the brain, to have a single body temperature. If an animal must keep ALL of its vital organs at the same temperature, and if one particular organ, such as the brain, is particularly sensitive to temperature, then the animal is obligated to maintain its body temperature at the temperature required by that organ.

79. (E) On the graph, the animal from the list with the highest relative metabolic rate is the cactus mouse. Only one animal has a higher metabolic rate—the shrew. The lowest metabolic rate included in this chart is the horse's.

80. (B) Because strain 2 cannot produce threonine on its own, it cannot grow unless it is supplemented.

81. (D) Strain three does not grow when it is supplemented with arginine and threonine; therefore, it obviously does not need them to grow. It does, however, need histidine to grow, and cannot produce it on its own, as evidenced by the fact that it will grow when histidine is supplemented.

82. (A) Mutations occur in the DNA and are passed on to the proteins that are translated via the mRNA.

83. (B) The mutation did not affect strain 4's ability to synthesize amino acids. This strain grows well regardless of whether or not its media is supplemented. It is likely that its mutation is a silent mutation.

84. (B) Graph 2 shows normal growth with an abrupt cutoff when the bacteriophage is introduced and begins to lyse bacteria.

85. (E) Graph 5 shows a lag for two weeks, during which period the media does not have enough nutrients, followed by an exponential surge in growth when nutrients are added.

86. (C) Graph 3 shows exponential growth—until the bacteria use up most of their nutrients and reach a plateau.

87. (A) The labeled oxygen from the carbon dioxide shows up in the plant-formed carbohydrates; therefore, this must be the source of the oxygen for carbohydrate synthesis.

88. (B) The labeled oxygen from the water is not incorporated into the starch, and, therefore, must be released as water is split to produce the hydrogen molecules for carbohydrate synthesis.

89. (D) Like water, the plant would incorporate the hydrogen molecules into the carbohydrates and release SO_4.

90. (D) Chlorophyll captures light energy and passes it to the cytochrome complexes that produce ATP and NADPH during the light reaction.

91. (B) The light reaction occurs in the thylakoid membrane of the chloroplasts, while the dark reaction occurs in the stroma of the chloroplasts.

92. (A) The Calvin cycle is known as the dark reaction because it does not use light directly. Instead, it utilizes ATP and NADPH produced from the light reaction as energy.

93. (D) Primary consumers eat producers; in this population, insects eat shrubs.

94. (E) The decomposers that break down dead and decaying material are not presented in this pyramid. An example of a decomposer would be a fungus that feeds on fallen trees.

95. (B) Because organisms at upper levels derive their food energy from lower levels of the food chain, and because energy is wasted from one level to the next, smaller and smaller biomasses can be supported as the food chain is ascended. Biomass is continuously lost through wastes and other metabolic processes.

CHAPTER THREE

CELLULAR AND MOLECULAR BIOLOGY

All living things are made up of cells. That's why you're likely to see four to five questions on the SAT II that directly relate to cells, and many more questions that assume you understand cell structure. Every function in biology involves a process that occurs within cells or at the interface between cells. Therefore, to understand biology, you need to appreciate the structure and function of the different parts of the cell and the plasma membrane, as well as the processes by which cells reproduce themselves and manufacture energy to enable organismal growth.

Cell Structure and Organization

Cell Theory

The cell was not discovered or studied in detail until the development of the microscope in the seventeenth century. Since then, much more has been learned, and a unifying theory known as the *cell theory* has been proposed. Matthias Schleiden and Theodor Schwann came up with the cell theory in 1838, while Rudolph Virchow discovered in 1855 that cells arise only from other cells.

The cell theory may be summarized as follows:

- All living things are composed of cells.
- The cell is the basic functional unit of life.
- Cells arise only from preexisting cells.
- Cells carry genetic information in the form of DNA. This genetic material is passed from parent cell to daughter cell.

Cells can be structurally categorized into two distinct groups, prokaryotic and eukaryotic.

Prokaryotic Cells

Our first group of cells encompasses bacteria and cyanobacteria (blue-green algae), unicellular organisms with a simple cell structure. These organisms have an outer cell membrane, but do not contain any membrane-bound organelles. They have no true nucleus; their genetic material consists of a single circular molecule of DNA concentrated in an area of the cell called the nucleoid region.

Prokaryotes may also contain plasmids, consisting of a few genes. Plasmids replicate independently and often incorporate genes that allow the prokaryotes to survive adverse conditions. Meanwhile, bacteria have a cell wall, cell membrane, cytoplasm, ribosomes, and, sometimes, flagella that are used for locomotion. Respiration occurs at the cell membrane.

Don't Mix These Up on Test Day

Prokaryotes are bacteria with no nucleus, no membrane-bound organelles, ribosomes, and cell walls made up of peptidoglycans.

Eukaryotes have a nucleus, membrane-bound organelles, and ribosomes. Examples include protists, fungi, plants, and animals. Fungi and plant eukaryotic cells have cell walls made of cellulose.

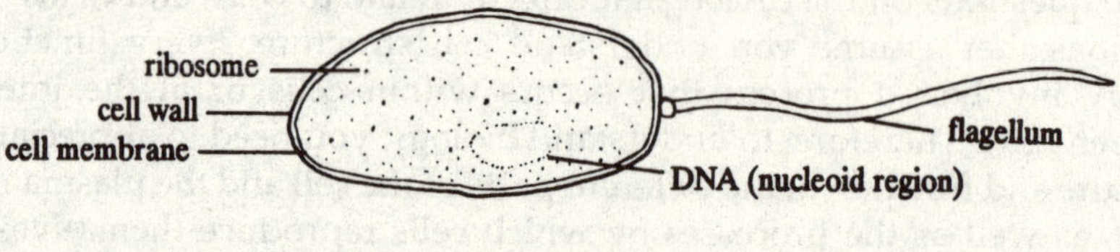

Prokaryotic Cell

Eukaryotic Cells

All multicellular organisms (like you, or a tree, or a mushroom) and all non-bacteria unicellular organisms—such as amoeba and paramecia—are composed of eukaryotic cells. A typical cell has a cell membrane and contains cytoplasm, which contains organelles suspended in a semifluid medium called the cytosol. The cell's genetic material consists of linear strands of DNA organized into chromosomes and located within a membrane-enclosed organelle called the nucleus.

Although both animal and plant cells are eukaryotic, they differ in a number of ways. For example, plant cells have a cell wall and chloroplasts, while animal cells do not. Centrioles, located in the centrosome area, are found in animal cells but not in plant cells.

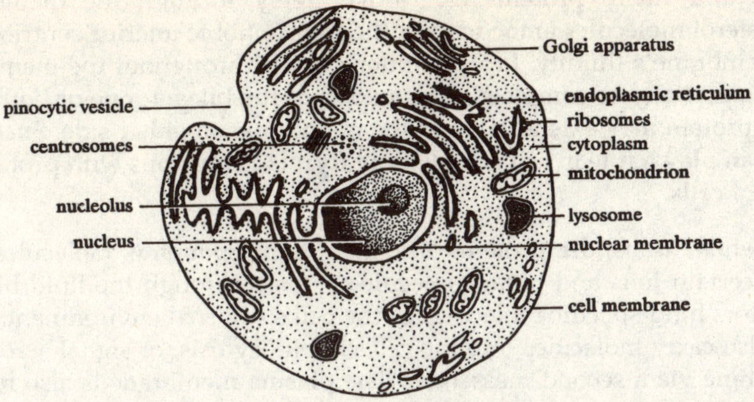

Eukaryotic Cell

Summary of Cell Properties:

Structure	Nucleus	Genetic Material	Cell Wall	Cell Membrane
Eukaryote	Yes	DNA	Yes/No	Yes
Prokaryote	No	DNA	Yes	Yes

Structure	Membrane Organelles	Ribosomes
Eukaryote	Yes	Yes
Prokaryote	No	Yes*

*Ribosomes in prokaryotes are smaller and have a different subunit composition than those in eukaryotes.

Organelles

Eukaryotic cells have specialized structures called organelles, which carry out particular functions for the cell. Organelles include the plasma membrane, nucleus, ribosomes, endoplasmic reticulum, Golgi apparatus, lysosomes, microbodies, vacuoles, mitochondria, chloroplasts, and the cytoskeleton.

Plasma Membrane

The plasma membrane encloses the cell and exhibits selective permeability; it regulates the passage of materials into and out of the cell. According to the generally accepted fluid mosaic model, the cell membrane consists of a phospholipid bilayer in which proteins are embedded. Phospholipids have both a hydrophilic phosphoric acid region and a hydrophobic fatty acid region. In a lipid bilayer, the hydrophilic regions are found on the exterior surfaces of the membrane, while the hydrophobic regions are found on the interior of the membrane.

Study Tip

On Test Day, you should have the function of every organelle down cold. If you do, it's likely that you will be rewarded with a higher score!

Lipids and many proteins can move freely through the membrane. Cholesterol molecules embedded in the hydrophobic interior contribute to the membrane's fluidity. Proteins interspersed throughout the membrane may be partially or completely embedded in the bilayer; one or both ends of the protein may extend beyond the membrane on either side. Such proteins can play a role in cell adhesion by forming functions with proteins in adjacent cells.

Meanwhile, transport proteins are membrane-spanning molecules that allow certain ions and polar molecules to pass through the lipid bilayer. Receptors bind specific molecules in the cell's external environment. They may also carry molecules into the cell via pinocytosis, or signal across the membrane via a second messenger. The plasma membrane is also readily permeable to oxygen and small polar molecules such as water. Small charged molecules are usually able to cross the plasma membrane through protein channels, while some larger molecules need carrier proteins.

History of the Cell Membrane

As early as 1895, scientists began to create models of cell membranes. But it wasn't until 1972 that Singer and Nicolson drew attention to the cell's phospholipid bilayer, and the structure of the cell membrane was discovered.

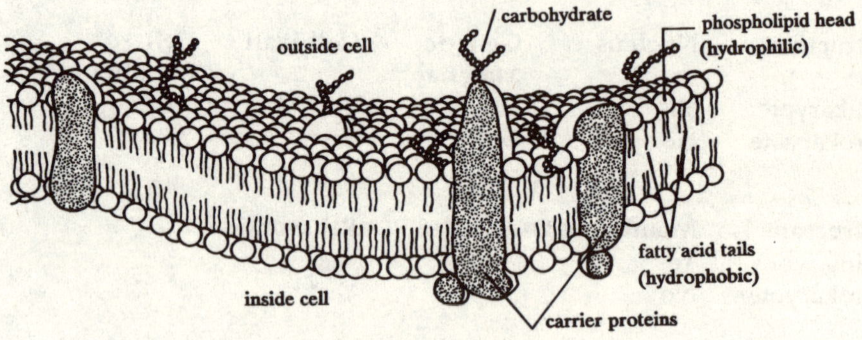

Fluid Mosaic Model

Nucleus

One of the most central organelles of the cell is the *nucleus*, which controls the activity of the cell, including cell division. It is surrounded by a nuclear membrane that maintains a nuclear environment distinct from that of the cytoplasm. There are also, however, nuclear pores in this membrane that allow selective two-way exchange of materials between the nucleus and cytoplasm. The nucleus contains *DNA*, which is complexed with structural proteins called histones to form *chromosomes*. These short, stubby rods contain the genetic material of the cell. Meanwhile, a dense structure within the nucleus in which ribosomal RNA (rRNA) synthesis occurs is known as the *nucleolus*.

Ribosomes

Ribosomes are the sites of protein production and are synthesized by the nucleolus. They consist of two subunits, one large and one small; each subunit is composed of rRNA and proteins. Free ribosomes are found in the cytoplasm, while bound ribosomes line the outer membrane of the endoplasmic reticulum (ER).

Endoplasmic Reticulum

The endoplasmic reticulum (ER) is a network of membrane-enclosed spaces connected at points with the nuclear membrane. If this network has ribosomes lining its outer surface, it is termed rough endoplasmic reticulum (RER); without ribosomes, it is known as smooth endoplasmic reticulum (SER). ER is involved in the transport of material throughout the cells, especially that material destined to be secreted from the cell. SER is involved in lipid synthesis and the detoxification of drugs and poisons, while RER is involved in protein synthesis. Proteins synthesized by the bound ribosomes cross into the cisternae of the RER, where they undergo chemical modification. The proteins then cross into SER, where they are secreted into cytoplasmic vesicles and transported to the Golgi apparatus.

Golgi Apparatus

This apparatus is a stack of membrane-enclosed sacs. It receives vesicles and their contents from SER and modifies them (through glycosylation, the process of adding sugar, for example). Next, it repackages them into vesicles and distributes them. It is an area that is particularly active in the distribution of newly synthesized material to the cell surface. Secretory vesicles, produced by the Golgi, release their contents to the cell's exterior by the process of exocytosis.

Lysosomes

Lysosomes contain hydrolytic enzymes involved in intracellular digestion. Maximally effective at a pH of 5, it is vital that these enzymes are enclosed within the lysosome, which has an acidic environment distinct from the neutral pH of the cytosol. Lysosomes fuse with endocytic vacuoles, breaking down material ingested by the cells. They also aid in renewing a cell's own components by breaking them down and releasing their molecular building blocks into the cytosol for reuse.

A cell in an injured or dying tissue may rupture the lysosome membrane and release its hydrolytic enzymes to digest its own cellular contents. This is referred to as autolysis, and is not common in adult organisms. In developing organisms, however, cells that are not going to become part of the adult organism often "commit suicide" in a process known as apoptosis.

Microbodies

Microbodies can be characterized as specialized containers for metabolic reaction. The two most common types of microbodies are peroxisomes and glyoxysomes. Peroxisomes contain oxidative enzymes that catalyze a class of reactions in which hydrogen peroxide is produced through the transfer of hydrogen from a substrate to oxygen. These microbodies break fats

A Closer Look

Endoplasmic refers to something that is within the cytoplasm, while *reticulum* is derived from a latin word that means "network."

Mnemonic

Remember:
The **r**oughness of the endoplasmic reticulum is determined by **r**ibosomes!

Think of It This Way

It may help to visualize the *Golgi apparatus* as the warehouse of the cell, a place where proteins are packaged for shipment.

The *lysosome* serves as the "stomach" of the cell—both the lysosome and the stomach have acidic pHs.

down into small molecules that can be used for fuel; they are also used in the liver to detoxify compounds, such as alcohol, that may be harmful to the body. Glyoxysomes, on the other hand, are usually found in the fat tissue of germinating seedlings. They are used by the seedling to convert fats into sugars until the seedling is mature enough to produce its own supply of sugars through photosynthesis.

Vacuoles

These membrane-enclosed sacs within the cell are formed after endocytosis and can fuse with a lysosome to digest their contents. Contractile vacuoles in freshwater protists pump excess water out of the cell. Plant cells have a large central vacuole called the tonoplast that is part of their endomembrane system. In plants, the tonoplast functions as a place to store organic compounds, such as proteins, and inorganic ions, such as potassium and chloride.

Quick Quiz

What do *mitochondria* and *chloroplasts* do?

Answer:

They are the organelles that convert energy into forms that the cell can use.

Mitochondria

These sites of aerobic respiration within the cell are important suppliers of energy. Each mitochondrion is bound by an outer and inner phospholipid bilayer membrane. The outer membrane is smooth and acts as a sieve, allowing molecules through on the basis of their size. The area between the inner and outer membranes is known as the intermembrane space. The inner membrane has many convolutions called cristae, as well as a high protein content that includes the proteins of the electron transport chain. The area bounded by the inner membrane is known as the mitochondrial matrix, and is the site of many of the reactions in cell respiration.

Mitochondria are somewhat unusual in that they are semiautonomous; that is, they contain their own circular DNA and ribosomes, which enable them to produce some of their own proteins. In addition, they are able to self-replicate through binary fission. They are believed to have been early prokaryotic cells that evolved a symbiotic relationship with the ancestors of eukaryotes.

Chloroplasts

Chloroplasts are found only in algal and plant cells. With the help of one of their primary components, *chlorophyll*, they function as the site of photosynthesis. They contain their own DNA and ribosomes and exhibit the same semiautonomy as mitochondria. They are also believed to have evolved via symbiosis.

Cytoskeleton

The cell gains mechanical support, maintains its shape, and carries out cell motility functions with the help of another important organelle, the cytoskeleton. The cytoskeleton is composed of microtubules, microfilaments, and intermediate fibers.

Microtubules. These hollow rods are made up of polymerized tubulins that radiate throughout the cells and provide it with support. They also provide a framework for organelle movement within the cell. Centrioles, which direct the separation of chromosomes during cell division, are composed of microtubules.

Cilia and Flagella. Cilia and flagella are specialized arrangements of microtubules that extend from certain cells and are involved in cell motility.

Microfilaments. Cell movement and support are maintained in part through the action of solid rods composed of actin subunits; these are termed microfilaments. Muscle contraction, for example, is based on the interaction of actin with myosin in muscle cells. Microfilaments move materials across the plasma membrane; they are active, for instance, in the contraction phase of cell division and in amoeboid movement.

Intermediate Fibers. These structures are a collection of fibers involved in the maintenance of cytoskeletal integrity. Their diameters fall between those of microtubules and microfilaments.

Membrane Trafficking

It is crucial for a cell to control what enters and exits it. In order to preserve this control, cells have developed the following mechanisms:

Permeability

Traffic through the membrane is extensive, but the membrane is selectively permeable; substances do not cross its barrier indiscriminately. A cell is also able to retain many small molecules and exclude others. In addition, different molecules move through the plasma membrane at different rates. The hydrophobic core impedes transport of hydrophilic ions and polar molecules. Hydrophobic molecules such as hydrocarbons and oxygen can cross easily. If two molecules are equally soluble, then the smaller molecule will cross the plasma membrane faster. Small, polar, uncharged molecules can pass through easily, but the lipid bilayer is not very permeable to large, uncharged polar molecules like glucose. It is also relatively impermeable to all ions, even small ones such as H^+ and Na^+.

Transport Proteins

Hydrophilic substances avoid contact with the lipid bilayer by passing through transport proteins. There are three types of transport proteins: *uniport*, *symport*, and *antiport*. Uniport proteins carry a single solute across the membrane. Symport proteins translocate two different solutes simultaneously in the same direction; transport occurs only if both solutes bind to the proteins. Antiport proteins exchange two solutes by transporting one into the cell and the other out of the cell.

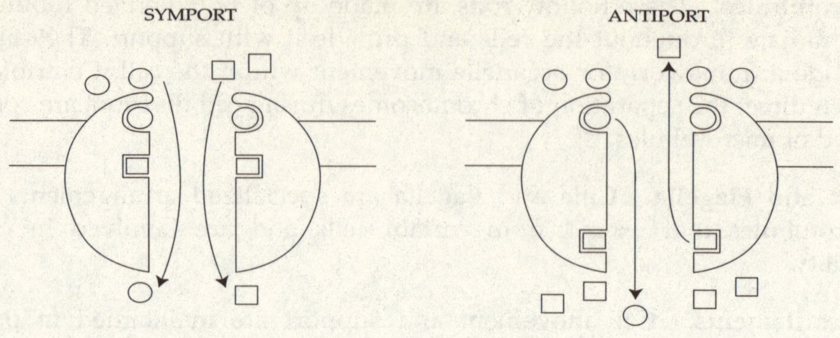

Symport and Antiport

Diffusion/Passive Transport

This is the net movement of dissolved particles down their concentration gradients, from a region of higher concentration to a region of lower concentration.

Facilitated Diffusion

Meanwhile, the net movement of dissolved particles down their concentration gradient—with the help of carrier molecules—is known as facilitated diffusion. This process does not require energy.

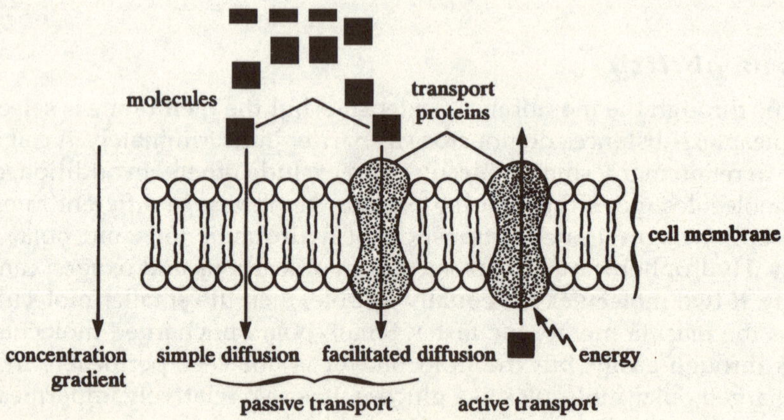

Movement Across Membranes

Osmosis

The process known as osmosis is the simple diffusion of water from a region of lower solute concentration to a region of higher solute concentration. If a membrane is impermeable to a particular solute, then water will flow across the membrane until the differences in the solute concentration have been equilibrated. Differences in the concentration of substances to which the membrane is impermeable affect the direction of osmosis.

Think of It This Way

It may help to visualize *diffusion* as the biological equivalent of a ball rolling down a hill. In *active transport*, on the other hand, energy must be expended—this process is the biological equivalent of pushing a ball up a hill.

Don't Mix These Up on Test Day

Passive transport:

- Moves with the gradient
- Requires no carrier
- Requires no energy

Facilitated diffusion:

- Moves with the gradient
- Requires a carrier
- Requires no energy

Active transport:

- Moves against the gradient
- Requires a carrier
- Requires energy

When the cytoplasm of the cell has a lower solute concentration than the extracellular medium, the medium is said to be *hypertonic* to the cell; water will flow out, causing the cell to shrink. On the other hand, when the cytoplasm of a cell has a higher solute concentration than the extracellular medium, the medium is *hypotonic* to the cell, and water will flow in, causing the cell to swell. If too much water flows in, the cell may lyse. Red blood cells, for example, lyse when put into distilled water. Finally, when solute concentrations are equal inside and outside, the cell and the medium are said to be *isotonic*. There is no net flow of water in either direction.

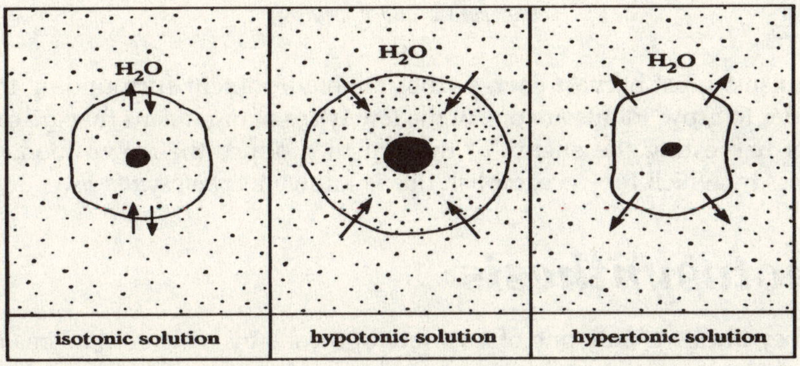

Osmosis

Don't Mix These Up on Test Day

In a *hypotonic* solution, a cell will swell until it looks like an O.

In a *hypertonic* solution, it will shrivel.

In an *isotonic* solution, it will remain the same size.

Active Transport

Active transport is the net movement of dissolved particles against their concentration gradient with the help of transport proteins. This process requires energy, and is necessary to maintain membrane potentials in specialized cells such as neurons.

Endocytosis/Exocytosis

Endocytosis is a process in which the cell membrane is invaginated, forming a vesicle that contains extracellular medium. Meanwhile, *pinocytosis* is the ingestion of liquids or small particles, while *phagocytosis* is the term assigned to the engulfing of large particles. In the latter, articles may first bind to receptors on the cell membrane before being engulfed.

These processes differ from *exocytosis*, which occurs when a vesicle with the cell fuses with the cell membrane and releases its contents to the outside. This fusion of the vesicle with the cell membrane can play an important role in cell growth and intercellular signaling. In both endocytosis and exocytosis, the material involved never actually crosses the cell membrane.

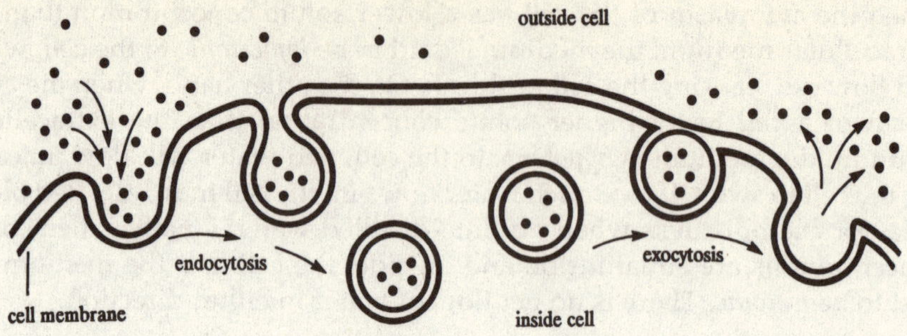

Endocytosis and Exocytosis

Organisms must harvest energy from the environment surrounding them in order to grow. Plants are one of the few types of organisms that are capable of harvesting the energy of the sun to produce their own food. The process by which they accomplish this is known as *photosynthesis*.

Photosynthesis

Photosynthesis is only one of a variety of means by which organisms harvest energy from the environment. Some simple bacteria, for example, use *chemosynthesis* to obtain energy for the manufacture of organic materials, oxidizing compounds of nitrogen, sulfur, or iron. The small amount of energy released by this oxidation is sufficient for the formation of glucose to support these bacteria—for example, nitrifying bacteria oxidize ammonia and nitrites to nitrates. Plants utilize these nitrates to make proteins, and the energy obtained from this oxidation to make glucose. This is in contrast to organisms known as *heterotrophs,* which must utilize other organisms as their energy source.

One of these energy sources is plants, which absorb light from the sun and use it to make the organic compounds that provide energy to other cells. Plants are *autotrophs,* meaning that they are self-feeders, sustaining themselves without utilizing other organisms. They are also *photoautotrophic,* using light as energy to synthesize carbohydrates, proteins, lipids, and other organic compounds. In this way, photosynthesis winds up either directly or indirectly nourishing almost all living things.

The actual process of photosynthesis occurs in the *chloroplast*, a highly organized plastid containing *chlorophyll* pigment. In many algae, the chlorophyll is distributed throughout the cytoplasm and there are no chloroplasts. Chlorophyll is a very complex molecule containing over 100 atoms. It is related to hemoglobin in that both contain metallic ions; hemoglobin, however, has iron at its center, while chlorophyll has magnesium.

Chloroplasts are found mainly in the cells of the mesophyll, the green tissue in the interior of the leaf. CO_2 enters and O_2 exits by way of the stomata, a pore in the leaf's surface. An envelope of two membranes bounds the stroma, the dense fluid within the chloroplasts. An elaborate system of thy-

Why Is It Green?

Chlorophyll is green because green is the only wavelength of light this pigment does NOT absorb. Instead, it reflects green light. The wavelengths of light principally absorbed and used by chlorophyll are in the red and blue regions of the visible spectrum.

lakoid membranes segregates the stroma from another compartment, the thylakoid space.

In some places, thylakoid sacs are layered in dense stacks called grana. Chlorophyll is found in the thylakoid membranes. Thylakoid functions to convert light energy, initially to chemical energy. The steps that convert chemical energy to allow the conversion of CO_2 into sugars occur in the stroma.

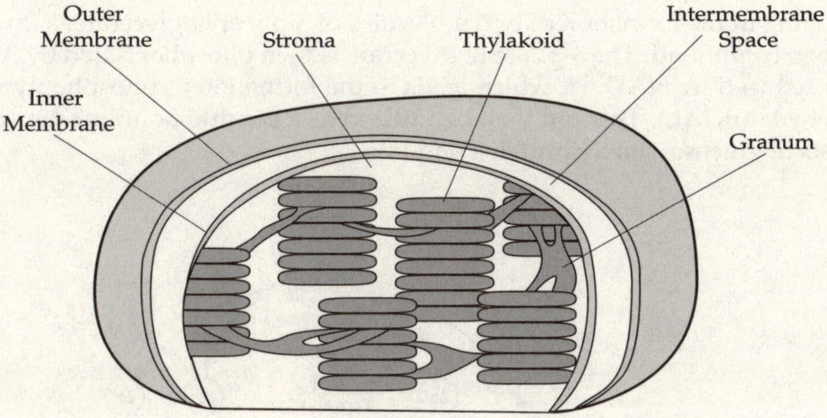

Chlorophyll

Photosynthesis can be summarized with this equation:

$$6CO_2 + 12H_2O + uv \text{ light} \longrightarrow C_6H_{12}O_6 + 6O_2 + 6H_2O$$

Photosynthesis involves the reduction of CO_2 to a carbohydrate. It can be characterized as the reverse of respiration, in that reduction occurs instead of oxidation. There are two steps in photosynthesis: light reactions and the Calvin cycle.

Light Reactions

When photons of light strike chlorophyll molecules, those molecules transfer the energy of the photons of light to their electrons. These electrons are raised to an "excited" higher energy level. The energy of these electrons is transferred to other components of the system and utilized in the following reactions:

- Breakdown of water (photolysis) into hydrogen ions and oxygen gas molecules:

$$H_2O \longrightarrow 2H^+ + \frac{1}{2} O_2$$

- ATP formation from ADP
- NADPH formation from NADP

Calvin Cycle

This cycle, also known as the "dark cycle," "fixes" CO_2 into organic material. It reduces the fixed carbon to carbohydrates through the addition of electrons. NADPH and ATP formed in the light reaction are utilized in the synthesis of carbohydrates in the dark reactions of the Calvin cycle. These light and dark reactions are fully separate. The NADPH provides the reducing power for the reduction of CO_2 to carbohydrate, and the air provides the carbon dioxide. CO_2 first combines with, or "is fixed to," ribulose, a five-carbon sugar. The resulting six-carbon compound is promptly split, resulting in the formation of two molecules of 3-phosphoglycerate, a three-carbon compound. The 3-phosphoglycerate is then phosphorylated by ATP and reduced by NADPH, which leads to the formation of phosphoglyceraldehyde (PGAL). This can then be utilized as a starting point for the synthesis of glucose, starch, proteins and fats.

Quick Quiz

What does the *Calvin cycle* do?

(A) It converts oxygen to energy.

(B) It transfers the energy of light photons to components of chlorophyll molecules.

(C) It "fixes" CO2 into organic material.

(Answer = (C))

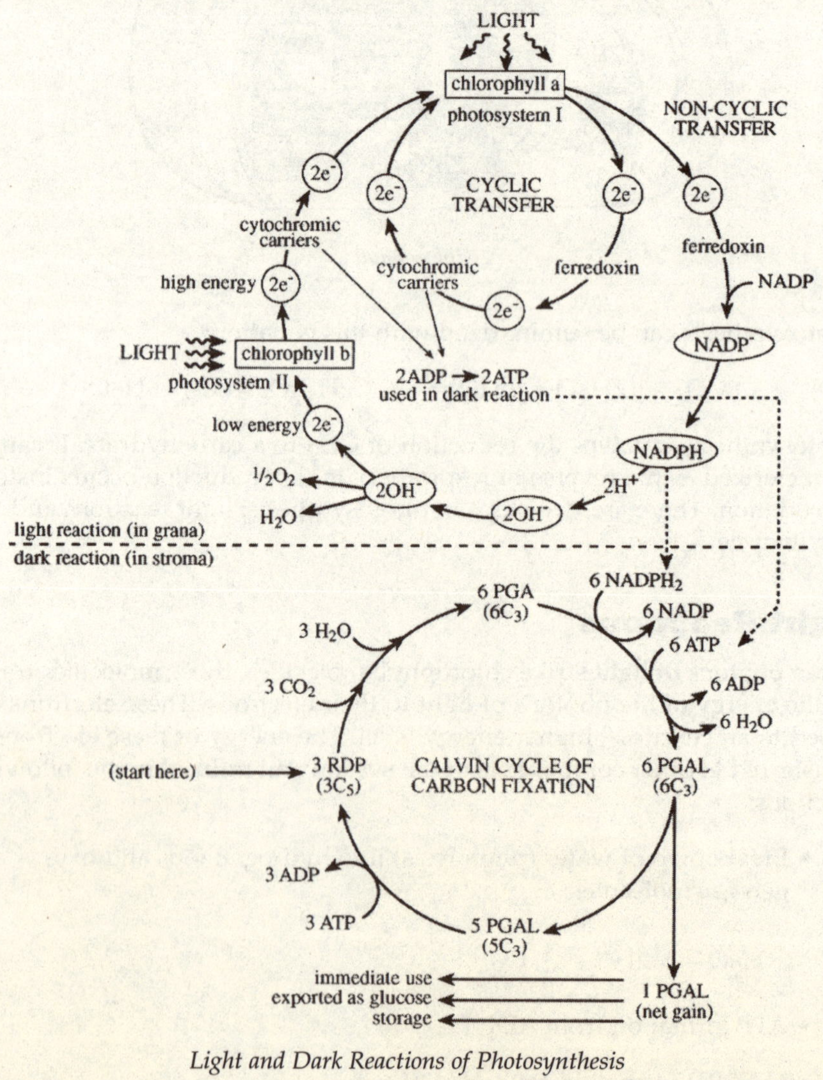

Light and Dark Reactions of Photosynthesis

The Leaf and Root Systems

The higher multicellular green plants have specialized organs, leaves, which are the principal sites of photosynthesis. In order to engage in photosynthesis, *leaves* have adapted in various ways.

First of all, the leaf has a *waxy cuticle* on top to conserve water. Its *upper epidermis*, the top layer of cells, has no openings, an adaptation which is also intended to inhibit water from being released. Another photosynthetic adaptation, the *palisade layer*, is the term given to elongated cells that are spread over a large surface area and contain chloroplasts. They are directly under the upper epidermis and are well exposed to light.

The leaf also possesses a *spongy layer*, where stomates open into air spaces that contact an internal moist surface of loosely packed spongy layer cells. Spongy cells contain chloroplasts. As in animals, this moist surface is necessary for diffusion of gases into and out of cells in both photosynthesis and respiration. Air spaces in leaves increase the surface area available for gas diffusion by the cells, allowing gaseous interchange of CO_2, H_2O, and O_2.

The lower epidermis of the leaf is punctuated by *stomate openings*, which further regulate the loss of water through transpiration and permit diffusion of carbon dioxide, water vapor, and oxygen between the leaf and the atmosphere. The size of these stomate openings is controlled by *guard cells*. These cells open during the day to admit CO_2 for photosynthesis and close at night to limit loss of water vapor through transpiration.

One explanation for the mechanism by which the guard cells open and close is as follows: During the day, the guard cells, which contain chloroplasts, produce glucose. High glucose content in the cells causes them to swell up via osmosis. This condition is known as *turgor*. Because the inner wall of the guard cell is thickened, the swelling produces a curvature of the opening between the guard cells, and the stomate opening increases. At night, photosynthesis ceases, cell turgor decreases, and the stomate opening closes. During a drought, the stomates will also close during the day to prevent loss of water by transpiration. In this case, photosynthesis ceases because of a lack of CO_2.

Finally, vascular bundles in plants serve as a circulatory system. *Veins* containing xylem and phloem bring water to the leaf from the roots (xylem) and carry manufactured food out of the leaf (phloem).

Solar Powered!

Plants take the energy of the sun and convert it into usable energy. Think of them as solar powered!

Mnemonic

Remember:

Xylems transport water, and **ph**loems transport **f**ood!

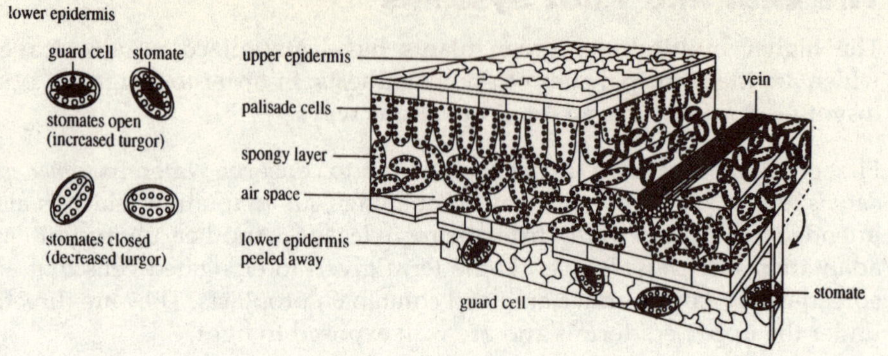

Leaf

The *root*, meanwhile, has specialized epidermal cells with thin-walled projections called root hairs. They provide increased surface area for absorption of water and minerals by diffusion and active transport.

Sugars for Energy

Only photosynthetic organisms can produce their own sugars. All other organisms must consume sugars to produce energy through cellular respiration.

Cellular Respiration

Organisms that cannot produce energy via photosynthesis rely on cellular respiration. In cellular respiration, food energy is transformed into energy that can be actively used by the cell in order to perform its various functions. To be more specific, glucose is broken down to release energy and to produce ATP, a high energy compound. This vital process is initiated with the series of reactions known as glycolysis. After this occurs, respiration can continue anaerobically (in the absence of oxygen) or aerobically (in the presence of oxygen).

Glycolysis

This is a series of reactions that leads to the oxidative breakdown of glucose into two molecules of pyruvate (the ionized form of pyruvic acid), the production of ATP, and the reduction of NAD^+ into NADH. This process occurs in the cytoplasm and is mediated by specific enzymes. For one molecule of glucose, two molecules of pyruvate are obtained. During this sequence of reactions, 2ATP are used and 4ATP are generated. Thus, net production amounts to 2ATP.

Glycolysis occurs in both prokaryotes and eukaryotes. It does not release much of the energy of glucose; at this stage, energy is still being held in high-energy C-H bonds. The chemical reaction occurs as follows:

$$\text{Glucose} + 2\text{NAD+} + 2\text{ATP} \longrightarrow 2\ \text{Pyruvate} + 2\text{NADH} + 4\text{ATP}$$

Pyruvate degradation can proceed in one of two directions: anaerobically, as fermentation, or aerobically, through the Krebs cycle followed by the electron transport chain.

Anaerobic Respiration

This process is also known as *fermentation*, a broad term covering the process by which NAD^+ is regenerated in cells in order to enable glycolysis to continue in the absence of O_2. Cells respond to high levels of NADH (which builds up when oxygen levels are low) by going into anaerobic respiration, which produces NAD^+ by reducing pyruvate into ethanol or lactic acid. Two additional ATP molecules are also gained during this process.

Certain organisms, such as yeast, practice *ethanol fermentation*; this is in fact the process by which beer and wine are produced. Humans, on the other hand, utilize *lactic acid fermentation* when oxygen levels are low in their bodies.

Ethanol Fermentation

Ethanol or alcohol fermentation commonly occurs only in yeast and in some bacteria. Pyruvate produced in glycolysis is decarboxylated to become acetaldehyde, which is then reduced by the NADH generated in glycolysis to yield ethanol. NAD^+ is regenerated, and glycolysis can continue.

Lactic Acid Fermentation

This type of fermentation takes place in certain fungi and bacteria and in human muscle cells during strenuous activity. When the oxygen supply to muscle cells lags behind the rate of glucose catabolism, the pyruvate generated is reduced to lactic acid. As in alcohol fermentation, the NAD^+ used in glycolysis must be regenerated when pyruvate is reduced.

In humans, lactic acid may accumulate in muscles during exercise, causing a decrease in blood pH that leads to muscle fatigue. Once the oxygen supply has been replenished, the lactic acid is oxidized back to pyruvate and enters cellular respiration. The amount of oxygen needed for this conversion is known as oxygen debt.

Ouch!

Lactic acid buildup, a product of lactic acid fermentation, is what makes your muscles sore after a good workout.

Aerobic Respiration

Aerobic respiration is far more efficient than anaerobic respiration. It harvests all the energy of glucose that has not yet been released through glycolysis, while anaerobic respiration releases only part of this energy. Aerobic respiratory processes include the Krebs cyle and the electron transport chain.

Krebs Cycle

The Krebs cycle, also known as the *citric acid cycle*, is an aerobic process that fully harvests the energy of glucose. It takes place in the mitochondrial matrix. This cycle begins when pyruvate is transported from the cytoplasm into the mitochondrial matrix, where it is decarboxylated. It loses a CO_2 molecule, and the acetyl group that remains is transferred to coenzyme A to form Acetyl CoA. Also, NAD^+ is reduced to NADH. This sequence of events is summarized in the following reaction:

$$\text{Pyruvate (3C)} + \text{Coenzyme A} \xrightarrow{NAD^+ \rightarrow NADH + H^+} \text{Acetyl CoA (2C)}$$

This cycle begins when the two-carbon acetyl group from acetyl CoA combines with oxaloacetate, a four-carbon molecule, to form the six-carbon citrate. Through a complicated series of reactions, two carbon dioxide molecules (CO_2) are released, and oxaloacetate is regenerated for use in another turn of the cycle. For each turn of the citric acid cycle, one ATP molecule is produced by substrate level phosphorylation. In addition, electrons are transferred to NAD^+ and FAD, generating NADH and $FADH_2$ respectively. These coenzymes then transport the electrons to the electron transport chain, where more ATP is produced via oxidative phosphorylation.

Studying the cycle, we can do some bookkeeping; keep in mind that for each molecule of glucose, two pyruvate molecules are decarboxylated and channeled into the citric acid cycle.

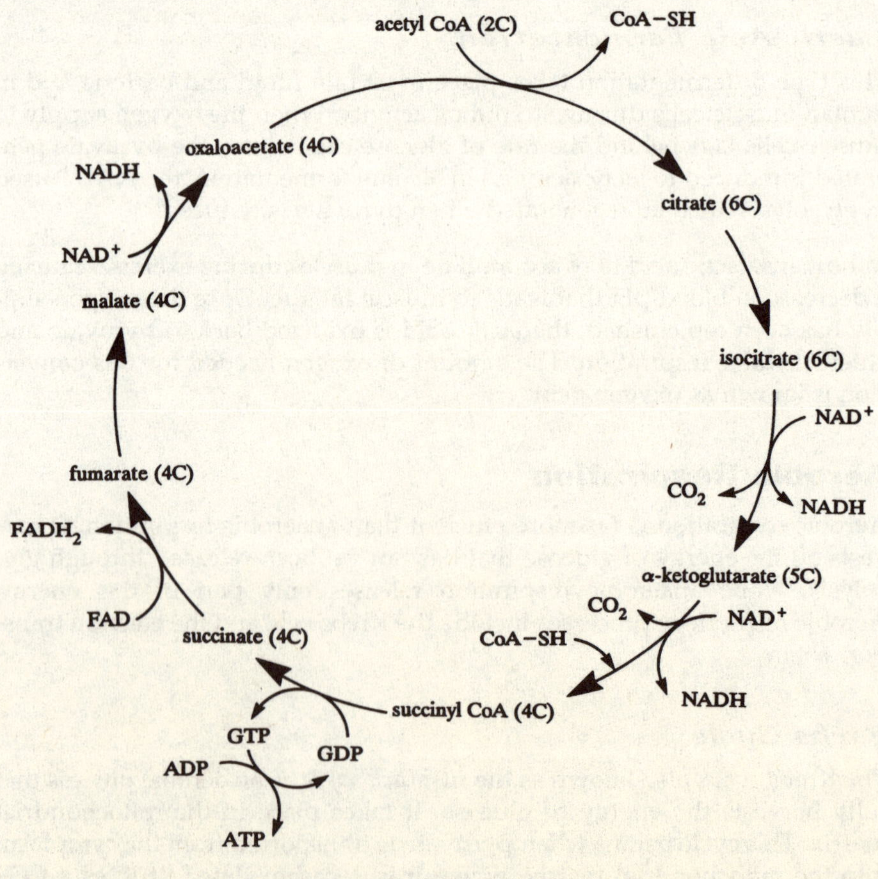

Krebs Cycle (Citric Acid Cycle)

A summary of the Krebs cycle reaction is provided below:

$$2 \times 3NADH \longrightarrow 6NADH$$
$$2 \times 1FADH_2 \longrightarrow 2FADH_2$$
$$2 \times 1ATP \longrightarrow 2ATP$$

The net reaction is:

$$2 \text{ Acetyl CoA} + 6NAD^+ + 2FAD + 4H_2O \longrightarrow$$
$$4CO_2 + 6NADH + 2FADH_2 + 2ATP + 4H^+ + 2CoA$$

Electron Transport Chain

The electron transport chain passes the hydrogen released by the Krebs cycle on to compounds called hydrogen acceptors, releasing energy in the process. It is a complex carrier mechanism located on the inside of the inner mitochondrial membrane. During oxidative phosphorylation, ATP is produced when high-energy potential electrons are transferred from NADH and $FADH_2$ to oxygen by a series of carrier molecules located in the inner mitochondrial membrane. As the electrons are transferred from carrier to carrier, free energy is released and then used to form ATP.

Most of the molecules of the electron transport chain are *cytochromes*, electron carriers that resemble hemoglobin in the structure of their active sites. The functional unit of the cytochrome contains a central iron atom, which is capable of undergoing a reversible redox reaction; that is, it can be alternatively reduced and oxidized.

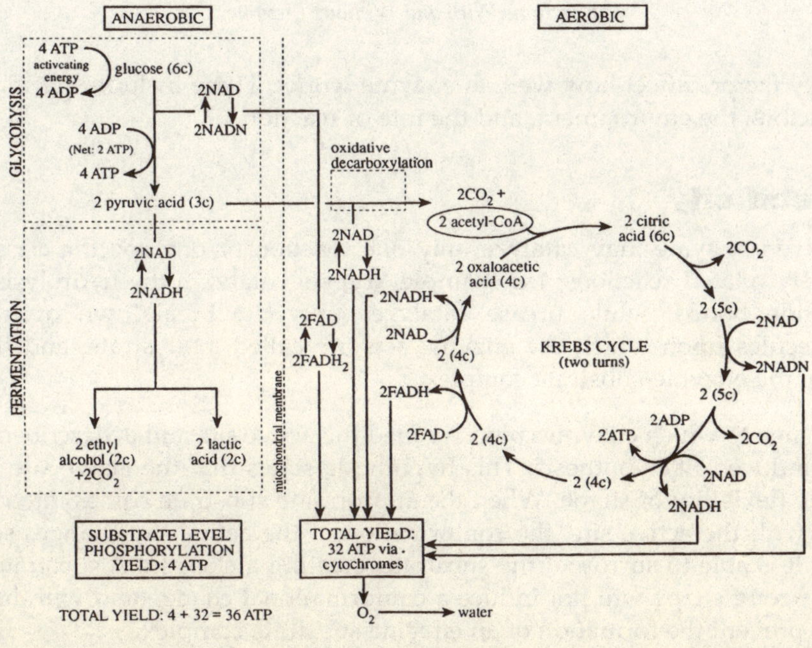

Total Metabolism

Don't Mix These Up on Test Day

Where Do They Take Place?

Glycolysis and fermentation take place in the cytoplasm.

The Krebs cycle takes place in the mitochondrial matrix.

The electron transport chain takes place in the inner mitochondrial membrane.

Study Tip

You don't have to memorize every detail of cellular respiration for the SAT II: Biology test, but you do need to be able to recognize the process.

In a Nutshell

Enzymes:

- Lower the activation rate of a reaction
- Increase the rate of the reaction
- Do not affect the overall energy change of the reaction
- Are not changed or consumed in the reaction

Enzymes

Enzymes *catalyze* and therefore accelerate reactions by reducing the activation energy necessary for these reactions to proceed. Reactions may occur without enzymes, although they do so at a much slower rate. Enzymes do not change equilibrium, but they do change the rate of a reaction. They are neither consumed nor changed, and the reactions they catalyze are usually reversible. The activation energy of a reaction with an enzyme is lower than an enzyme-free reaction, while its overall free energy is the same.

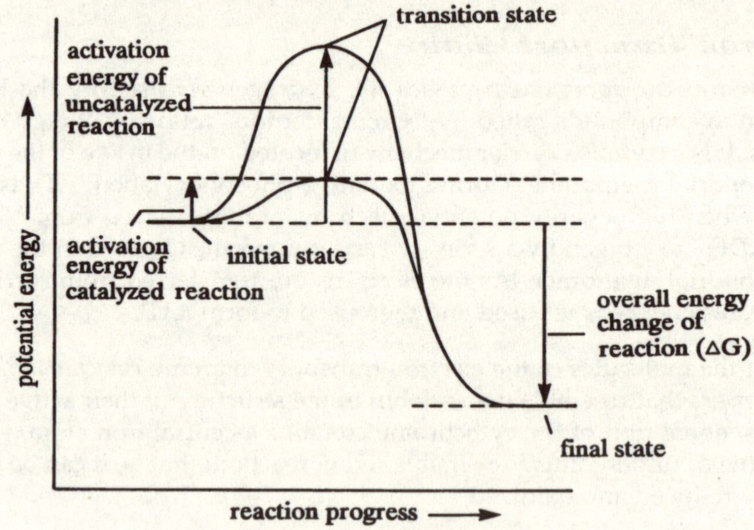

Reactions With and Without Enzymes

Many factors affect how well an enzyme works. These include specificity, cofactors, the environment, and the rate of reaction.

Specificity

A given enzyme may catalyze only one reaction or one specific class of closely related reactions; for example, trypsin catalyzes the hydrolysis of peptide bonds, while urease catalyzes only the breakdown of urea. Molecules upon which the enzyme acts are called a substrate, and they form the enzyme-substrate complex.

The area to which enzymes bind is called the active site and is described by the induced fit hypothesis. This hypothesis states that the active site has some flexibility of shape. When the appropriate substrate comes into contact with the active site, the conformation of the active site changes such that it is able to surround the substrate, creating a close fit. A substrate of the wrong shape will not induce a conformational change and will therefore prevent the formation of an enzyme substrate complex.

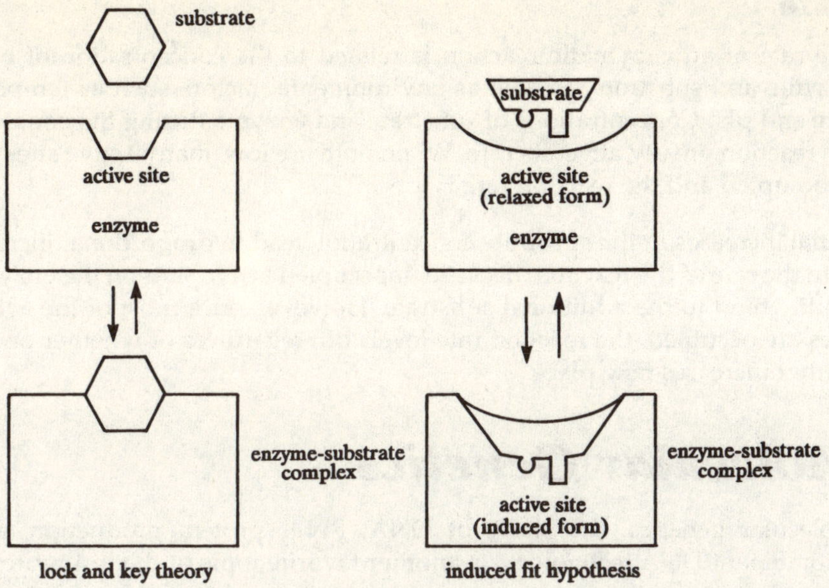

Models for Ensyme-Substrate Interactions

Cofactors

Many enzymes require incorporation of nonprotein molecules to become active. These cofactors can aid in binding or stabilizing the enzyme in an active conformation.

Make Sure You Get Your Vitamins

Many cofactors are vitamins. This is why vitamin deficiencies often cause severe diseases like scurvy. Scurvy, caused by a lack of Vitamin C, used to be commonplace on board ships, where no fresh food was available.

Environment

Rates of enzyme-catalyzed reactions tend to double for every 10°C increase in temperature until their optimal temperature is reached. For most enzymes in the human body, the optimal temperature is 37°C. At higher temperatures, enzymes become denatured as their three-dimensional structure is destroyed. Each enzyme also has an optimal pH, above and below which enzymatic activity declines. Human enzyme activity typically reaches its maximum level at around 7.2, the pH of the body. An exception is pepsin, which works best in the highly acidic conditions of the stomach.

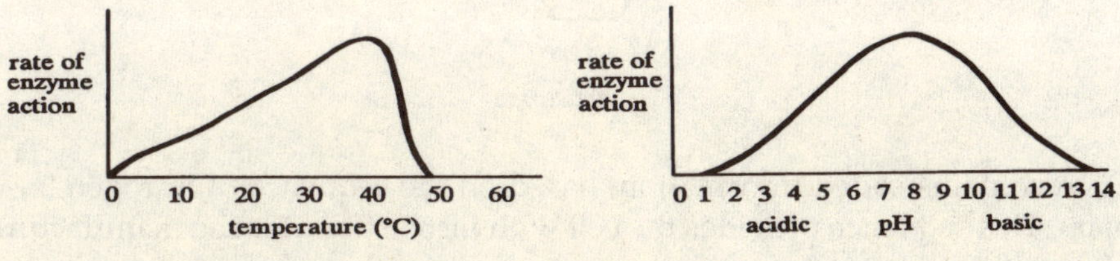

Effects of Temperature and pH on Enzyme Activity

Focus on Rate

The rate of an enzyme-related reaction may be changed by:

- Concentration
- Temperature
- pH

Rate

The rate of an enzymatic reaction is related to the concentration of both enzyme and substrate, as well as environmental factors such as temperature and pH. Concentrations of substrate and enzyme during the course of the reaction greatly affect its rate. When both are low, many active sites are unoccupied and the reaction rate is low.

Initial increases in the substrate concentration lead to proportional increases in the rate of the reaction, because unoccupied active sites on the enzyme readily bind to the additional substrate. However, once most of the active sites are occupied, the reaction rate levels off, regardless of whether or not further increases take place.

Molecular Genetics

Molecular genetics, the study of DNA, RNA, protein production, and recombinant DNA technology, is another favorite topic of the test writers.

DNA

DNA, or deoxyribonucleic acid, is an essential component of the cell. Found in the nucleus, it contains the genetic information of the organism. Its structure, characteristics, and most important functions are described in detail below.

Structure of DNA

The basic unit of DNA is the *nucleotide*, which is composed of deoxyribose (a sugar) bonded to both a phosphate group and a nitrogenous base. There are two types of bases: the double-ringed *purines* and the single-ringed *pyrimidines*. The purines in DNA are adenine (A) and guanine (G), and the pyrimidines are cytosine (C) and thymine (T).

Now You Know Who to Blame

DNA, the molecule of heritability, is what actually causes you to end up with your mother's nose or your father's ears.

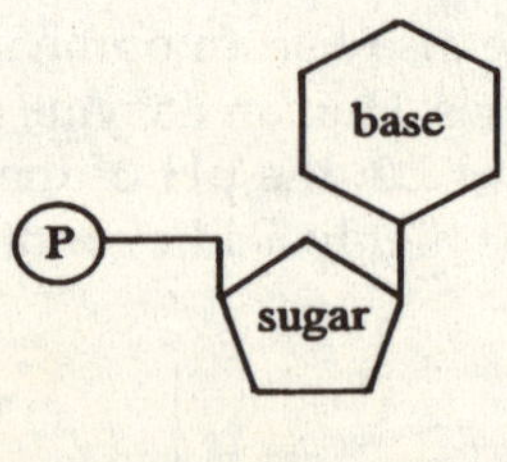

Nucleotide

DNA contains coded information, based on the sequence of nitrogen base pairs. This sequence provides the cell with instructions for the manufacture of specific proteins (particularly enzymes, but also many structural proteins), which in turn control the complete structural and functional characteristics of the cell.

DNA possesses the ability to self-replicate; in other words, DNA's informational molecule is capable of reproducing itself. This ability is key to the reproduction of cells and, indeed, entire organisms. Since the coded sequence of nitrogen bases determines the kinds of proteins found in organisms and thus their distinct nature, DNA is also the basis for heredity. Self-replication of the coded sequence allows the sequence to be passed on from generation to generation.

One of DNA's most significant properties is that it is mutable. Under certain conditions, a sequence of nitrogen base pairs (the code) may be altered (in what is known as a gene mutation). This changes the amino acid sequence and with it the nature of the proteins that are synthesized. The corresponding characteristics of the cells and the organism as a whole are altered. These changes continue through all succeeding molecular generations of DNA and generations of the organism. Such stable gene changes are what permit evolution to occur.

In a Nutshell

DNA:

- Is self-replicating
- Is mutable
- Provides blueprints for every protein humans and other organisms make

Watson-Crick Model

Scientists James D. Watson and Francis H. C. Crick described the arrangement of nucleotides in DNA in 1953. According to their model, DNA occurs as a double strand of thousands of nucleotides in the shape of a twisted ladder or double helix. The backbone or upright portions of this ladder are made up of alternating deoxyribose sugars and connecting phosphates. The nitrogen bases extend laterally from the sugar units of the backbone. Those from the opposing chains are linked in pairs by weak hydrogen bonds, forming the "rungs" of the ladder.

Nitrogen bases can pair in two ways: guanine-cytosine and thymine-adenine. There are no other possible combinations in normal DNA. These combinations form the basis for replication and utilization of the genetic code.

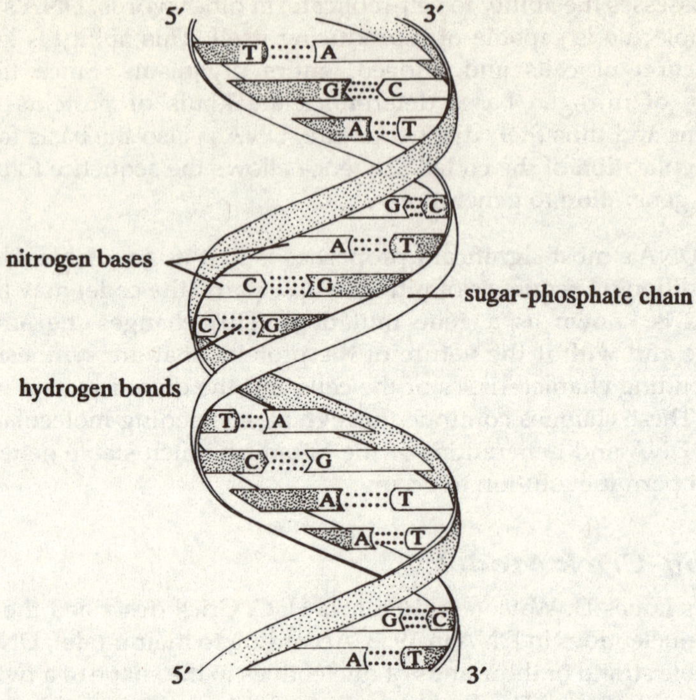

DNA Molecule

DNA Replication

DNA must be replicated in order to pass on its vital instructions for the organism. A double-stranded DNA molecule splits into two separate strands between the base pairs by breaking hydrogen bonds. The free nitrogen-base ends select complementary free nucleotides, according to the base-pairing rules (T-A and G-C), to form two double strands. This replication is *semiconservative*, as half of the original DNA is present in each of the new or daughter double-stranded DNA molecules.

Snowball Effect

Mutations in the DNA repair mechanism enzymes lead to a much higher mutation rate in the organism as a whole.

DNA Repair

In order to faithfully maintain the genetic information encoded in the DNA, a mechanism for repair and proofreading is needed. DNA is constantly exposed to mutagenizing compounds such as *uv* light, radiation, or chemicals. Scientists have identified more than 50 different types of DNA repair enzymes. Sometimes the damage can be directly reversed, although at other times the repair must take advantage of the base pairing of DNA to correct its mistakes.

The Genetic Code

The genetic code is a four-letter code made up of the DNA nitrogen bases A, T, G, and C. Each chromosome is a long DNA molecule with thousands of these bases as letters. Each group of three bases along the molecule specifies (codes for) a particular amino acid. For example, GGC codes for the amino acid we know as alanine, AAC codes for valine, and so forth. Sixty-four three-letter combinations may form from four letters ($4 \times 4 \times 4$).

Since there are 64 code names for the 20 amino acids present in the cytoplasm, there is room for synonyms—with several three-letter combinations specifying the same amino acid. This is known as a degenerate genetic code. Read on for a description of the mechanism by which the coded messages leave the nucleus and specified amino acids are combined at the ribosomes in an orderly pattern to synthesize protein.

RNA

Like DNA, RNA (ribonucleic acid) is a chain of nucleotides. However, it is usually single stranded; the sugar is ribose, and the four bases are adenine, guanine, cytosine, and uracil. RNA carries the messages dictated by the DNA to the cytoplasm. The chart below details the major differences between DNA and RNA:

DNA-Unique Features	RNA-Unique Features
Double stranded except when replicating	Nearly always single stranded
Deoxyribose sugar in the nucleotides	Ribose sugar in nucleotides
Thymine base forms a thymine-adenine base pair (T-A)	Uracil base instead of thymine. The base pair is uracil-adenine (U-A)
Capable of self-replication	Does not normally replicate (except in the case of some viruses)
Only one type of DNA per organism. This DNA acts as the original source of information, acting like a master record. Its information is copied onto RNA molecules.	Three types of RNA (mRNA, tRNA, rRNA)

Translator

RNA acts as a "translator," decoding the protein messages found in DNA.

Types of RNA

RNA takes information from the DNA and allows the cell to decode it. All RNA molecules are involved in some aspect of information carrying for specific protein synthesis. All of the three types of RNA are single stranded and have a uracil, as opposed to a thymine, base. None are capable of replication.

Don't Mix These Up on Test Day

There are three types of RNA, each with its own function:

- *mRNA*, or messenger RNA, carries messages.
- *tRNA*, or transfer RNA, carries amino acids.
- *rRNA*, or ribosomal RNA, is a structural component of ribosomes.

mRNA. mRNA, or messenger RNA, carries codes from DNA in the nucleus to ribosomes in the cytoplasm. It is manufactured at the chromosome from a DNA template in much the same way as DNA is replicated. The mRNA is assembled from individual ribonucleotide units that line up according to the base-pairing rules, along one of the strands (the "sense strand") of the DNA double helix. The result is a long, filamentous strand of mRNA that is complementary to the parent DNA strand for a given nucleotide sequence (unlike during replication, the entire DNA strand is not represented in the mRNA product).

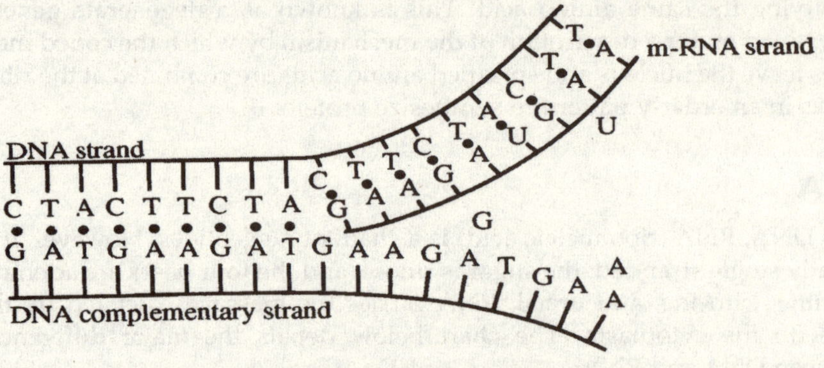

Messenger RNA (mRNA)

Observe that the DNA code sequence acts as a template along which specific molecules of ribonucleotides attach themselves in the correct sequence to form the polynucleotide mRNA. The specific DNA code is imprinted on the mRNA in the form of corresponding or complementary nitrogen bases, comparable to the way a photographic negative shows light areas as dark areas (negative).

The mRNA has the "inverted" complementary, or negative, codes of the original master on DNA. For example, since the DNA code for the amino acid valine is AAC, the mRNA for valine is the complementary UUG. These negative codes on mRNA are the actual working blueprints for protein synthesis.

Once the mRNA has received the message (coded bases), it separates from the DNA strand and leaves the nucleus. Each molecule of mRNA becomes associated with one or more spherically shaped ribosomes, found attached to the walls of the endoplasmic reticulum (RER) or floating freely in the cytoplasm. The mRNA is now ready to receive the amino acids brought to it by tRNA.

tRNA. tRNA, or transfer RNA, is a smaller type of RNA, functioning as a carrier of amino acid molecules. Since there are 20 kinds of amino acids in the cytoplasm, there are at least 20 different kinds of tRNA, all in the cytoplasm. Actually, there are nearly 64 types of tRNA, one for each amino acid triplet (excepting "nonsense" sequences).

Like mRNA, tRNA is coded for by DNA, but, unlike mRNA, it is a comparatively short ribonucleotide polymer instead of a long filament. For this reason, it has a significantly smaller number of nitrogen bases. The tRNA molecule also has two ends. One end contains a three-nucleotide sequence, the *anticodon*. This is complementary to the codon in mRNA regions specifying particular amino acids. The other end is specialized for carrying the corresponding amino acid.

rRNA. rRNA, or ribosomal RNA, is a structural component of the ribosomes and is associated with numerous proteins in the ribosome. It is produced in a region of the nucleus known as the nucleolus.

Protein Synthesis

Combinations of amino acids build up complex proteins in a process known as protein synthesis. Protein synthesis can be divided into three phases. First of all, mRNA is produced during transcription; then post-transcriptional RNA processing occurs, followed by a number of stages that can be grouped together under the term translation.

Transcription

The first stage of protein synthesis is termed *transcription*. This is the process by which information coded in the base sequence of DNA is transcribed into a strand of mRNA. The strand of mRNA is synthesized from a DNA template in a process similar to DNA replication. The DNA helix unwinds at the point of transcription, and synthesis occurs using only one DNA strand as a template. The base-pairing rules are the same as for DNA, with U replacing T (G bonds with C; A bonds with U).

mRNA is synthesized by the enzyme *RNA polymerase*, which must bind to sites on the DNA called promoters in order for RNA synthesis to begin. Synthesis continues until the polymerase encounters a termination sequence, which signals RNA polymerase to stop transcription, thus allowing the DNA helix to reform. The mRNA strand is then processed and leaves the nucleus through nuclear pores.

Post-Transcriptional RNA Processing

Most eukaryotic DNA does not code for proteins; noncoding or garbage sequences are found between coding sequences. A typical gene consists of several coding sequences called exons. These are interrupted by noncoding sequences called introns. During processing, the introns are cleaved and removed, while the exons are spliced to form an mRNA molecule coding for a single polypeptide. Processing occurs within the nucleus, and is also necessary for tRNA and rRNA production. As this process is concluded, mRNA moves out of the nucleus through pores in the nuclear membrane termed nuclear pores.

Link to Genetics

Protein synthesis is linked to genetics. You can think of DNA as being responsible for the genotype of the organism, while the proteins produced in protein synthesis are responsible for the organism's phenotype.

Quick Quiz

Match the numbered stages of protein synthesis with the correct lettered descriptions below.

1. post-transcriptional RNA processing
2. translation
3. transcription

(A) the stage during which mRNA is produced and the information coded in the base sequence of DNA is transcribed into a strand of mRNA

(B) the stage during which mRNA, tRNA, and rRNA are processed

(C) the final stage of protein synthesis,during which the genetic code of nucleotide sequences is translated into a sequence of amino acids

Answers:
1. = (B)
2. = (C)
3. = (A)

Translation

The remaining stages of protein synthesis occur in the cytoplasm and are called *translation*, since the genetic code of nucleotide sequences is translated into a sequence of amino acids. The mRNA attaches itself along the surface of one or more ribosomes with the help of the rRNA.

The cytoplasm contains tRNA units. Both ends of the tRNA molecule play critical roles in protein synthesis. One end, the carrier end, picks up (with the aid of an enzyme and ATP) an amino acid from the cytoplasm. This amino acid is attached to a specified tRNA by an aminoacyl tRNA synthetase that is highly specific for both the amino acid and that tRNA. It is not yet known exactly how these enzymes recognize specific tRNAs, although the nucleotide sequence anticodon at the nucleotide end, not the carrier end, is probably influential.

The carrier end for all the tRNA molecules consists of a CCA sequence. The recognition end has an exposed nitrogen-base triplet that represents the positive (DNA-like) code for that particular amino acid. For example, the tRNA carrying the triplet code CUU at its recognition end will have the nitrogen-base structure at the carrier end that enables it to pick up the amino acid glutamic acid.

With the amino acid attached, the tRNA is ready to move to a ribosome. When the ribosome detects a particular triplet on the mRNA, it is receptive to the particular tRNA molecule that carries the appropriate amino acid. Thus, if the ribosome detects a GAA on the mRNA, it will call for the tRNA associated with GAA's complementary triplet CUU, which carries glutamic acid.

The positive CUU recognition triplet of tRNA then forms a temporary bond with the complementary triplet GAA on the mRNA. This complementary base-pair recognition results in the glutamic acid being deposited at that point in the growing amino acid sequence. Note that in the diagram below, glutamic acid is now in close proximity with aspartic acid, an amino acid previously deposited on the ribosome as governed by its tRNA carrier, coded AUC. Specific enzymes link the correctly stationed glutamic acid and aspartic acid molecules in a dehydration-synthesis reaction to form a peptide bond (and thus a dipeptide). The "uncharged" tRNA with the aspartic amino acid code then returns to the cytoplasm to pick up more molecules of aspartic acid.

The ribosome then proceeds to the next triplet on the mRNA. The tRNA with the appropriate amino acid at the carrier end (glycine, in the diagram below) forms a temporary bond with the mRNA; again, the two amino acids unite in a dehydration synthesis reaction, and a longer polypeptide forms. At the same time, the CUU tRNA breaks away from both the mRNA and its amino acid (glutamic acid), and floats off into the cytoplasm.

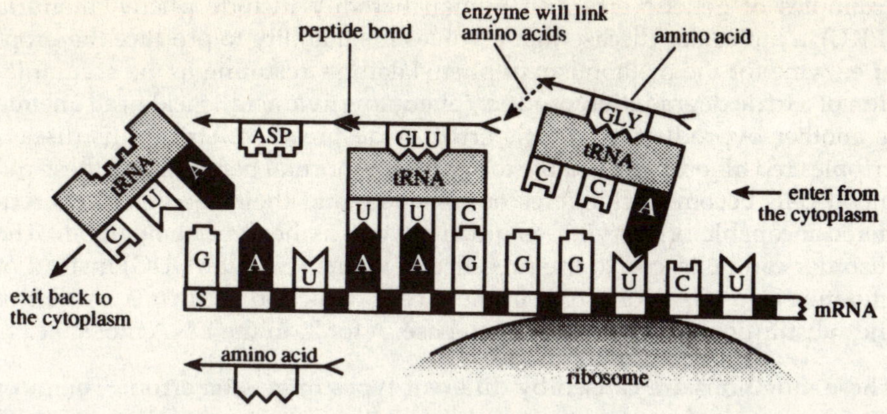

Protein Synthesis

The ribosome moves over again, detects another nitrogen triplet on the mRNA, and searches for another appropriate tRNA with its amino acid (glycine, for example). The chain of amino acids gets longer and longer. When the ribosome reaches the end of an mRNA filament, the polypeptide chain has been completed, built according to gene-determined instructions. It then floats off into the cytoplasm to do its job as a functional enzyme or structural protein. Frequently, there are many ribosomes that associate with a single mRNA molecule to form a structure known as a polysome; synthesis of the same protein occurs simultaneously at each of the ribosomes.

Following the release of the protein from the ribosome, the protein immediately assumes its characteristic native conformation. This conformation is determined by the primary sequence of amino acids. Furthermore, the polypeptide chains can form intramolecular and intermolecular cross-bridges with disulfide bonds. The result is a complex, intertwined, functional protein.

Sometimes protein synthesis malfunctions. DNA molecules do not always replicate perfectly. Errors may be introduced, or environmental mutagens can cause changes in the DNA sequence. These *mutations* are then reflected in altered protein formations in the resulting molecules.

Mutations

In a gene mutation, nitrogen bases are added, deleted, or substituted, in the process creating different genes; inappropriate amino acids are inserted into polypeptide chains, and a mutated protein is produced. Hence a mutation is a genetic "error," with the wrong base or no base on the DNA at the particular position in question.

Examples of genetic errors in human heredity include phenylketonuria (PKU), a molecular disease that involves the inability to produce the proper enzyme for the metabolism of phenylalanine, resulting in the accumulation of a toxic degradation product (phenylpyruvic acid). Sickle-cell anemia is another byproduct of genetic error. Widespread in Africa, this disease cripples red blood cells. Unable to synthesize normal hemoglobin, these red blood cells become sickle (crescent) shaped, and their hemoglobin is rendered incapable of carrying as much oxygen as healthy hemoglobin. The disorder can be traced to the presence of valine (GUA or GUG) instead of glutamic acid (GAA or GAG) in these cells, which in its turn is a result of the substitution of just one nitrogen base, A for T, in the DNA molecule.

These mutations are caused by different types of genetic errors. The most common types of mutations are point mutations and frameshift mutations.

Point Mutations

A point mutation occurs when a single nucleotide base is substituted by another nucleotide base. If the substitution occurs in a noncoding region, or if the substitution is transcribed into a codon that codes for the same amino acid, there will be no change in the amino acid sequence (a "silent" mutation). If the substitution changes the sequence, the result can range from insignificant to lethal, depending on the effect the substitution has on the protein.

Frame Shift Mutations

Base-pair insertions involve the addition of nucleotides, while base-pair deletions involve a loss of these nucleotides. Such mutations usually have serious effects on the protein coded for, since nucleotides are read as a series of triplets. The addition or loss of nucleotide(s) (except in multiples of three) will change the reading frame of the mRNA, and is known as a *frameshift mutation*. The proteins that are formed in the wake of such mutations, if synthesized at all, will most likely be nonfunctional.

Recombinant DNA Technology

Recombinant DNA technology allows for the manipulation of genetic material. This technology has enabled scientists to research the mutations associated with Alzheimers and cystic fibrosis—and has even led to the cloning of sheep.

Hello, Dolly

Recombinant DNA technology forms the scientific basis for cloning, as in the recent cloning of Dolly the sheep.

Some tools that have proven particularly effective at isolating and expressing genes of interest include:

- *Restriction enzymes:* These cut DNA at specific nucleotide sequences.

- *Vectors:* The most common type of vector is a circular DNA strand called a plasmid that can be manipulated to carry particular genes. It can be inserted into either prokaryotic or eukaryotic cells to alter the genetic expression of those cells. Other types of vectors include genetically engineered viruses, such as adenovirus and vaccinia.

- *Sequencing:* This utilizes restriction enzymes to cut the long DNA into shorter, more manageable pieces. The DNA is then analyzed and run on an electrophoretic gel; this gel will separate strands of DNA differing in length by only a single nucleotide. Through this method, the exact sequence of the DNA can be determined.

All of these techniques are currently being implemented in the Human Genome Project, which is attempting to map the entire human genome. This concludes our review of molecular genetics.

Biological Chemistry

You are made up of many different elements; your skin is not identical to your hair, which in its turn is vastly different from your lungs and neurons. Yet each of these parts of your body are made up of the same building blocks.

Living things are made up mostly of carbon, hydrogen, oxygen, nitrogen, sulfur, and phosphorous. There are also trace amounts of magnesium, iodine, iron, calcium, and other minerals in the elements of protoplasm, the substance of life.

Compounds are elements united by chemical bonds, such as H_2O, CO_2, and $C_6H_{12}O_6$, and atoms are the units of elements. Molecules are the units of compounds, such that two or more atoms combined make up a molecule.

The chemical compounds in living matter can be classified as inorganic compounds that do not contain carbon, such as NaCl or HCl, or as bio-organic compounds that are part of or are made by living things, and which contain carbon. The principle bio-organic compounds are carbohydrates, lipids, proteins, and nucleic acids.

Carbohydrates (Sugars and Starches)

Carbohydrates contain carbon, hydrogen, and oxygen, and are characterized by the way in which they maintain a hydrogen-to-oxygen ratio of 2:1. Examples of carbohydrates are glucose (a monosaccharide), maltose and sucrose (disaccharides), and glycogen, starch, and cellulose (polysaccharides that are very insoluble in aqueous solutions).

Power Food

Carbohydrates have 4 kcal of energy per gram. This is why athletes stock up on carbohydrates before a big competition—and why normal people who don't burn much energy should avoid eating too many of them.

Structure

Monosaccharides, such as glucose, have a molecular formula of $C_6H_{12}O_6$ and a hexagonal structure consisting of six carbons, with an H and an OH bound to each carbon.

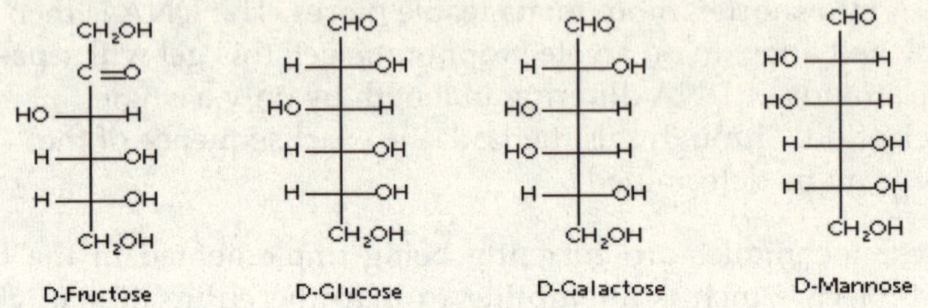

Monosaccharides

Disaccharides (for example, maltose) consist of two glucose molecules put together through dehydration synthesis whereby the removal of water forms a bond between the two molecules.

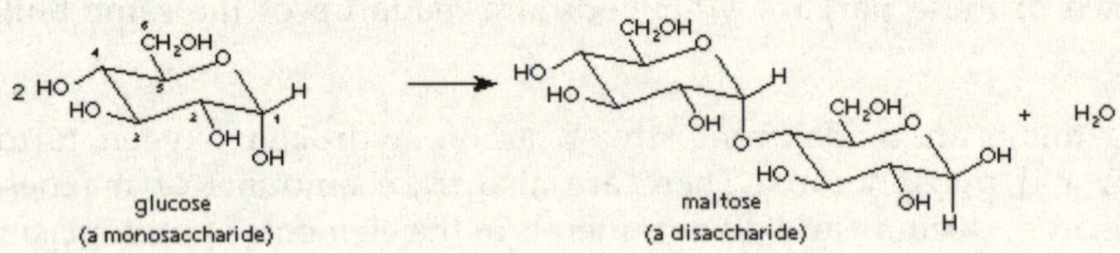

Disaccharides

Polysaccharides like starch, cellulose, and glycogen are chains of repeating monosaccharides of a particular type, usually glucose. These molecules are too large to be soluble in water and do not pass through membranes.

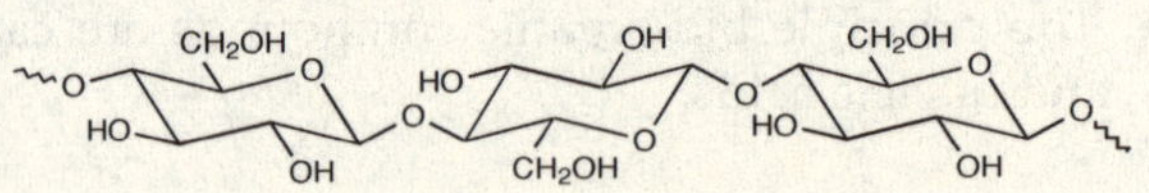

Cellulose, a 1,4'-β-D-Glucose polymer

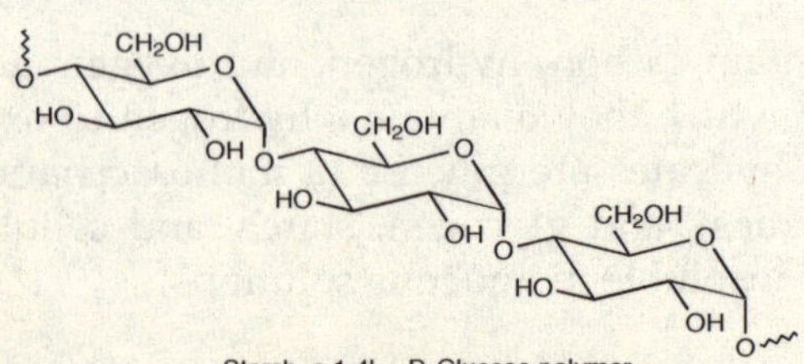

Starch, a 1,4'-α-D-Glucose polymer

Polysaccharides

Dehydration and Hydrolysis

By removing water through dehydration, smaller units can form an O bond between themselves to synthesize a large polymer. By adding water through hydrolysis, large polymers can break down to reform the smaller molecules.

High Energy

Lipids pack even more energy than carbohydrates—9 kcal of energy per gram.

Lipids (Fats and Oils)

Characteristics

Lipids are also made up of carbon, hydrogen, and oxygen; however, their hydrogen-to-oxygen ratio is much greater than carbohydrates' 2:1. Lipids may be characterized as three fatty acid molecules bonded to one glycerol molecule.

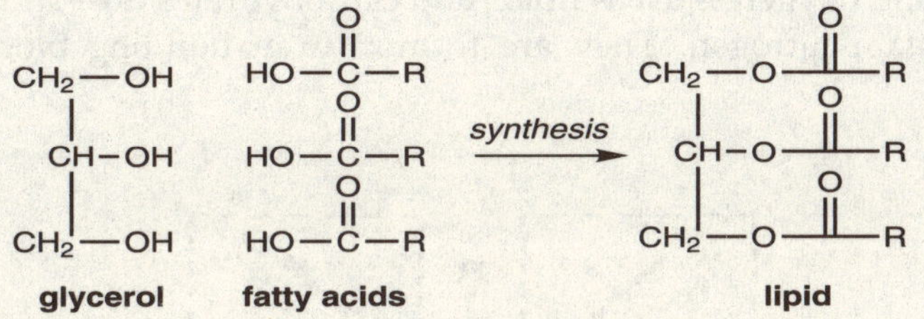

Synthesis of Lipids

Three dehydration reactions are needed to form one fat molecule. Lipids, unlike carbohydrates, do not form polymers.

Lipid Derivatives

Lipids are the chief means of food storage in animals, since lipids release more energy per gram weight than any other class of biological compounds. And as major components of fatty adipose tissue, they also provide insulation and protection against injury.

Phospholipids contain glycerol, two fatty acids, a phosphate group, and a nitrogen-containing alcohol like lecithin or cephalin. Lecithin is a major constituent of cell membranes, and cephalin is found in brain, nerves, and neural tissue.

Waxes are esters of fatty acids and alcohols. They are found as protective coatings on skin, fur, and leaves of higher plants, and on the cuticle of the exoskeleton of many insects in the form of lanolin. Meanwhile, steroids have three fused cyclohexane rings and one fused cyclopentane ring. Examples of this lipid derivative include cholesterol, sex hormones, and corticosteroids.

The fatty, acidlike carbon chains termed carotenoids contain conjugated double bonds and carry six-membered carbon rings at each end. As pigments, they produce red, yellow, orange, and brown colors in plants and animals. Two carotenoid subgroups are the carotenes and the xanthophylls.

Finally, porphyrins are also known as tetrapyrroles and contain four pyrrole rings joined together. They are often complexed around a metal. Some examples include heme and cytochromes, which utilize iron, and chlorophyll, which utilizes magnesium.

Don't Skimp on Proteins

Proteins are vital components of the cell, and are involved in almost all of its functions. They are a necessary part of every healthy diet. Like carbohydrates, proteins yield 4 kcal of energy per gram.

Proteins

Characteristics

Proteins make up 50 percent of the dry weight of a cell and are vital to almost everything a cell does. Proteins contain the elements of carbon, hydrogen, and oxygen—just as lipids and carbohydrates do—and they also always contain nitrogen. They are formed from building blocks called *amino acids*.

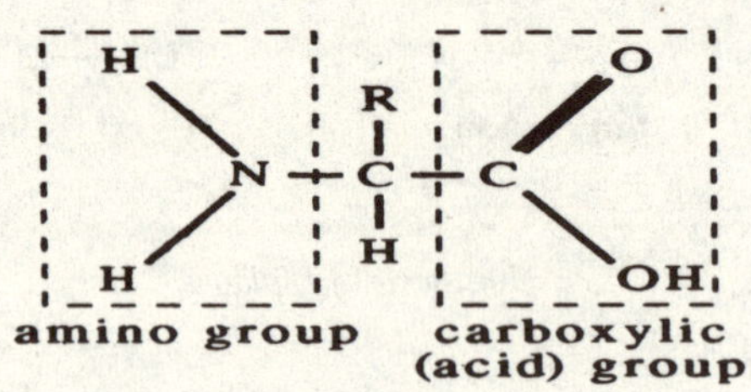

Amino Acid

Amino acids unite through dehydration synthesis to form peptide (C-N) bonds. Chains of such bonds produce a chain of amino acids called a *polypeptide*, another polymer.

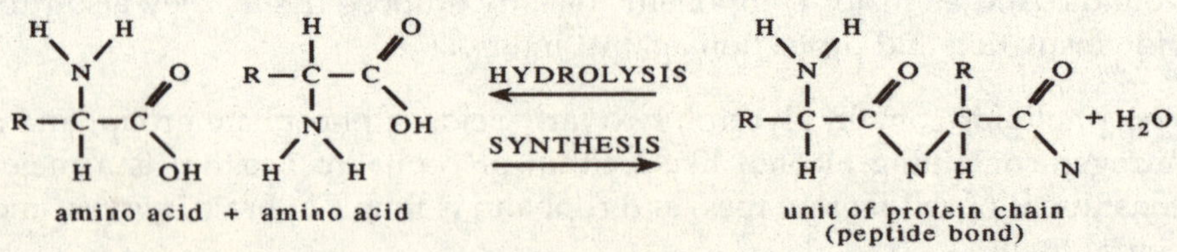

Polypeptide

Classification

Proteins can be classified on the basis of structure. Simple proteins consist entirely of amino acids. Albumins and globulins are primarily globular in nature and are typically functional proteins, such as carriers or enzymes.

Meanwhile, schleroproteins are fibrous in nature and are structural proteins. One example is collagen.

Conjugated proteins, on the other hand, contain a simple protein portion, plus at least one nonprotein fraction. Lipoproteins are bound to a lipid; mucoproteins are bound to a carbohydrate. The proteins that are bound to pigmented molecules are known as chromoproteins. Finally, metalloproteins complex around a metal ion, while nucleoproteins like histones are bound to nucleic acids.

Functions

Various types of proteins and their functions are listed in the table below.

Type of Protein	Function	Examples
Hormonal	Chemical messengers	Insulin, glucagon
Transport	Transport of other substances	Hemoglobin, carrier proteins
Structural	Physical support	Collagen
Contractile	Movement	Actin/myosin
Antibodies	Defense	Immunoglobulins, interferons
Enzymes	Biological catalysts	Amylase, lipase, ATPase

Levels of Structure

Proteins are structured in such a way that different relationships are formed between the original, unique sequence of amino acids (known as the *primary level of structure*) and more complex, three-dimensional compounds. This basic primary level of structure is built upon to form secondary, tertiary, and quaternary levels of structure.

The *secondary level of structure* repeatedly coils or folds segments of the chain in patterns that contribute to the overall conformation of the protein. Secondary structure is caused by hydrogen bonds at regular intervals along the polypeptide backbone; alpha helices and beta pleated sheets are numbered among the most common types of these structures.

The *tertiary level of structure*, meanwhile, can be characterized as irregular contortions due to bonding between side chains of the various amino acids. Tertiary structure can be reinforced by strong covalent bonds, called disulfide bridges, between cysteines. It can also be modular, with two or more globular regions called domains connected by relatively flexible regions of the polypeptide chain.

In a Nutshell

Protein Levels of Structure:

- *Primary*: a linear sequence of amino acids
- *Secondary*: coils or folds of the protein
- *Tertiary*: the interactions between different areas of the protein
- *Quaternary*: the interactions of one protein with another

The final *quaternary level of structure* of the protein results from the relationship between the different polypeptides, called subunits. For example, some proteins, especially cell surface receptors, are made up of two proteins that interact through quaternary structures to produce specific macromolecules.

Functional Groups

Functional groups are attached to carbon skeletons and give compounds their functionality. These regions are commonly involved in chemical reactions.

- *Hydroxyl (OH).* Hydroxyl groups are contained in polar compounds known as alcohols, such as ethanol.
- *Carbonyl (C=O).* Carbonyl groups are polar groups contained in aldehydes and ketones, including formaldehyde.
- *Carboxyl (COOH).* Carboxylic groups are polar groups contained in carboxylic acids. They lose their H^+ ions to form acids like acetic acid (vinegar).
- *Amino (NH_2).* These are polar groups contained in amines, such as methylamine.
- *Sulfhydryl (SH).* Sulfhydryl groups help stabilize the structure of proteins. They are contained in thiols, such as mercaptoethanol.
- *Phosphates (PO_4^{2-}).* Found in organic phosphates like glycerol phosphate, these groups store energy that can be passed from one molecule to another by the transfer of a phosphate group.

Water

Like the earth's surface, your body is mostly water. Cells contain about 70–95 percent water. The molecule as a whole is neutral, but it does have partial positive and negative charges, making it polar.

One of a Kind

Water is the only compound that exists in the earth's natural environment as a solid, a liquid, and a gas.

Special Properties of Water

Water possesses a number of distinctive properties. It is cohesive, implying that due to hydrogen bonding, its molecules stick together. It is also characterized by surface tension. At the interface between water and air, there is an ordered arrangement of water molecules; hydrogen molecules are bonded to one another and to the water below to make the water act as though it were coated with a film.

Water's high specific heat means that a slight change in water's temperature is accompanied by the absorption or release of a relatively large amount of heat. And it also possesses a high heat of vaporization, the quantity of heat a liquid must absorb for one gram of it to be converted from the liquid to the gaseous state. Hydrogen bonding contributes to this propensity by restraining molecules; much heat is required to break the hydrogen bonds.

Another property that can be traced to hydrogen bonding is water's ability to expand as it freezes. Water is one of the few substances that is less dense as a solid than it is a liquid. Finally, charged regions of polar water molecules are attracted to ions and polar compounds, making water an effective medium for complex chemical reactions in organisms.

Aqueous Solutions

Most chemical reactions occur in water with solutes. A number of factors are important in these reactions: solute concentration, acids, bases, and pH.

Solute Concentration. A mole is the number of grams of a substance that equals its molecular weight in daltons. Molecular weight is the sum of the weights of all the atoms in a molecule; thus, the molecular weight of NaCl is 28. Therefore, to obtain one mole of NaCl, 28 grams must be weighed out. One mole of one substance has the same number of molecules as a mole of another substance. To make a one-liter solution with one mole of NaCl, we would weigh out 28 grams of NaCl and completely dissolve it in water. Then we would bring the total volume to one liter and this would be a one-molar (1M) solution of NaCl.

Acids, Bases, and pH. A hydrogen atom shared between two water molecules shifts from one molecule to the other. The hydrogen atom leaves its electron behind and is now a hydrogen ion, a single proton with a charge of 1^+. The water molecule that lost this ion is a hydroxide ion with a charge of –1 (OH^-). The protons binds to the second water molecule at one of its unshared orbitals, making that molecule a hydronium ion (H^{30+}). You can think of this as a dissociation of water into hydrogen and hydroxide ions.

Equation: $H_20 \rightarrow H^+ + OH-$

This is a reversible reaction, and the concentration of water molecules greatly exceeds that of the ions.

Acids and Bases. Since dissociation of water results in one H^+ and one OH^-, the concentration will be equal in pure water. The concentration of each is 10^{-7} M. This means that there is only one ten-millionth of a mole of hydrogen ions per liter of pure water, and an equal number of hydroxide ions. Imbalance of the concentration of these two ions is a result of acids or bases when they are dissolved in water. For example, when HCl, an acid, which is a substance that increases the H^+ concentration in water, is added to water, hydrogen ions dissociate from the chloride ions:

Equation: $HCl \rightarrow H^+ + Cl^-$

Now there are two sources of H^+ in the solution, resulting in more H^+ ions than OH^- ions. Such a solution is known as an acidic solution. A substance

Acid Rain

Acid rain forms when industrial wastes lower the pH of rain water. It has a pH that is twenty five times more acidic than normal rain. Acid rain causes considerable damage to the environment.

that reduces the hydrogen ion concentration in a solution is known as a base. This disassociation of the base will result in more OH^- ions than H^+ ions. An example of this is NaOH, a base that dissociates into sodium and hydroxide ions.

Equation: $NaOH \longrightarrow Na^+ + OH^-$

The OH^- ions will combine with the H^+ ions from the dissociation of water, leaving more OH^-. The result is a basic solution. Other bases reduce H^+ ion concentration directly by accepting H^+ ions into themselves. Ammonia is an example of a base that will bind a hydrogen ion from the solution.

Equation: $NH_3 + H_2O \longrightarrow NH_4^+ + OH^-$

Either case will result in a reduction of the H^+ concentration. Solutions with a relatively high concentration of OH^- are called basic solutions.

pH Scale. In any solution, the product of the H^+ and the OH^- concentrations is constant at 10^{-14} M. This is written as:

$$[H^+][OH^-] = 10^{-14}\ M$$

where brackets indicate molar concentration. A neutral solution at room temperature has H^+ and OH^- concentrations of 10^{-7}, so the product is 10^{-14}. When acid is added to increase H^+ ion concentration to 10^{-5}, OH^- will decline to 10^{-9}. Therefore, an acid not only adds hydrogen ions to a solution but also removes hydroxide ions, because of the tendency for H^+ ion to combine with OH^- to form water. A base has the opposite effect by removing hydrogen ions from the solution.

Because these concentrations can vary so much, scientists have developed the pH scale. The pH of a solution is defined as the negative logarithm of the hydrogen ion concentration expressed in moles per liter:

$$pH = -\log[H^+]$$

For a neutral solution, H^+ ion concentration is 10^{-7}, so $-\log[H^+] = -(-7) = 7$. The pH declines as H^+ ion concentration increases. A pH value of less than 7 denotes an acidic solution, and the lower the number the more acidic the solution. The pH for basic solutions is above 7. Human physiologic pH, meanwhile, must be maintained at around 7. If your pH level falls below or rises above this level, your ability to carry gases is affected, and the enzymes in your body will not work as well.

Now that you've worked through this chapter, you should have a good handle on cellular and molecular biology. Let's test your understanding of the topics we've covered with a short quiz.

Cellular and Molecular Biology Quiz

1. Which of the following is NOT a distinction between plants and animals?

   (A) Plants contain cell walls made of cellulose.
   (B) Plants have intermediate larval stages.
   (C) Plants are extensively branched.
   (D) Animals are generally heterotrophic and motile.
   (E) all of the above

2. Which of the following is NOT an organelle?

   (A) nucleus
   (B) Golgi apparatus
   (C) lysosome
   (D) chlorophyll
   (E) chloroplast

3. Which is correctly associated?

   (A) RNA: thymine
   (B) DNA: uracil
   (C) RNA: replication
   (D) mRNA: picks up amino acids
   (E) RNA: ribose sugars

4. What is the best evidence that genes control synthesis of proteins?

   (A) Proteins are macromolecules.
   (B) RNA directs amino acid synthesis.
   (C) The amino acid sequence of polypeptides is changed by gene mutation.
   (D) DNA serves as a template for RNA.
   (E) mRNA is found in the ribosome.

5. The tRNA code for the amino acid valine is AAC. What is the mRNA code for valine?

   (A) TTG
   (B) GGU
   (C) CCA
   (D) CCG
   (E) UUG

6. The source of oxygen given off in photosynthesis is

   (A) water
   (B) carbon dioxide
   (C) glucose
   (D) starch
   (E) chlorophyll

7. Which of the following is NOT a lipid derivative?

   (A) waxes
   (B) steroids
   (C) carotenoids
   (D) albumins
   (E) lecithin

GO ON TO THE NEXT PAGE

8. Which is NOT a characteristic of proteins?

(A) They can self-replicate.
(B) They can act as hormones.
(C) They can catalyze chemical reactions.
(D) They act in cell membrane trafficking.
(E) They can bind foreign materials.

9. The rough endoplasmic reticulum differs from the smooth endoplasmic reticulum due to the presence of

(A) lysosomes
(B) ribosomes
(C) mitochondria
(D) Golgi apparati
(E) histones

10. Which of the following is found in eukaryotes but not in prokaryotes?

(A) ribosomal RNA
(B) plasma membrane
(C) nuclear membrane
(D) ribosomes
(E) none of the above

11. Which of the following statements regarding photosynthesis is NOT true?

(A) The light cycle occurs only during exposure to light.
(B) The dark cycle occurs only in the absence of light.
(C) ATP is produced during the light cycle.
(D) During the dark cycle, sugars are produced.
(E) Red and blue light are optimal for photosynthetic function.

12. All of the following organelles are membrane bound except the

(A) rough endoplasmic reticulum
(B) Golgi apparatus
(C) nucleus
(D) mitochondria
(E) ribosome

13. Which of the following statements about the Krebs cycle is NOT true?

(A) The Krebs cycle occurs in the matrix of the mitochondrion.
(B) The Krebs cycle is linked to glycolysis by pyruvate.
(C) The Krebs cycle is the single greatest direct source of ATP in the cell.
(D) Citrate is an intermediate in the Krebs cycle.
(E) The Krebs cycle produces nucleotides such as NADH and $FADH_2$.

GO ON TO THE NEXT PAGE

14. Cells that are involved in active transport, such as cells of the intestinal epithelium, utilize large quantities of ATP. In such cells, there are

    (A) high levels of adenylate cyclase activity
    (B) many polyribosomes
    (C) many mitochondria
    (D) high levels of DNA synthesis
    (E) many lysosomes

15. The process by which a cell engulfs large particulate matter is called

    (A) pinocytosis
    (B) exocytosis
    (C) cytokinesis
    (D) phagocytosis
    (E) osmosis

16. The basis for the pairing of the two strands of DNA in the double helix is

    (A) covalent bonding
    (B) ionic bonding
    (C) hydrogen bonding
    (D) hydrophobic interactions
    (E) hydrophilic interactions

17. Which of the following statements about enzymes is NOT true?

    (A) The activity of enzymes is unaffected by genetic mutation.
    (B) Enzymes may interact with nonprotein molecules in order to engage in biological activity.
    (C) Enzymes optimally operate at a particular pH.
    (D) Enzymes optimally operate at a particular temperature.
    (E) Enzymes are almost always proteins.

18. Which of the following is a correct association?

    (A) mitochondria: transports materials from the nucleus to the cytoplasm
    (B) lysosome: digestive enzymes for intracellular use
    (C) endoplasmic reticulum: selective barrier for the cell
    (D) ribosome: electron transport chain
    (E) polysome: group of Golgi complexes

GO ON TO THE NEXT PAGE

19. If O_{18}-labeled glucose is given to a rat, where will the label first appear?

    (A) in exhaled O_2
    (B) in exhaled CO_2
    (C) in exhaled H_2O
    (D) in plasma H_2O
    (E) in intracellular H_2O

20. Which of the following is in a different chemical category than the others?

    (A) cytosine
    (B) thymine
    (C) arginine
    (D) guanine
    (E) uracil

Answers and Explanations to Cellular and Molecular Biology Quiz

1. **(B)** Animals generally go through a larval stage—a developmental stage between the fertilized egg and the adult (examples include the metamorphosis of tadpoles and caterpillars). Plants do not pass through intermediate larval stages. Another distinction between plants and animals is that plants are typically photosynthetic and sessile, while animals are generally heterotrophic and motile. Also, plant structure is adapted for maximum exposure to light, air, and soil by extensive branching; animals, meanwhile, are adapted for minimum surface exposure. They are extremely compact. Finally, plant cells contain cell walls composed of cellulose, while animal cells do not.

2. **(D)** Chlorophyll is an essential component of an organelle, the chloroplast, but it is not an organelle in its own right.

3. **(E)** RNA is made up of a ribose sugar bound to a phosphate group, which is then bound to one of the four bases. It is also characterized by having uracil as one of its bases rather than thymine (as in (A)). DNA, meanwhile, utilizes thymine rather than uracil (B). As for (C), RNA does not replicate in eukaryotic cells. Although RNA will synthesize DNA in some retroviruses, this is not known as the replication of DNA. In (D), tRNA actually picks up the amino acids and brings them to the ribosome, while the mRNA carries the message of the protein that is to be produced to the ribosome. Neither are associated with ribose sugars.

4. **(C)** A point or frameshift mutation in a gene will be evidenced by a corresponding mutation in the protein, such as a change in the amino acid sequence of polypeptides. (A), on the other hand, has nothing to do with how DNA might control the synthesis of a protein, and (B) is false because DNA actually directs amino acid synthesis by serving as the template that determines the sequence of the mRNA. (D) and (E) are both true, but are not the best answers; they could be nothing more than coincidental occurrences and do not fully support the statement.

5. **(E)** mRNA includes a coded base sequence called a codon. tRNA, which carries the amino acids to the mRNA and the ribosome, has a complimentary strand called the anticodon. Therefore, a tRNA with an anticodon of AAC should match the mRNA codon UUG. Remember that in RNA, adenine bonds with uracil, and guanine bonds with cytosine.

6. **(A)** In the light reaction, light splits H_2O into excited electrons, H^+ and O_2. The excited electrons go on to form ATP and the H^+ electrons are incorporated into the carbohydrates produced during the dark reaction. O_2 is released into the environment as a waste product of this reaction. In (B), CO_2 donates the carbon and the oxygen required for carbohydrate formation in the dark reaction. (C) and (D) are end products of photosynthesis, and (E) chlorophyll is involved in the initial capture of sunlight.

7. **(D)** Albumins are not lipid derivatives; they are globular proteins that act as carriers, enzymes, and perform a variety of other functions in the body. Waxes (A) are esters of fatty acids and monohydroxylic alcohols. They are found as protective coatings on skin, fur, leaves of higher plants, and on the cuticle of the exoskeleton of many insects. Steroids (B) such as cholesterol and estrogen have three fused cyclohexane rings and one fused cyclopentane ring. Carotenoids (C) are fatty, acidlike carbon chains containing conjugated double bonds and carrying six-membered carbon rings at each end. These compounds are pigments, and produce red, yellow, orange, and brown colors in plants and animals. Finally, lecithin (E) is an example of a phospholipid. Phospholipids contain glycerol, two fatty acids, a phosphate group, and nitrogen-containing alcohol.

8. (A) DNA is the only molecule capable of self-replication. Proteins may function as hormones (chemical messengers), enzymes (catalysts of chemical reactions), structural proteins (providers of physical support), transport proteins (carriers of important materials), and antibodies (binders of foreign particles).

9. (B) Ribosomes, the site of protein production, give the rough ER its characteristic appearance. The function of the ER is to transport proteins around the cells or to export them out of the cell. Sections of the ER are lined with ribosomes, where proteins are produced and then transported to the appropriate areas. (A) Lysosomes are membrane-bound organelles in the cytoplasm with very low pHs. Filled with proteolytic enzymes, they are the site of degradation in the cell. As for (C), mitochondria, another type of membrane-bound organelle, serve as the site of cellular respiration. The Golgi apparatus (D) is also a membrane-bound organelle, and follows the ER in the production of proteins. It is here that proteins are glycosylated, post-translationally modified, and packaged. In the final choice, (E), histones are proteins in the nucleus which bind to DNA like "beads on a string."

10. (C) One basic difference between eukaryotes and prokaryotes is that eukaryotes possess membrane-bound organelles, while prokaryotes do not. Prokaryotic respiratory and photosynthetic systems are not bound by membranes, and neither is the prokaryotic nucleus. Both of the cells types in (A) and (B) have ribosomal RNA (although it differs in composition and size) and both have plasma membranes.

11. (B) Photosynthesis occurs in two steps. Step one is the light reaction in which visible light, especially that in the red/blue wavelengths, produces ATP and $NADPH_2$ through the splitting of CO_2 and H_2O. O_2 is produced during the splitting of water. During the dark cycle, the second step, carbohydrates such as glucose are synthesized whenever $NADPH_2$, ATP, and CO_2 are present, and this will occur regardless of the presence or absence of light.

12. (E) The ribosome, found in both prokaryotes and eukaroytes, is located in the endoplasmic reticulum and is involved in polypeptide synthesis. Membrane-bound organelles, on the other hand, are found only in eukaryotes. These include the ER, Golgi apparatus, mitochondria, nuclei, and lysosomes.

13. (C) The single greatest direct source of ATP in the cell is the electron transport chain, not the Krebs cycle. The Krebs cycle does occur in the matrix of the mitochondria (A), however, and oxidative phosphorylation (the electron transport chain) does occur in the inner membrane of the mitochondria known as the cristae. Citrate is an intermediate in this cycle (D). The Krebs cycle only forms two ATP directly; all of the other ATP that form during this cycle are produced when NADH and $FADH_2$ donate their electrons to the electron transport chain.

14. (C) Cells that are involved in active transport, such as the epithelial cells of the intestine, will require large amounts of ATP. If a cell utilizes large amounts of ATP, it must produce it in many mitochondria. In (A), high levels of adenylate cyclase activity are found in cells that are the target cells for hormone activation. Many polyribosomes (B), meanwhile, are found in cells that have a high level of protein synthesis. As for (D), high levels of DNA synthesis are found in cells that undergo rapid reproduction and mitosis, and many lysosomes would be found in phagocytic cells, enabling them to digest the foreign material they have endocytosed (E).

15. (D) Phagocytosis is the process of engulfing large matter, such as a bacterium. Meanwhile, pinocytosis (A) is the process of taking in small amounts of liquid, and exocytosis (B) is the term given to the release of proteins from the cell. In (C), cytokinesis is the division of the cytoplasm during mitosis, while osmosis is the movement of water from an area of lower solute to an area of higher solute concentration (E).

16. (C) DNA is a double-stranded helix, composed of the purines adenine and guanine and the pyrimidines cytosine and thymidine. Adenine binds with thymidine, while guanine binds cytosine via weak hydrogen bonds. These weak bonds enable the helices to separate easily to facilitate DNA replication. (A) Covalent bonding is characterized by shared electron pairs, while (B) ionic bonding is characterized by electron transfer. These are both strong forms of intermolecular bonds. In the last two choices, (D) hydrophobic interactions are attractive forces between nonpolar molecules, while (E) hydrophilic interactions are attractive forces between polar molecules.

17. (A) Enzymes are biological catalysts usually composed of proteins. They work at an optimal temperature and pH, typically around the physiological temperature of 56°C and a pH of 7.2. At higher temperatures, the proteins will denature and lose their function. They will often have to interact with cofactors such as vitamins or ions for optimal activity. Mutation will also affect the DNA sequence coding for these proteins, leading to an altered polypeptide and, often, a change in spatial configuration, which leads in its turn to a change in function.

18. (B) Lysosomes are membrane-bound organelles containing digestive enzymes. Typically, they have a low pH. (A) Mitochondria, on the other hand, are involved in cellular respiration, while (C) the ER transports polypeptides around the cell and to the Golgi apparatus for packaging. Finally, (D) the ribosome is the site of protein synthesis, and (E) polysomes are groups of ribosomes that make large quantities of a particular polypeptide.

19. (B) When radioactively labeled glucose is given to a rat, it enters respiration, beginning with glycolysis and proceeding through the Krebs cycle and the electron transport chain. The glucose is broken down into pyruvic acid during glycolysis, at which point no CO_2 is formed. As pyruvic acid forms acetyl coenzyme A, CO_2 is released, and this CO_2 may contain the labeled oxygen. Later, as acetyl coenzyme A enters the Krebs cycle and the Krebs cycle reactions occur, more CO_2 is released. This CO_2 may also contain the labeled oxygen. (A) is incorrect because we generally do not exhale oxygen. The only oxygen exhaled is that which is not absorbed into the alveoli. (C), (D), and (E) are all incorrect for the following reasons: The water produced by respiration is formed from electrons, which are sent down the electron transport chain in the mitochondria and combined with hydrogen ions and molecular oxygen, which is taken in through the lungs. This molecular oxygen is the final electron acceptor of the chain. Therefore, the oxygen in respiratory waste product water comes from the oxygen that is inhaled into the alveoli. It does not come from the oxygen that is part of the glucose molecule that is respired.

20. (C) Of the compounds listed, (A), (B), (D), and (E) are all nitrogenous bases. They are present in such compounds as RNA and DNA. The nitrogenous bases guanine and adenine are purines, and the nitrogenous bases cytosine, thymine, and uracil are pyrimidines. (C), arginine, is an amino acid and not a nitrogenous base. Amino acids are the building blocks for proteins.

CHAPTER FOUR

ECOLOGY

We are not alone in the world; we depend on other organisms for survival. If plants did not harvest the energy of the sun and transform it into carbohydrates, we and our livestock would have nothing to eat. And if fungi and other decomposers did not break down dead organisms, we would be maneuvering around dinosaur corpses today.

Ecology is the study of the interactions between organisms and their environment. The environment encompasses all that is external to the organism and is necessary for its existence. An organism's environment consists of the physical (*abiotic*) environment and the living (*biotic*) environment. The physical environment comprises all nonliving, extrinsic things, and the biotic environment encompasses all living things that directly or indirectly influence the life of the organism. Abiotic factors include temperature, light, water, and topography, while the relationship between that organism and other organisms is numbered among biotic factors.

Don't Mix These Up on Test Day

The *abiotic environment* encompasses:

- Temperature
- Light
- Water
- Oxygen supply
- Soil

The *biotic environment* encompasses:

- Organisms and their relationships with other organisms

Energy Flow

All living things require energy to carry on their life functions. The complex pathways involved in the transfer of energy through the living components of the ecosystem (or biotic community) may be mapped in the form of a food chain or food web, as each organism affects one or more other organisms.

Food Chain

Energy from the sun enters living systems via the photosynthetic production of glucose by green plants. Within the food chain, energy is transferred from the original sources in green plants through a series of organisms, with repeated stages of consumption and, finally, decomposition. The important players in this food chain are producers, primary consumers, secondary consumers, tertiary consumers, and decomposers.

The Chain of Life

Producers are consumed by primary consumers, who are consumed by secondary consumers, who are consumed by tertiary consumers—and all of these organisms are broken down by decomposers.

Producers

Autotrophic green plants and chemosynthetic bacteria are characterized as producers. They utilize the energy of the sun and simple raw materials (carbon dioxide, water, and minerals) to manufacture carbohydrates, proteins, and lipids. The radiant energy of the sun is captured and stored in the C-H bond. Producers always form the initial step in any food chain.

Consumers

Primary consumers are animals that consume green plants (herbivores)—for example, the cow, the grasshopper, and the elephant. Secondary consumers, meanwhile, are carnivores that consume primary consumers—for example, frogs, tigers, and dragonflies. Finally, *tertiary consumers* feed on secondary consumers; examples include snakes that eat frogs.

Decomposers

Finally, at the end of the chain we find saprophytic organisms and organisms of decay, which include bacteria and fungi. Producers and consumers concentrate and organize materials of the environment into complex living substances. Hence living things give off wastes during their life and, eventually, die. Bacteria and fungi decompose the organic wastes and dead tissue into simpler compounds such as nitrates and phosphates, which are returned to the environment to be used again by living organisms. These processes are demonstrated in food webs and nutrient cycles (involving nitrogen, carbon, and water).

Food Web

The food chain can be expanded into the food web, which more accurately depicts what happens in nature. Almost every species is consumed by one or more other species, some of which are on different food chain levels. The result is a series of branches among all the food chains of a community that together form a web.

The greater the number of pathways in a community food web, the more stable that community is. For example, owls eat rabbits. If rabbits in a given area died off because of disease, there would be more vegetation available to mice; the mice would thrive and provide substitute food for the owls. Meanwhile, the decimated rabbit population would have a better chance of recovering while the owls concentrated on the mice, and the various populations involved would eventually return to their former levels.

Food Pyramids

Without a constant supply of energy from the sun, an ecosystem would soon run down. As food is transferred from one level of the food chain to the next, a transfer of energy occurs. Every energy transfer involves a loss of energy. In addition to the energy lost in this transfer, life forms at each level of the food chain utilize some of the energy they obtain from the food for their own metabolism (that is, to support life functions), and lose some additional energy in the form of heat. An energy pyramid is thus a fundamental property of all ecosystems, at all levels.

Pyramids of Energy

Each member of a food chain utilizes some of the energy it obtains from its food for its own metabolism (life functions) and loses some additional energy in the form of heat. Since this means a loss of energy at each feeding level, the producer organism at the base of the pyramid contains the greatest amount of energy. Less energy is available to the primary consumer and still less to the secondary and tertiary consumers. The smallest amount of available energy can therefore be found at the top of the pyramid.

Pyramids of Mass

Since organisms at the upper levels of the food chain derive their energy from organisms at lower levels, and since energy is wasted from one level to the next, smaller and smaller biomasses can be supported as a pyramid of energy is ascended. As a mass of living substances passes along a food chain from producer to primary consumer to secondary consumer, some mass is lost through the expulsion of wastes, the oxidation of foods, and other metabolic processes. As the pyramid of mass is ascended, each level in the pyramid is reduced to approximately 10 percent of the mass of the level beneath it.

Pyramids of Numbers

Consumer organisms that are higher in the food chain are usually larger and heavier than those lower down. Since the lower organisms have a greater total mass, there must be a greater number of lower level organisms (a large bass eats tiny minnows, but eats many of them). With the greatest number of organisms at the base (the product level) and the smallest number at the top (the final consumer level), we have a pyramid of numbers.

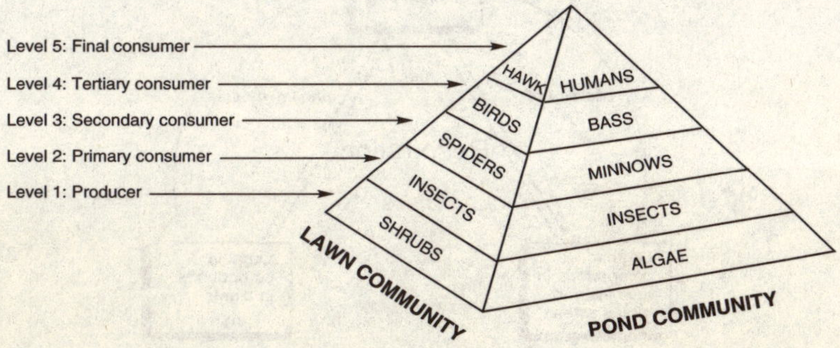

Pyramid of Numbers

Eat or Be Eaten

Organisms play different roles in the food pyramid at different times. Where in the food pyramid would you be if you ate a plate of french fries? How about if you ate a hamburger? Or if a lion ate you?

The Higher You Go, the More You Lose

As a general rule, energy, mass, and numbers are lost as a food pyramid is ascended.

Note that since other factors, such as the generation time and the size of the organisms, must be considered, the pyramids of numbers and biomass do not apply to all levels at all times (unlike the pyramid of energy). Thus bacteria, which could be regarded as the highest level in the food chain, are greater in biomass and numbers than all other organisms. This is due to the small size and short generation time of the bacteria. In general, as we ascend the pyramid, we find decreasing energy content and mass along with fewer numbers of organisms.

Nutrient Cycles

Certain crucial elements, such as nitrogen and carbon, must be recycled from decaying organisms so that the next generation can utilize them. Material is cycled and recycled between organisms and their environments, passing from inorganic forms to organic forms and then back to the inorganic forms. Many of these cycles are accomplished largely through the activities of scavengers (like hyenas and vultures) and decomposers (saprophytes, such as bacteria and fungi).

Waste Not, Want Not

Because raw materials are finite in quantity here on Earth, nutrients like carbon and nitrogen must continually be recycled.

Carbon Cycle

This cycle commences as gaseous CO_2 enters the living world when plants take it in and use it to produce glucose via photosynthesis. At this stage, the carbon atoms in CO_2 are bonded to hydrogen and other carbon atoms. The plants uses the glucose to make starch, proteins, and fat.

Next, animals eats plants and use the digested nutrients to form carbohydrates, fats, and proteins characteristic of the species. Part of these organic compounds is used as fuel in respiration in plants and animals. The metabolically produced CO_2 is then released to the air. Aside from expelled wastes, the rest of the organic carbon remains locked within an organism until its death, at which time decaying processes return the CO_2 to the air.

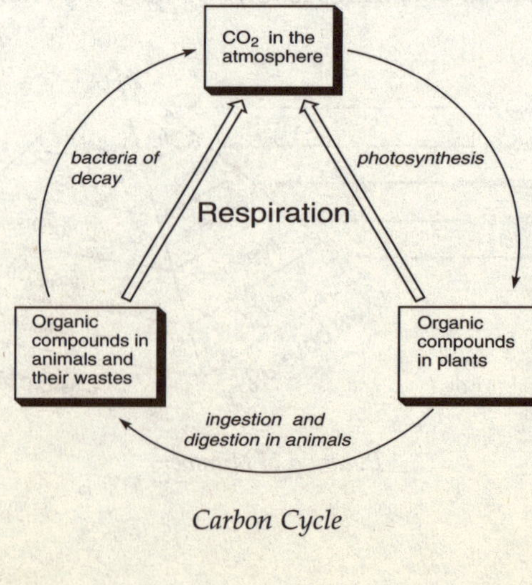

Carbon Cycle

Nitrogen Cycle

Nitrogen is an essential element of amino acids and nucleic acids, which are the building blocks for all living things (as well as viruses). Since there is a finite amount of nitrogen on the earth, it is important that it be recovered and reused.

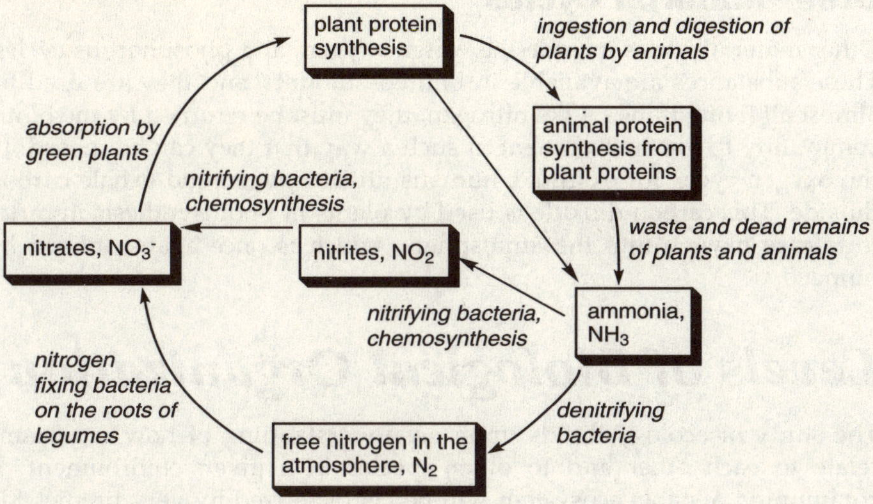

Nitrogen Cycle

The following bullets refer to the figure of the nitrogen cycle above.

- Elemental (free) nitrogen, at the bottom of the figure, is chemically inert and cannot be used by most organisms. Lightning and nitrogen-fixing bacteria in the roots of legumes change the nitrogen to usable, soluble nitrates.
- The nitrates are absorbed by plants and are used to synthesize nucleic acids and plant proteins.
- Animals eat the plants and synthesize specific animal proteins from the plant proteins. Both plants and animals give off wastes and, eventually, die.
- The nitrogen locked up in the wastes and dead tissue is released by the process of decay, which converts the proteins into ammonia.
- Two fates await the ammonia (NH_3): Part of it is nitrified to nitrites by chemosynthetic bacteria and then to usable nitrates by nitrifying bacteria. The rest of the ammonia is denitrified. This means that the ammonia is broken down to release free nitrogen, which returns us to the beginning of the cycle.

Well Hidden

Pull up the roof of a legume, and you'll see many large nodules. These serve as the site of nitrogen fixation for symbiotic bacteria.

Note that there are four kinds of bacteria: decaying, nitrifying, denitrifying, and nitrogen-fixing. The bacteria have no use for the excretory ammonia, nitrites, nitrates, and nitrogen they produce. These materials are essential, however, for the existence of other living organisms.

Other Material Cycles

Other material cycles include the water, oxygen, and phosphorous cycles. These substances are available in limited amounts, and they are used by almost all living things. Like nitrogen, they must be returned by the biotic community to the environment in such a way that they can be reused. In the oxygen cycle, for example, humans utilize oxygen and exhale carbon dioxide. This carbon dioxide is used by plants in photosynthesis in order to release oxygen into the atmosphere, which is once again utilized by humans.

The Big Picture

Let's take a moment to assess the links between ecological levels of biological organization and the cellular biology we discussed in an earlier chapter. From smallest to largest, the levels of biological organization found on Earth are as follows:

Cell —> Tissue —> Organ —> Organism —> Population —> Community —> Ecosystem —> Biosphere

Levels of Biological Organization

The study of ecology builds upon an understanding of how organisms relate to each other and to other species in a given environment or community. A stable ecosystem will be characterized by very predictable relationships, while an unstable ecosystem's interrelationships will be far more volatile. These relationships may be understood in terms of different biological levels of organization, from the organism level through the biosphere level.

The Organism Level

Smaller units of different levels of organization combine to make up organisms. The organism contains many organ systems, which in their turn encompass organs. Organs are formed from tissue, tissue from cells, cells from many different molecules, molecules from atoms, and atoms from subatomic particles.

The Population Level

A *population* is a group of organisms of the same species living together in a given location. A group of similar organisms that are capable of interbreeding are considered a species. Examples of populations include dandelions on a lawn, flies in a barn, minnows of a certain species in a pond, and lions in a savanna. Environmental factors such as nutrients, water, and sunlight limitations aid in maintaining populations at relatively constant levels.

The Communities Level

A community consists of populations of different plants and animal species interacting with each other in a given environment. The term *biotic community* is used to indicate only the different populations and not their environmental relations. An ecosystem includes the community and the environment. Generally, a community incorporates populations from all five kingdoms—protist, plant, fungus, monera, and animals—which all depend upon each other for survival.

The following are familiar examples of communities. In a given lawn community, dandelions, grasses, mushrooms, earthworms, nematodes, and bacteria may reside, while the pond community contains dragonflies, algae, minnows, and insect larvae. Forest communities often incorporate moss, trees, bacteria, lichens, ferns, deer, chipmunks, spiders, and foxes. And sea communities include fish, whales, and plankton.

The Ecosystem Level

Ecosystems give the communities a place to live. Ecosystems involve the interaction of living (biotic) communities with the nonliving (abiotic) environment. In studying the ecosystem, the biologist emphasizes the effects of the biotic community on the environment and vice versa. Examples of ecosystems include meadows, open sea, deep sea, forests, rocky shores, ponds, marshes, grasslands, and deserts.

The Biosphere Level

This level incorporates the portions of the planet that support life—that is, the atmosphere, the lithosphere (rock and soil surface), and the hydrosphere (the oceans). It is a relatively thin zone extending a few feet beneath the earth's surface, several miles down into the deepest sea, and several miles high into the atmosphere.

The Physical Environment

Physical elements like water, temperature, sunlight, oxygen supply, and substratum determine what kinds of animals and plants live in a given environment.

Water

This is the basic constituent of the internal environment of all living things— it makes up 95 percent of some cells. If water is not readily available, the organism may possess adaptations for the storage and conservation of it.

Temperature

Organisms must maintain their internal temperatures at an optimal level. Protoplasm is destroyed at temperatures below 0°C and at extremely high temperatures. Hence adaptations are necessary to protect organisms against these extremes, which we will describe in more detail in our discussion of homeostasis in the chapter on organismal biology. The temperature of a location depends upon its latitude and altitude. The same changes in habitat that occur as one approaches the colder polar regions also occur as one ascends toward the colder regions of a mountain top.

Sunlight

This serves as the ultimate source of energy for all organisms. Green plants must compete for sunlight in forests. To this end, they develop adaptations to capture as much sunlight as possible (including broad leaves, branching, greater height, and vine growth). In water, the photic zone—the top layer through which light can penetrate—is where all photosynthetic activity takes place. In the aphotic zone, only animal life and other heterotrophic life exist.

Plants That *Like* Pollution

Duckweed grows rapidly when exposed to pollution, "choking" other plant and animal life by utilizing all the available oxygen.

Oxygen Supply

Oxygen supply poses no problem for terrestrial life, since air is composed of approximately 20 percent oxygen. Aquatic plants and animals utilize oxygen dissolved in water, where oxygen is present only in parts per million. Pollution can significantly lower oxygen content in water, threatening aquatic life. It can also benefit certain organisms at the expense of others.

Substratum (Soil or Rock)

Substratum determines the nature of plant and animal life in the soil. Some soil factors include:

- *Acidity (pH).* Acid rain may make soil pH too low for most plant growth. Rhododendrons and pines, however, are more well suited to acidic soil.
- *Texture of soil and clay content.* These determine the quantity of water the soil can hold. Most plants grow well in loams that contain high percentages of each type of soil.
- *Minerals.* Nitrates, phosphates, and other minerals determine the type of vegetation soil will support. Beach sand has been leached of all minerals and is unable to support plant life.
- *Humus quantity.* This is determined by the amount of decaying plant and animal life in the soil.

Interactions Within the Ecosystem

Within an ecosystem, interactions take place among members of a community and between the members of the community and the environment. These interactions involve a flow of energy and a cycling and recycling of materials. Various concepts important to an understanding of these interactions are outlined below.

Survival of the Fittest

Two species occupying similar niches will do one of the following:

- Compete until one species is driven to extinction
- Develop superiority in different geographical settings, allowing one species to take over one area and the other to take over another
- Evolve in divergent directions

The Niche

The concept of *niche* denotes the functional role and position of an organism in the ecosystem. The niche is distinct from the habitat, which is the physical place where an organism lives. The niche incorporates this physical place, but also encompasses what the organism eats, where and how it obtains its food, which climatic factors it can tolerate and which are optimal, the nature of its parasites and predators, where and how it reproduces, and so forth. In sum, the niche embodies every aspect of an organism's existence.

It is implicit in the definition of niche that no two species can ever occupy the same niche. Organisms occupying the same niche compete for the same limited resources: food, water, light, oxygen, space, minerals, and reproductive sites. There may be many organisms in the niche, but they are all of the same species and thus have the same survival requirements. The niche is so specific that a species can be identified by the niche it occupies.

Species competing for similar niches share at least one resource in common. Therefore, they will compete for that resource. There are a number of possible outcomes of such competition. In one scenario, one species may be competitively superior to the other and drive it to extinction. Another possibility involves one species proving competitively superior in some regions, while the other may prove superior in other regions, under different environmental conditions. This would result in the elimination of one species in some places and the other species in other places. Finally, two species may rapidly evolve in divergent directions under the strong selection pressure resulting from intense competition. In this fashion the two species would rapidly evolve greater differences in their niches in order to ensure their survival.

Nutritional Interactions

Different types of organisms interact with their environments in different ways in order to obtain food. Autotrophs, for example, manufacture their own food and store their own energy. Green plants utilize the energy of the sun to manufacture food, while chemosynthetic bacteria obtain energy from the oxidation of inorganic sulfur iron and nitrogen compounds. On the other hand, heterotrophs (herbivores, carnivores, and omnivores) cannot synthesize their own food and must depend upon autotrophs or other heterotrophs in the ecosystem to obtain food and energy.

Herbivores consume only plants or plant foods. Due to the toughness of cellulose-containing plant tissues, herbivores possess ingestive structures for crushing and grinding these tissues and for extracting plant fluids. Their long digestive tracts offer more surface area and time for digestion. Much of the food they consume is indigestible; thus, symbiotic bacteria capable of digesting cellulose cohabit the digestive tracts of herbivores. Thus, many herbivores, such as cows and horses, have hooves instead of toes to enable faster movement on the grasslands. They have incisors adapted for cutting and molars adapted for grinding their food. Examples of herbivores include cows, deer, rabbits, horses, zebras, and beavers. Insects or other invertebrates can also be herbivores.

Carnivores, meanwhile, eat other animals exclusively. In general, carnivores possess pointed teeth and fanglike canine teeth for tearing flesh. They have relatively short digestive tracts due to the easy digestibility of meat. *Omnivores*, on the other hand, are animals that eat both plants and animals. Their digestive tracts enable them to digest and absorb both plant matter and meat.

Interspecific Interactions

A community is not simply a collection of different species living within the same area. It is an integrated system of species that are dependent upon one another to one extent or another. The major types of interspecific interactions are discussed below.

Symbiosis

Symbionts live together in an intimate, often permanent association that may or may not be beneficial to them. Some symbiotic relationships are obligatory—that is, one or both organisms cannot survive without the other. Types of symbiotic relationships are generally classified according to the benefits the symbionts receive. Symbiotic relationships include commensalism, mutualism, and parasitism.

Don't Mix These Up on Test Day

These types of symbiosis come up frequently on the SAT II: Biology test:

- *Commensalism* is a +/0 relationship in which one organism benefits and the other is unaffected.
- *Mutualism* is a +/+ relationship in which both organisms benefit.
- *Parasitism* is a +/– relationship in which one organism benefits and the other is harmed.

Commensalism. In this relationship, one organism is benefited by the association and the other is not affected (this is symbolized as +/0). The host neither discourages nor fosters the relationship. The remora (sharksucker), for example, attaches itself by a hold-fast device to the underside of a shark. Through this association, the remora obtains the food the shark discards, wide geographic dispersal, and protection from enemies. The shark is totally indifferent to the association. A similar association links the barnacle and the whale. The barnacle is a sessile crustacean that attaches to the whale and obtains wider feeding opportunities through its host's migrations.

Mutualism. This is a symbiotic relationship from which both organisms derive some benefit. In the instance of the tick bird and rhinoceros, the rhinoceros aids the bird through the provision of food in the form of parasites on its skin. The bird in its turn aids the rhinoceros by removing the parasites and by warning the rhinoceros of danger when it suddenly flies away.

A more intimate mutualistic association exists between a fungus and an algae in the form of the lichen. Lichens are found on rocks and tree barks. The green algae produces food for itself and the fungus by photosynthesis. Meshes of fungal threads support the algae and conserve rain water. Thus, the fungus provides water, respiratory carbon dioxide, and nitrogenous wastes for the algae, all of which are needed for photosynthesis and protein synthesis. Lichens are significant in that they were the first organism capable of establishing a terrestrial existence; they are pioneer organisms in the order of ecological succession on bare rock.

Nitrogen-fixing bacteria and legumes also engage in mutualism. Nitrogen-fixing bacteria invade the roots of legumes and infected cells grow to form root nodules. In the nodule, the legume provides nutrients for the bacteria, and the bacteria fixes nitrogen (by changing it to soluble nitrate, a mineral essential for protein synthesis by the plant). These bacteria are a major source of usable nitrogen, which is needed by all plants and animals.

Protozoa and termites work together in a similar fashion. Termites chew and ingest wood, but are unable to digest its cellulose. Protozoa in the digestive tract of the termite secrete an enzyme that is capable of digesting cellulose, and both organisms share the carbohydrates. In this manner, the protozoa are guaranteed protection and a steady food supply, while the termite obtains nourishment from the ingested wood. Likewise, in the case of intestinal bacteria and humans, bacteria utilize some of the food material not fully digested by humans and, in turn, manufacture vitamin B_{12}.

Parasitism. A parasite takes from the host but gives nothing in return; thus, the parasite benefits at the expense of the host (+/–). Examples of parasites include leeches, ticks, and sea lampreys. Parasitism exists when competition for food is most intense. Few autotrophs (green plants) exist as parasites (mistletoe is an exception). In general, parasites flourish among organisms like bacteria, fungi, and animals.

Ectoparasites cling to the exterior surface of the host with suckers or clamps, bore through the skin, and suck out juices. Endoparasites, on the other hand, live within the host. In order to gain entry into the host, they must break down formidable defenses, including skin, digestive juices, antibodies, and white blood cells. Parasites possess special adaptations to overcome these defenses.

Parasitism is advantageous and efficient, since the parasite lives with a minimum expenditure of energy. Parasites may even be parasitic on other parasites. Thus, a mammal may have parasitic worms, which in turn are parasitized by bacteria, which in turn are victims of bacteriophages.

A prominent example of a parasitical relationship is that between the virus and its host cell. All viruses are parasites. They contain nucleic acids surrounded by a protein coat, and are nonfunctional outside their host cells. As

Parasites Exercise Restraint

It is not to a parasite's advantage to maximize its food intake and to severely injure its host. If the host dies, the parasite will lose its free ride and be forced to fend for itself, a situation that it would prefer to avoid.

viral nucleic acid enters the host, the virus takes over the host cell functions and redirects them into replication of the virus. In consequence, the life functions of this host cell slow down or cease.

Disease bacteria function in the same way with their animal hosts. Most bacteria are either chemosynthetic or saprophytic (bacteria of decay). Diphtheria bacteria are parasitic upon humans, anthrax on sheep, and tuberculosis on cows or humans. Most disease fungi, meanwhile, are saprophytic. Ringworm, however, a fungus that attacks the foot, is an exception; it is parasitic on humans.

The tapeworm-human relationship is a particularly good example of parasitism. Tapeworms can live inside their hosts' intestines for many years, growing longer and longer. It is interesting to note that successful parasites do not kill their hosts, as this would, counterproductively, lead to death of the parasite itself. The more dangerous the parasite is to its host, the less chance it has of ultimate survival.

Predation

Predators are free-living organisms (carnivores and herbivores) that feed on other living organisms. The effects of predators on their prey vary; the predator might severely limit the numbers of its prey and even go so far as to cause its extinction. On the other hand, the predator might barely impact the numbers of prey, if the predator is scarce or commonly utilizes another food source. In many cases, the predator aids in controlling the numbers of the prey, as the lynx does with the snow hare, but not to the extent of endangering the existence of the prey population. Thus the predator helps to regulate the size of the prey population. Predatory-prey relationships evolve toward such a balance. Examples of predators include hawks, lions, humans, and the Venus flytrap, a carnivorous plant.

Good Thing Someone's Cleaning This Place Up!

If the earth had no saprophytes to consume decaying matter, we'd still be climbing over the carcasses of dinosaurs and other animals that died millions of years ago.

Saprophytism

Saprophytes include those protists and fungi that decompose (digest) dead organic matter externally and absorb the nutrients. Saprophytes decompose organic materials; for this reason, they constitute a vital link in the cycling of material within the ecosystem. Mold, mushrooms, bacteria of decay, and slime molds are all examples of saprophytes.

Scavengers

These animals consume animals that have died naturally or been killed by others. They therefore need no adaptations for killing their prey. Decomposers, such as the bacteria of decay, can be considered scavengers, as can the vulture. The snapping turtle and the hyena are organisms that might be considered both scavengers and predators.

Intraspecific Interactions

Competition is not restricted to interspecific interactions. Individuals belonging to the same species utilize the same resources; if a particular resource is limited, these organisms must compete with one another. Members of the same species compete, but they must also cooperate. Intraspecific cooperation may be extensive (as in the formation of societies in animal species) or may be nearly nonexistent. Hence, within a species, relationships between individuals are influenced by both disruptive and cohesive forces. Competition (for food or a mate, for example) is the chief disruptive force, while cohesive forces include reproduction, protection from predators, and destructive weather.

Interactions Between Organisms and Their Environments

Animals have developed a variety of ways of coping with their environments, including osmoregulation and thermoregulation.

Osmoregulation

Osmoregulation may be defined as the ways in which organisms regulate their supply of water. Saltwater fish, for example, live in a hyperosmotic environment that causes them to lose water and take in salt. In constant danger of dehydration, they must compensate by constantly drinking and actively excreting salt across their gills. Freshwater fish, in contrast, live in a hypo-osmotic environment that causes intake of excess water and excessive salt loss. These fish correct this condition by drinking infrequently, absorbing salts through the gills, and excreting dilute urine.

Don't Mix These Up on Test Day

Osmoregulation is the regulation of water and salt levels within an organism.

Thermoregulation is the regulation of an organism's temperature.

On land, insects excrete solid uric acid crystals in order to conserve water, while desert animals possess adaptations for avoidance of desiccation. The camel can tolerate a wide range of body temperatures and possesses fat layers in areas that come into contact with solar radiation. The horned toad, meanwhile, has a thick, scaly skin. Other desert animals burrow in the sand during the day and search for food at night, thereby avoiding the intense heat that causes water loss.

As for plants, nondesert land plants possess waxy cuticles on leaf surfaces and stomates on their lower leaf surfaces only, and shed leaves in winter to avoid water loss. The water-conserving adaptations of desert plants include extensive root systems, fleshy stems to store water, spiny leaves to limit water loss, extra thick cuticles, and small numbers of stomates.

Thermoregulation

Organisms must also develop ways to regulate heat. Cellular respiration transfers only some of the energy derived from the oxidation of carbohydrates, fats, and proteins into the high-energy bonds of ATP. Roughly 60 percent of the total energy is not captured; most of this is transformed to heat. The vast majority of animals and plants are cold-blooded, or ectothermic—most of their heat energy escapes to the environment. Consequently, the body temperature of ectotherms, also known as poikilotherms, is very close to that of their surroundings. Since an organism's metabolism is closely tied to its body temperature, the activity of ectothermic animals such as snakes is radically affected by environmental temperature changes. As the temperature rises (within limits, since very high temperatures would be lethal), these organisms become more active; as temperatures fall, they become sluggish.

Some animals, notably mammals and birds, are endotherms; they are warm-blooded, or homeothermic. They have evolved physical mechanisms that allow them to make use of the heat produced as a consequence of respiration. Physical adaptations like fat, hair, and feathers actually retard heat loss. Homeotherms maintain constant body temperatures higher than the environment around them. Hence they are less dependent upon environmental temperature than poikilothermic animals, and are able to inhabit a comparatively greater range of variable conditions as a result.

Stability in the Ecosystem

Nature's Irony

In nature's irony, each level in the succession of a community changes the physical environment, making it a better place for organisms in the next stage of succession to live, but pushing out its own organisms in the process.

An ecosystem is self-sustaining; it will remain stable as long as both the physical environment (abiotic factors) and the biotic community are relatively stable. A stable ecosystem requires a constant energy source, a living system incorporating this energy into organic compounds, and, finally, a cycling of materials between the living system and its environment so that materials can be returned to the environment and renewed.

The most stable type of ecosystem is known as the *climax community*. This is the final stage of *ecological succession*.

The Climax Community and Ecological Succession

A climax community is the stable living (biotic) part of the ecosystem mentioned above. In this community, populations exist in balance with each other and with the environment. The type of climax community that is found in a particular ecosystem depends upon all the abiotic factors that characterize this ecosystem, such as rainfall, soil conditions, temperature, and shade.

Ecological succession is the orderly process by which one biotic community replaces or succeeds another until a climax community is established. Each community stage in an ecological succession is identified by a dominant species—the one that exerts control over the other species that are present. Thus, in a grassland community, grass is the dominant species.

Changes occur because each community that establishes itself changes the environment, making it more unfavorable for itself and more favorable for the community that is to succeed it. Successive communities are composed of populations that are able to exist under the new conditions. Finally, a stage arises in which a population alters the environment in such a way that the original conditions giving rise to that population are recreated. Replacement stops and we have our climax community, an ecological steady-state.

This climax community becomes a permanent part of the ecosystem, unless the abiotic factors are drastically altered by climatic or geologic upheavals. If such upheavals occur and the equilibrium is upset, new climax conditions are produced and a new series of successions is initiated, resulting in the establishment of new communities in the ecosystem.

An example of this would be ecological succession in a rocky, barren area. Consider an area in northeastern United States, barren, perhaps, as a result of a severe forest fire. Lichen would be a good candidate to be the first or pioneer organism to resettle this virgin area. Recall that a lichen is an association between an alga and a fungus that can live on a rocky surface. Acids produced by the lichen attack rock, breaking it down to form the first layers of soil. Since lichens thrive only on a solid surface, conditions at this stage are worse for the lichen but better for mosses. Airborne spores of mosses land on the soil and germinate. The result is a new *sere* (a series of ecological communities formed in ecological succession), with moss supplanting lichen as the dominant species in the community.

As the remains of the moss build up the soil still more, annual grasses and then perennial grasses with deeper roots become the dominant species. As time marches on, we find shrubs and then trees. The first trees will be the sun-loving gray birch and poplar. As more and more trees compete for the sun, the birch and poplar will be replaced by white pine and, finally, maples and beeches, which grow in deep shade.

The growth of maples and beeches produces the same conditions that originally favored their appearance. And so this community remains for a thousand years. In the final maple-beech community, you would also find foxes, deer, chipmunks, and plant-eating insects. These are all animals that would not have been found in the original barren rock terrain.

To summarize this example of ecological succession:

Lichen —> mosses —> annual grasses —> perennial grasses
—> shrubs —> sun-loving trees (poplar) —>
thick shade trees (hemlock, beech, maple)

Time elapsed: About 1,000 years

Elevation and Climax Communities

It is important to remember that the dominant species of the climax community is determined by such physical factors as temperature, nature of the soil, and rainfall. Thus the climax community at higher elevations in New

York state is hemlock-beech-maple, while at lower elevations, it is more often oak-hickory. In cold Maine, the climax community is dominated by the pine; in the wet areas of Wisconsin, by cypress; in sandy New Jersey, by pine; in Georgia, by oak, hickory, and pine; and on a cold, windy mountain top, by scrub oak.

We've discussed the progression of climax communities in terrestrial habitats. Now, let's turn to a community that starts with a pond:

- *Step 1—pond.* This pond contains plants such as algae and pondweed and animals such as protozoa, water insects, and small fish.
- *Step 2— shallow water.* The pond begins to fill in with reeds, cattails, and water lilies.
- *Step 3—moist land.* The former pond area is now filled with grass, herbs, shrubs, willow trees, frogs, and snakes.
- *Step 4— woodland.* Pine or oak becomes the dominant tree of the climax community.

The Cold Places

Starting from the equator, the further north or south you go, the colder and more barren it gets. The same holds true for altitude: High places are generally chillier and able to support less life than low places.

Biomes

The conditions in a particular terrestrial and climatic region select plants and animals possessing suitable adaptations for that particular region. Each geographic region is inhabited by a distinct community called a biome. These biomes also define major climates.

Terrestrial Biomes

The evolutionary origin of plants and animals can be traced to the seas. In order to survive on land, these organisms had to develop adaptations to face an environment characterized by a relative lack of water and a relative lack of food and supporting medium. The new environment was also distinguished by varying temperatures (as compared to the oceans, which have a relatively constant temperature), as well as by varying composition of the soil as compared to the definite salt compositions in the oceans.

Land biomes are characterized and named according to the climax vegetation of the region in which they are found. The climax vegetation is the vegetation that becomes dominant and stable after years of evolutionary development. Since plants are important as food producers, they determine the nature of the inhabiting animal population; hence the climax vegetation determines the climax animal population. There are eight types of terrestrial biomes that can be formed as a result of all these factors:

Tropical Forests

Tropical forests are characterized by high temperatures and high levels of rainfall. The climax community includes a dense growth of vegetation that does not shed its leaves. Vegetation like vines and epiphytes (plants growing on the other plants) and animals like monkeys, lizards, snakes, and birds inhabit the typical tropical forest, or rain forest. Trees grow closely together; sunlight barely reaches the forest floor. The floor is inhabited by saprophytes living off dead organic matter. Tropical rain forests are found in central Africa, Central America, the Amazon basin, and southeast Asia.

Savanna

The savanna (grassland) is characterized by low rainfall (usually 10–30 inches per year), although it gets considerably more rain than the desert biomes do. Grassland provides no protection for herbivorous mammals (such as bison, antelopes, cattle, and zebras) from carnivorous predators. That is why animals that do inhabit the savanna have generally developed long legs and hoofs, enabling them to run fast. Examples of savanna include the prairies east of the Rockies, the Steppes of the Ukraine, and the Pampas of Argentina.

Desert

The desert receives less than ten inches of rain per year, and this rain is concentrated within a few heavy cloudbursts. The growing season in the desert is restricted to those days after rain falls. Generally, small plants and animals inhabit the desert. Most desert plants (for example, cactus, sagebrush, and mesquite) conserve water actively; desert animals like the lizard, meanwhile, live in burrows. Few birds and mammals are found in the deserts, except those that have developed adaptations for maintaining constant body temperatures. Examples of desert biomes include the Sahara in Africa, the Mojave in the United States, and the Gobi in Asia.

You Won't Sweat Here

We generally think of deserts as very hot and dry. However, cold deserts also exist. Deserts develop in regions in which less than 30 cm of rain falls per year; temperature does not play any role in their formation.

Temperate Deciduous Forest

Temperate deciduous forests have cold winters, warm summers, and a moderate rainfall. Trees such as beech, maple, oaks, and willows shed their leaves during the cold winter months. Animals found in temperate deciduous forests include the deer, fox, woodchuck, and squirrel. Temperate deciduous forests are located in the northeastern and central eastern United States and in central Europe.

Temperate Coniferous Forest

Temperate coniferous forests are cold, dry, and inhabited by fir, pine, and spruce trees. Much of the vegetation here has evolved adaptations for water conservation—that is, needle-shaped leaves. These forests are found in the extreme northern part of the United States and in southern Canada.

Two Tundras

There are two types of *tundra*: arctic tundra, which is found just south of the North Pole, and alpine tundra, located above the tree line on high mountains all over the world.

Taiga

The taiga receives less rainfall than the temperate forests, has long, cold winters, and is inhabited by a single type of coniferous tree, the spruce. The forest floors in the taiga contain moss and lichens. Birds are the most common animal; however, the black bear, the wolf, and the moose are also found here. Taiga exists in the extreme northern parts of Canada and Russia.

Tundra

Tundra is a treeless, frozen plain located between the taiga and the northern icesheets. It has a very short summer and a very short growing season, during which time the ground becomes wet and marshy. Lichens, moss, polar bears, musk oxen, and arctic hens make their homes here.

Polar Region

The polar region is a frozen area with very few types of vegetation or terrestrial animals. Animals that do inhabit polar regions generally live near the polar oceans.

Terrestrial Biomes and Altitude

The sequence of biomes between the equator and the pole is comparable to the sequence of regions on mountains. The nature of those regions is determined by the same decisive factors—temperature and rainfall. The base of the mountain, for example, would resemble the biome of a temperate deciduous area. As one ascends the mountain, one would pass through a coniferous-like biome, then taigalike, tundralike, and polarlike biomes.

Aquatic Biomes

In addition to the eight terrestrial biomes, there are aquatic biomes, each with its own characteristic plants and animals. More than 70 percent of the earth's surface is covered by water, and most of the earth's plant and animal life is found there. As much as 90 percent of the earth's food and oxygen production (photosynthesis) takes place in the water. Aquatic biomes are classified according to criteria quite different from the criteria used to classify terrestrial biomes. Plants have little controlling influence in communities of aquatic biomes, as compared to their role in terrestrial biomes.

Aquatic areas are also the most stable ecosystems on Earth. The conditions affecting temperature, amount of available oxygen and carbon dioxide, and amount of suspended or dissolved materials are stable over very large areas, and show little tendency to change. For these reasons, aquatic food webs and aquatic communities tend to be balanced. There are two types of major aquatic biomes: marine and freshwater.

Marine Biomes

The oceans connect to form one continuous body of water that controls the earth's temperature by absorbing solar heat. Water has the distinctive ability to absorb and utilize large amounts of heat without undergoing a great temperature change. Marine biomes contain a relatively constant amount of nutrient materials and dissolved salts. Although ocean conditions are more uniform than those on land, distinct zones in the marine biomes do exist, including the intertidal zone, littoral zone, and pelagic zone.

Intertidal Zone. The intertidal zone is a region exposed at low tide that undergoes variations in temperature and periods of dryness. Populations in the intertidal zone include algae, sponges, clams, snails, sea urchins, sea stars (starfish), and crabs.

Littoral Zone. The littoral zone is a region on the continental shelf that contains ocean area with depths of up to 600 feet, and extends several hundred miles from the shores. Populations in littoral zone regions include algae, crabs, crustacea, and many different species of fish.

Pelagic Zone. The pelagic zone is typical of the open seas and can be divided into photic and aphotic zones. The *photic zone* is the sunlit layer of the open sea extending to a depth of 250–600 feet. It contains plankton—passively drifting masses of microscopic photosynthetic and heterotrophic organisms—and nekton—active swimmers such as fish, sharks, and whales that feed on plankton and smaller fish. The chief autotroph is the diatom, an alga.

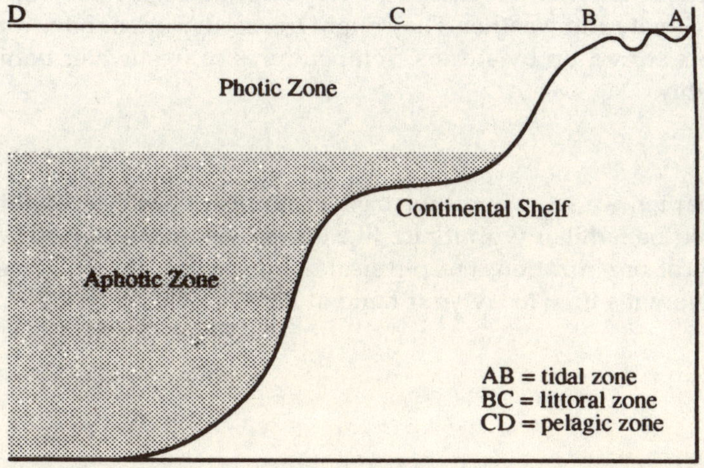

Aquatic Biomes

Meanwhile, the *aphotic zone* may be defined as the region beneath the photic zone with no sunlight and no photosynthesis; only heterotrophs can survive here. Deep-sea organisms in this zone have adaptations that enable them to survive in very cold water, high pressure, and complete darkness.

Not to Be Underestimated

Since nearly three quarters of the earth is covered by the oceans, marine biomes are extremely important. Ocean temperatures affect the planet's climate and wind patterns, while marine algae supply us with a large proportion of our oxygen.

The zone contains nekton and benthos—the crawling and sessile organisms. Some are scavengers and some are predators. The habitat of the aphotic zone is fiercely competitive.

Freshwater Biomes

Rivers, lakes, ponds, and marshes—the links between the oceans and land—contain freshwater. Rivers are the routes by which ancient marine organisms reached land and evolved terrestrial adaptations. Many forms failed to adapt to land and developed adaptations for freshwater. Others developed special adaptations suitable for both land and freshwater. As in marine biomes, factors affecting life in freshwater include temperature, transparency (illumination due to suspended mud particles), depth of water, available CO_2 and O_2, and most importantly, salt concentration.

Freshwater biomes differ from salt water biomes in three basic ways. Freshwater has a lower concentration of salt (greater concentration of water) than the cell, creating a diffusion gradient that results in the passage of water into the cell. Freshwater organisms have homeostatic mechanisms to maintain water balance by the regular removal of excess water. These mechanisms include the contractile vacuole of protozoa and excretory systems of fish. Plant cells have rigid cell walls, building up cell pressure (cell turgor) as waste passes in. This pressure counteracts the gradient pressure, stops the influx of water, and, as a result, establishes a water balance.

In rivers and streams, on the other hand, strong, swift currents have caused fish to develop strong muscles and plants to develop rootlike holdfasts. Finally, freshwater biomes (except for very large lakes) are affected by variations in climate and weather. They might freeze, dry up, or have mud from their floors stirred up by storms. Temperatures of freshwater bodies vary considerably.

In this chapter, we've covered the basic concepts of ecology. By this point, you should be familiar with topics like energy flow, nutrient cycles, levels of biological organization, the physical environment, the ecosystem, and biomes. Now it's time to try your hand at the following quiz.

Ecology Quiz

1. In a pond community, the greatest mass present would consist of

   (A) algae
   (B) insect
   (C) frogs
   (D) fish
   (E) fungi

2. Mutualism is illustrated by

   (A) lichens
   (B) tapeworms
   (C) bread mold
   (D) nematodes
   (E) epiphytes

3. A climax community

   (A) consists of only one species of life
   (B) is populated mainly by so-called pioneer organisms
   (C) is stable within a given climate
   (D) is independent of the environment
   (E) consists of decaying organic matter

4. In the pyramid of energy, the greatest amount of stored chemical bond energy is found in

   (A) primary producers
   (B) secondary producers
   (C) primary consumers
   (D) secondary consumers
   (E) tertiary consumers

5. Living in a close nutritional relationship with another organism in which one organism benefits while the other is neither harmed nor benefited is best defined as

   (A) symbiosis
   (B) mutualism
   (C) saprophytism
   (D) commensalism
   (E) parasitism

6. Digestion of cellulose by bacteria found in a termite's gut is an example of

   (A) mutualism
   (B) parasitism
   (C) saprophytism
   (D) commensalism
   (E) autotrophism

7. Which of the following fixes atmospheric N_2?

   (A) lightning
   (B) bacteria on the roots of legumes
   (C) *uv* light
   (D) two of the above
   (E) all of the above

8. Denitrifying bacteria

   (A) turn ammonia into NO_2 (nitrites)
   (B) turn ammonia into N_2
   (C) turn ammonia into NO_3 (nitrates)
   (D) do not use nitrogen in their life cycle
   (E) none of the above

GO ON TO THE NEXT PAGE

9. A stable ecosystem

   (A) requires a constant energy source
   (B) requires a living system
   (C) is self-sustaining
   (D) requires cycling of materials between the living system and the environment
   (E) all of the above

10. Which of the following would be a dominant species in a climax community?

   (A) mosses
   (B) sun-loving trees
   (C) shrubs
   (D) annual grasses
   (E) shade trees

11. Which of the following is a marine zone?

   (A) intertidal zone
   (B) littoral zone
   (C) pelagic zone
   (D) all of the above
   (E) two of the above

12. In lichen, algal cells produce glucose, while associated fungal cells obtain water and minerals and attach the lichen to a rock surface. This relationship is best described as

   (A) succession
   (B) commensalism
   (C) parasitism
   (D) mutualism
   (E) parallel evolution

13. A biome characterized by permafrost and located immediately north of the timberline and south of the permanent ice sheet would best be characterized as

   (A) taiga
   (B) arctic tundra
   (C) alpine tundra
   (D) coniferous forest
   (E) spruce/moose forest

14. In a food chain of grass —> prairie dog —> snake —> owl, the secondary consumer would be

   (A) grass
   (B) prairie dog
   (C) snake
   (D) owl
   (E) microbes of decay

15. In the northeastern United States, the final succession climax stage of a freshwater lake will be a

   (A) marsh
   (B) grassland
   (C) bog
   (D) estuary
   (E) deciduous forest

Answers and Explanations to Ecology Quiz

1. (A) In an ecology pyramid, the primary producers (photosynthetic or chemosynthetic organisms, such as algae) are always the largest population. (B) and (E) are primary consumers, while (D) and (C) are secondary consumers.

2. (A) Mutualism is a close nutritional relationship between two species in which both benefit. Lichen is the result of a mutualistic relationship between fungi and algae. The algae attach to a rock via rootlets that the fungi produce. Through these rootlets, the fungi receives water, nutrients, and an attachment, the lichen. The alga in its turn receives an attachment (lichen) to the rock, and can produce carbohydrates through photosynthetically utilizing the water and nutrients from the fungi's rootlets. This is considered a +/+ situation. Tapeworms (B) are parasites. The tapeworm gets nutrition from the host as the host loses it, a +/– situation. Saprophytes, such as bread mold (C), are organisms that feed on dead and decaying material, while nematodes (D) are either free-living saprophytes or parasites. Finally, (E) epiphytes may be defined as plants that live on the branches of other plants. In this manner they receive greater exposure to sunlight than they would normally have access to, exemplifying commensalism, a +/0 relationship.

3. (C) A climax community is the final community in a particular biome's succession. In the northeastern part of the United States, the climax community is the deciduous forest, while in the midwest it is the grasslands. (A) is incorrect because many species will live in a biome. Meanwhile, pioneer species (B) are the species that colonize a biome, such as lichen on rocks, and are therefore the earliest species in a biome. (D) The climax community is dependent on the environment, especially the climate; factors such as type of soil and amount of rainfall will determine what organisms will survive and thrive there. As for (E), dead and decaying matter are present in all communities, but are not the sole inhabitants of any of them.

4. (A) In the pyramid of energy, the greatest amount of energy is always found among primary producers. Primary producers are either plants or photosynthetic bacteria. Energy is lost with each level in the pyramid, as it is utilized for maintenance of the organism, movement, and warmth. Only a fraction of the energy produced becomes new tissue that can be harvested by the next level up in the pyramid. This should remind you of the general chemistry principle that with any exchange of energy, some is lost. Primary consumers are herbivores, while carnivores that ingest the herbivores are known as secondary consumers. They in their turn are preyed upon by tertiary consumers.

5. (D) In a commensal relationship, which is a form of a symbiotic relationship, two organisms live in close association with each other. One benefits from this association, while the other is neither harmed nor benefited (in what is sometimes described as a "+/0" relationship). An example of a commensal relationship is the epiphyte plant, which lives on the branches of rainforest trees, gaining the advantage of being closer to sunlight. Symbiosis (A) is a general term describing close nutritional relationships of all types, including mutualism, commensalism, and parasitism. For explanations of the concepts of mutualism (B), saprophytism (C), and parasitism (E), see explanation (2) above.

6. (A) Termites cannot actually digest cellulose, but protozoans in their digestive systems can. In return, these tiny organisms receive a home and food and water. The terms mutualism, parasitism (B), saprophytism (C), and commensalism (D) are all defined in the explanation to question 2 above. (E) Autotropism, meanwhile, describes self-feeders such as photosynthetic or chemosynthetic organisms.

7. (D) Elemental nitrogen (N_2) is chemically inert and cannot be used by most organisms. Lightning and nitrogen-fixing bacteria in the roots of legumes change the nitrogen to usable, soluble nitrates.

8. (B) Denitrifying bacteria break down NH_3 into N_2. Nitrifying bacteria turn ammonia into NO_2 (nitrites), while nitrogen-fixing bacteria turn N_2 into NO_3 (nitrates).

9. (E) A stable ecosystem is self-sustaining and will therefore remain stable in the presence of a relatively stable physical environment (abiotic factors) and a relatively stable biotic community. A stable ecosystem requires a constant energy source, a living system incorporating this energy into organic compounds, and a cycling of materials between the living system and the environment.

10. (E) The summary of ecological succession from a rocky barren area to a final climax community looks like this: lichen —> mosses —> annual grasses —> perennial grasses —> shrubs—> sun-loving trees —> thick shade trees (such as hemlock and beech). Therefore, shade trees would be the dominant species in the climax community.

11. (D) All of the zones listed may be classified as types of marine biomes. The intertidal zone (A) is a region exposed at low tides that undergoes variations in temperature and periods of dryness. Populations in the intertidal zones include algae, sponges, clams, snails, sea urchins, starfish, and crabs. Meanwhile, the littoral zone (B) is the term used for a region on the continental shelf that contains ocean area with depths of up to 600 feet and extends several hundred miles from the shores. Populations in this zone include algae, crabs, crustacea, and many different species of fish. Finally, the pelagic zone, typical of the open sea, is divided into the photic zone (the sunlit layer containing plankton and fish, sharks, or whales) and the aphotic zone (the sunless zone containing the crawling and sessile organisms).

12. (D) The relationship between algal cells and fungal cells is known as mutualism (defined in explanation (2) above, as are commensalism (B) and parasitism (C)). Ecological succession (A) is the orderly process by which one biotic community replaces or succeeds another until a climax community is established, while parallel evolution (E) occurs when groups within the branches of an evolutionary tree develop in similar ways when exposed to similar environments.

13. (B) The arctic tundra is a treeless, frozen plain found between the taiga lands and the northern ice sheets, as well as south of the permanent ice sheet. It has a very short summer and growing season. Lichens, moss, polar bears, musk oxen, and arctic hens are found here. The taiga (A), on the other hand, receive less rainfall than the temperate forests and have long, cold winters. These forests are made up entirely of spruce trees, and the most common animals are the moose, the black bear, the wolf, and birds.

14. (C) The snake, a secondary consumer, consumes the prairie dog, a primary consumer. For definitions of the terms named in this question, see explanation (4).

15. (E) The northeastern part of the United States would have the climax community of a temperate deciduous forest. These forests have cold winters, warm summers, and moderate rainfall. They are populated by deer, fox, woodchucks, and squirrels, and their trees (beech, maple, oaks, and willows) shed their leaves during the cold winter months.

CHAPTER FIVE

CLASSICAL GENETICS

One of the newer fields of science is classical genetics, the branch of biology that deals with heredity and variation in organisms. Classical genetics tries to determine how traits are inherited in organisms through performing test crosses. The foundations for this field were laid by the monk Gregor Mendel, who in the mid 19th century performed a series of experiments to determine inheritance in garden pea plants.

Mendelian Genetics

Around 1865, based on his observations of the seven characteristics of the garden pea, Gregor Mendel developed the basic principles of genetics—dominance, segregation, and independent assortment. Although Mendel formulated these principles, he was unable to propose any mechanism for hereditary patterns, since he knew nothing about chromosomes or genes. Hence his paper was largely ignored until the early 1900s.

After Mendel's work was rediscovered, Thomas H. Morgan tied the principles of genetics to the chromosome theory. He linked specific genes (traits) to specific chromosomes visible in the salivary glands of *Drosophila melanogaster*, the fruit fly. Morgan brought to light the giant chromosomes, at least 100 times the size of normal chromosomes, that are found in the fruit fly's salivary glands. These chromosomes are banded, and the bands are thought to coincide with gene locations. Morgan also described sex-linked genes.

The fruit fly is a highly suitable organism for research. With its short life cycle, it reproduces often and in large number, providing large sample sizes. Its chromosomes are large and easily recognizable in size and shape. They are also few in number (eight chromosomes/four pairs of chromosomes). Finally, mutations occur relatively frequently in this organism.

Perfect Peas

It is suspected that Mendel might have doctored the results of his experiments, because the data from his pea crosses is almost too perfect.

Flashback

Remember that in protein transcription, which we discussed in our chapter on cellular and molecular biology, DNA sequence was responsible for genotype, while the protein produced in the course of protein synthesis was responsible for phenotype.

Definitions

Before we begin our discussion of the Mendelian laws, let's get a few essential definitions out of the way.

- An *allele* is one of a pair of genes.
- A *dominant* trait is a hereditary trait that is expressed even when combined with another genes.
- A *recessive* trait is a hereditary trait that is hidden by a dominant, contrasting allele.
- *Homozygous* indicates that an individual is genetically pure for a given trait, that is, that this individual possesses two identical alleles (*TT*).
- *Heterozygous* describes *hybrid* individuals, who have two contrasting alleles for a given trait (*Tt*).
- *Phenotype* is the outward appearance of an individual with respect to a given trait, that is, the expressed characteristic (for example, tall = *TT* or *Tt*).
- *Genotype* is the genetic characteristics of an individual with respect to a given trait (*Tt* = hybrid tall).

Mendelian Laws

As a result of Mendel's experiments with the fruit fly and other organisms, it is known that:

- Chromosomes contain the genetic or heredity determinant in linear sequence.
- Each gene has a definite position in the sequence of genes on each chromosome.
- Homologous chromosome pairs have genes in the same position in the gene sequence.
- The paired genes, known as alleles, are present on two corresponding chromosomes. The dominant gene is customarily represented by a capital letter, usually the first letter of the name of the dominant trait. The recessive gene is shown by the same letter in lowercase. Thus, if yellow is dominant and green is recessive, the allelic genes would be represented as *Y* (dominant allelic gene for yellow) and *y* (recessive allelic gene for green).
- Paired alleles can be homozygous (or *pure*, with identical genes for one trait), as in *YY* or *yy*, or heterozygous (hybrid, with contrasting genes for one trait), as in *Yy*.

The observations listed above led to the formulation of the following Mendelian laws.

Dominance of Phenotypic Traits

When two individuals with contrasting pure traits are crossed, only one of these traits appears in the offspring. The trait that shows up is the dominant trait, while the trait that remains hidden is the recessive trait. Because of the dominant character of one allelic gene over the other, we must distinguish between phenotype and genotype of the individual. Recall that the phenotype is the characteristic that is observed—that is, the expressed trait. It is determined by both the genotype (the genetic makeup of the allelic pair) and the environment.

Let's take a look at three different combinations of *Y* and *y* alleles and the ways in which their genotypes and phenotypes compare and contrast with each other.

Genes	**Genotype**	**Phenotype**
YY	Homozygous Yellow	Yellow
Yy	Heterozygous Yellow	Yellow
yy	Homozygous Green	Green

Note that *Yy* is just as yellow as *YY*. One "dose" of *Y* is just as strong as two "doses."

Now let's cross a homozygous tall plant with a short one:

Parents = Homozygous tall (*TT*) + Homozygous short (*tt*)

A *Punnett square* can be used to illustrate the cross:

P (parents) *TT* × *tt*
Gametes *T* *T* *t* *t*

Fertilization:

	T	*T*
t	*Tt*	*Tt*
t	*Tt*	*Tt*

Punnett Square

A *Punnett square*, as shown in the text at left, is a useful tool. It provides a quick way to determine the probable traits of offspring produced from particular crosses.

Mendel's Laws

Mendel's laws are:

- Law of Segregation
- Law of Independent Assortment

The dominance of phenotypic traits is also sometimes referred to as one of Mendel's laws, the "Law of Dominance."

Results of the cross:

F_1 offspring are the offspring of the first generation. They are 100 percent hybrid (heterozygous tall). Note that even though one parent was short, all offspring are tall. The recessive trait disappears in the F_1 generation.

Law of Segregation

When two heterozygous dominant individuals are crossed, the recessive trait reappears in about 25 percent of the second F_2 generation. In meiosis, the heterozygous gene pairs separate out in random assortment. During fertilization, new chromosome combinations (and therefore new allelic combinations) are produced. Thus, two recessive alleles that have separated from dominant alleles can form a recessive pair.

F_1 cross *Tt* x *Tt*:

	T	*t*
T	*TT*	*Tt*
t	*Tt*	*tt*

Results of the cross.

F_2 generation: *TT*, *Tt*, *Tt*, *tt* (recessive phenotype reappears)
Phenotype ratio = 3 tall: 1 short
Genotype ratio = 1 pure tall: 2 hybrid tall : 1 short

Law of Independent Assortment

When dihybrids are crossed, the genes for each trait are distributed independently of the genes for all other traits. The chromosomes carrying these genes segregate independently in the disjunctive phase of the first meiotic division. This law holds true only if the genes are not linked on the same chromosome. Obviously, if alleles for different traits are on the same chromosome, they will all stay together (except after recombination).

Example of a dihybrid cross in which tall and yellow are both dihybrids:

	TY	*Ty*	*tY*	*ty*
TY	*TTYY*	*TTYy*	*TtYY*	*TtYy*
Ty	*TTYy*	*Ttyy*	*TtYy*	*Ttyy*
tYty	*TtYY*	*TtYy*	*ttYY*	*ttYy*
TtYy	*TtYy*	*Ttyy*	*ttYy*	*ttyy*

Results of the cross:

Phenotype ratio:

9 tall yellow ($\frac{9}{16}$): 3 tall green ($\frac{3}{16}$) : 3 short yellow ($\frac{3}{16}$) : 1 ($\frac{1}{16}$) short green

The simplest approach to an independent assortment problem is to consider each of the genes separately, determine the predicted Mendelian ratios for each of the traits alone, and then use the laws of probability to combine these. For example, in the cross above, the predicted Mendelian phenotype ratios are $\frac{3}{4}$ for tall and $\frac{1}{4}$ for green. The probability of observing these phenotypes together is the product of their independent probabilities—that is, $\frac{3}{4} \times \frac{1}{4}$, or $\frac{3}{16}$.

Three to One

Note that in the dihybrid cross under discussion here, each trait assorts individually in a 3:1 ratio, as is generally the case in a monohybrid cross. There are 9 tall yellow and 3 tall green for a total of 12 tall. There are also 3 short yellow and 1 short green, which amounts to a total of 4 short. Hence both the tall:short ratio (12:4) and the yellow:green ratio are 3:1.

Inheritance Patterns

Ethical restraints forbid geneticists to perform test crosses in human populations. Instead, they must rely on examining matings that have already occurred, using tools such as pedigrees. A *pedigree* is a family tree depicting the inheritance of a particular genetic trait over several generations. By convention, males are indicated by squares, and females by circles. Matings are indicated by horizontal lines, and descendants are listed below. Individuals affected by the trait are generally shaded, while unaffected individuals are unshaded. When carriers of sex-linked traits have been identified (typically, female heterozygotes), they are usually half shaded in family traits.

The following pedigrees illustrate two types of heritable traits, recessive disorders and sex-linked disorders. When analyzing a pedigree, look for individuals with the recessive phenotype. Such individuals have only one possible genotype—homozygous recessive. Matings between them and the dominant phenotype behave as test crosses; the ratio of phenotypes among the offspring allows deduction of the dominant genotype. In any case in which only males are affected, sex-linkage should be suspected.

Study Tip

When faced with a difficult pedigree on your SAT II: Biology exam, remember that recessive phenotypes can only have one possible genotype—homozygous recessive.

Recessive disorder. Note how the trait skips a generation in the autonomal recessive disorder depicted in the figure below. Albinism is an example of this form of disorder.

Quick Quiz

Match each of the following numbered traits with the correct lettered definition.

1. recessive trait
2. sex-linked recessive trait
3. dominant trait

(A) appears in every generation

(B) skips generations

(C) is almost always found in males

Answers:

1. = (B)
2. = (C)
3. = (A)

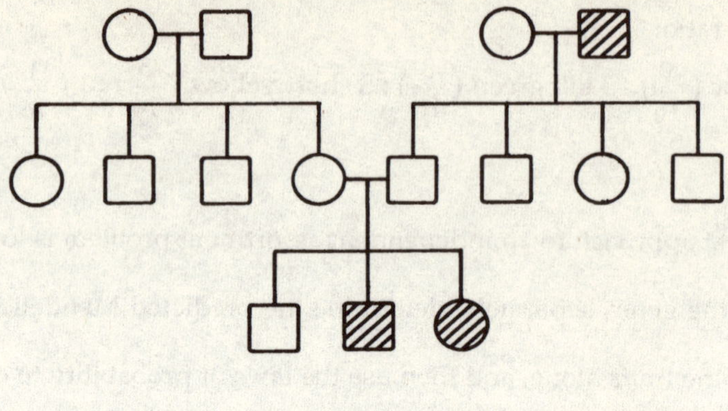

Recessive Disorder

Sex-linked disorders. Gender skewing is evident in this type of disorder, which includes traits such as hemophilia.

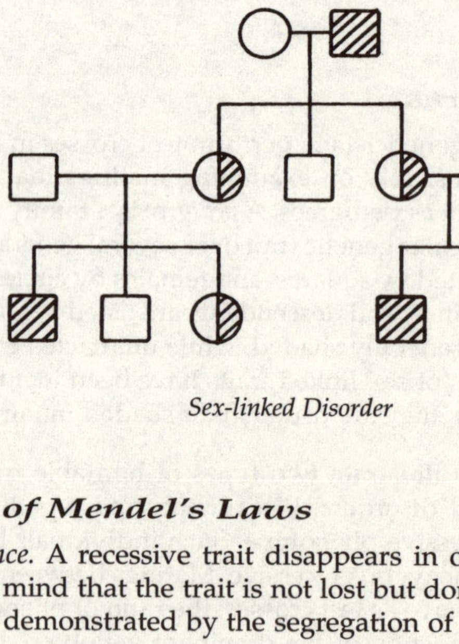

Sex-linked Disorder

Summary of Mendel's Laws

- *Dominance.* A recessive trait disappears in one (F_1) generation. Keep in mind that the trait is not lost but dominated phenotypically, as demonstrated by the segregation of subsequent generations.
- *Law of Segregation.* A recessive trait reappears after being masked for one generation by association with a dominant trait.
- *Independent Assortment of Genes (if not linked on the same chromosome).* Four types of gametes are formed by the dihybrid, due to the disjunction of homologous chromosomes. Nonlinked genes are thus inherited independently.

- *Pedigrees.* Pedigrees depict the inheritance of a particular genetic trait over several generations. They can be used to study Mendel's Laws in action.

The study of classical genetics requires an understanding of meiosis, the mechanism of gamete formation. Mendel knew that alleles were inherited from each parent, and that these alleles were somehow linked to the various characteristics he studied in his peas, but it was not until meiosis was truly elucidated that the mechanisms behind heredity were understood.

Meiosis

As we have seen, sexual reproduction differs from asexual reproduction in that two parents are involved and the end result of their union is genetically unique offspring. This process occurs via the fusion of two gametes, specialized sex cells produced by each parent. *Meiosis* is the process whereby these sex cells are produced.

As in mitosis, the gametocyte's chromostomes are replicated during the S phase of the cell cycle, and the centrioles replicate at some point during interphase. The first round of division (meiosis I) produces two intermediate daughter cells. The second round of division (meiosis II), similar to mitosis, involves the separation of the sister chromatids, resulting in four genetically distinct haploid gametes. In this way, a diploid cell produces haploid daughter cells.

Each meiotic division has the same four stages as mitosis, although it goes through each of them twice (except for DNA replication). The stages of meiosis are detailed in the following paragraphs.

Interphase I

Gametocyte chromosomes are replicated during the S phase of the cell cycle, while the centrioles replicate at some point during interphase.

Prophase I

During this stage, chromatin condenses into chromosomes, the spindle apparatus forms, and the nucleoli and nuclear membrane disappear. Homologous chromosomes (chromosomes that code for the same traits, one inherited from each parent), come together and intertwine in a process called *synapsis*. Since at this stage each chromosome consists of two sister chromatids, each synaptic pair of homologous chromosomes contains four chromatids, and is therefore often called a *tetrad*.

Sometimes chromatids of homologous chromosomes break at corresponding points and exchange equivalent pieces of DNA; this process is called *crossing over*. Note that crossing over occurs between homologous chromosomes and not between sister chromatids of the same chromosomes. The

Don't Mix These Up on Test Day

Mitosis:

- Produces $2n$ cells from $2n$ cells
- Occurs in all dividing cells
- Does not involve the pairing up of homologous chromosomes
- Does not involve crossing over

Meiosis:

- Produces n cells from $2n$ cells
- Occurs only in sex cells (gametocytes)
- Involves the pairing up of homologous chromosomes at the metaphase plate, forming tetrads
- Involves crossing over

chromatids involved are left with an altered but structurally complete set of genes.

The chromosomes remain joined at points called chiasmata where the crossing over occurred. Such genetic recombination can "unlink" linked genes, thereby increasing the variety of genetic combinations that can be produced via gametogenesis. Recombination among chromosomes results in increased genetic diversity within a species. Note that sister chromatids are no longer identical after recombination has occurred.

Metaphase I

Homologous pairs (tetrads) align at the equatorial plane, and each pair attaches to a separate spindle fiber by its kinetochore.

Anaphase I

Homologous pairs separate and are pulled to opposite poles of the cell. This process is called *disjunction,* and it accounts for a fundamental Mendelian law. During disjunction, each chromosome of paternal origin separates (or disjoins) from its homologue of maternal origin, and either chromosome can end up in either daughter cell. Thus, the distribution of homologous chromosomes to the two intermediate daughter cells is random with respect to parental origin. Each daughter cell will have a unique pool of alleles provided by a random mixture of maternal and paternal origin. These genes may code for alternative forms of a given trait.

Telophase I and Cytokinesis

A nuclear membrane forms around each new nucleus. At this point, each chromosome still consists of sister chromatids joined at the centromere. The cell divides through cytokinesis into two daughter cells, each of which receives a nucleus containing the haploid number of chromosomes. Between cell divisions there may be a short rest period, or interkinesis, during which the chromosomes partially uncoil.

Prophase II

The centrioles migrate to opposite poles and the spindle apparatus forms.

Metaphase II

The chromosomes line up along an equatorial plane. The centromeres divide, separating the chromosomes into pairs of sister chromatids.

Anaphase II

The sister chromatids are pulled to opposite poles by the spindle fibers.

Deadly Mutations

If, during Anaphase I or II of meiosis, homologous chromosomes or sister chromatids fail to separate (in what is termed *nondisjunction*), one of the resulting gametes will have two copies of a particular chromosome and the other gamete will have none. Subsequently, during fertilization, the resulting gamete may have one too many (47) or one too few (45) copies of the chromosome in question. Few of these mutated gametes survive. Those that do encounter difficulties associated with conditions like Down's syndrome (Trisomy 21).

Telophase II

Finally, a nuclear membrane forms around each new haploid nucleus. Cytokinesis follows and two daughter cells are formed. Thus, by the time meiosis is completed, four haploid daughter cells are produced per gametocyte. In females, only one of these three becomes a functional gamete.

The random distribution of chromosomes in meiosis, coupled with crossing over in prophase I, enables an individual to produce gametes with many different genetic combinations. This is why sexual reproduction produces genetic variability in offspring, as opposed to asexual reproduction, which produces identical offspring. The possibility of so many different genetic combinations is believed to increase the capability of a species to evolve and adapt to a changing environment.

In a Nutshell

Meiosis I involves the reduction of chromosome number.

Meiosis II is more similar in nature to mitosis; chromosome number remains unchanged.

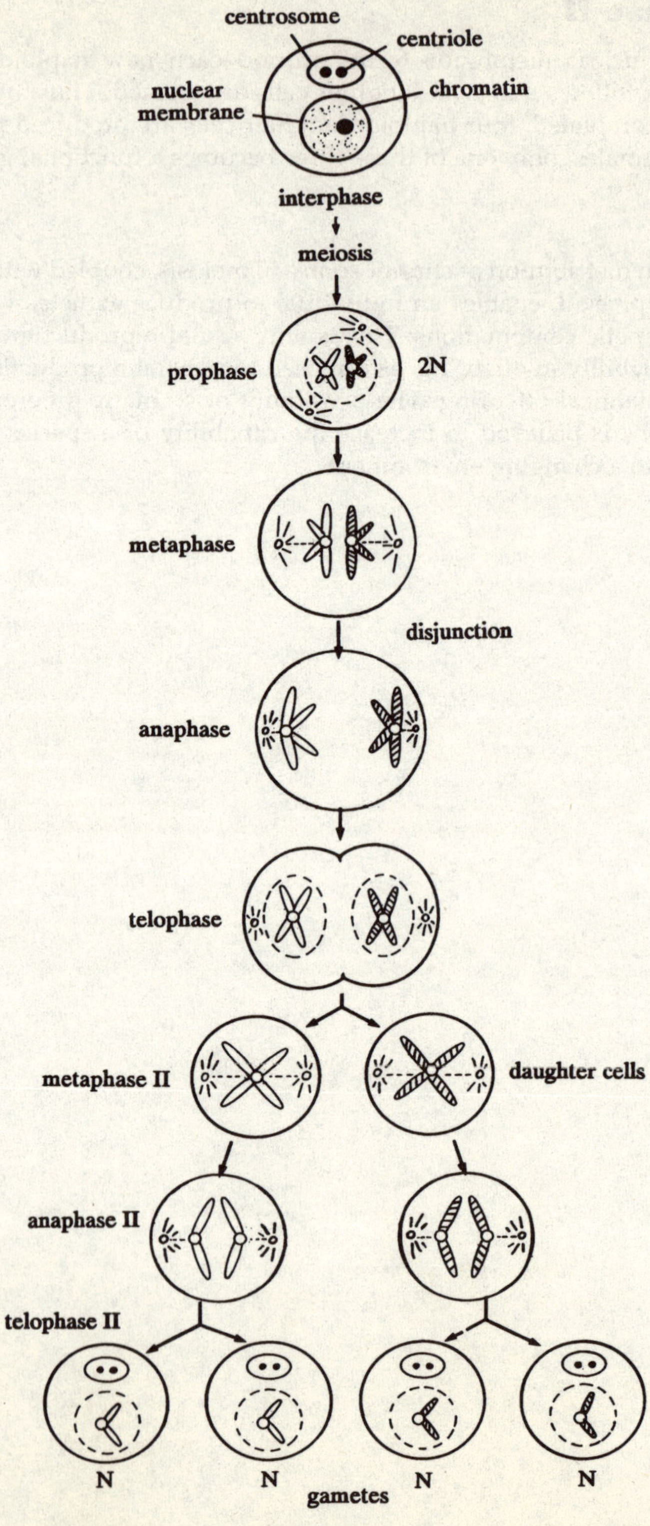

Meiosis

Non-Mendelian Inheritance Patterns

While Mendel's laws hold true in many cases, these laws cannot explain the results of certain crosses. Sometimes an allele is only incompletely dominant or, perhaps, codominant. The genetics that enable the human species to have two genders would also not be possible under Mendel's laws.

Don't Mix These Up on Test Day

In *incomplete dominance*, two traits are blended together; both are partially expressed, and neither dominates.

In *codominance*, both traits are fully expressed; neither dominates.

Incomplete Dominance

Incomplete dominance is a blending of the effects of contrasting alleles. Both alleles are expressed partially, neither dominating the other.

An example of incomplete dominance is found in the four-o'clock plant and in the snapdragon flower. When a red flower (*RR*) is crossed with a white flower (*WW*), a pink blend (*RW*) is created. When two pink flowers are crossed, the yield is 25 percent red, 50 percent pink, and 25 percent white (phenotypic and genotypic ratio 1:2:1).

Sex Determination

Most organisms have two types of chromosomes: *autosomes*, which determine most of the organism's body characteristics, and *sex chromosomes*, which determine the sex of the organism. Humans have 22 pairs of autosomes and one pair of sex chromosomes. The sex chromosomes are known as *X* or *Y*. In humans, *XX* is present in females and *XY* in males. The *Y* chromosome carries very few genes. Sex is determined at the time of fertilization by the type of sperm fertilizing the egg, since all normal eggs contain *X* chromosomes only. If the sperm carries an *X* chromosome, the offspring will be female; if the sperm carries a *Y* chromosome, the offspring will be male.

This process is illustrated in the Punnett square below:

	X	*Y*
X	*XX*	*XY*
X	*XX*	*XY*

The ratio of the sex of the offspring is 1:1.

Sex Linkage

Genes for certain traits, such as color blindness or hemophilia, are located on the *X* chromosomes. Hence these genes are linked with the genes controlling sex determination. These genes seem to have no corresponding allele on the Y chromosome, with the result that the X chromosome contributed by the mother is the sole determinant of the character of the offspring. Genes determining two afflictions—hemophilia and red-green color-blindness—are sex-linked (on the X chromosome). They are recessive, implying that they can be hidden by a dominant normal allele on another X-chromosome in a female. For this reason, the female with two X chromosomes may carry, but will rarely exhibit, these afflictions. The male, on the other hand, with his "noninformational" Y chromosome, has no dominant allele to mask the recessive gene on his X chromosome. As a consequence of this, males exhibit sex-linked traits much more frequently than females do.

Cross 1: Let's see what happens when we cross a hemophilia-carrying female and a normal male:

$XX_h \times XY$:

	X	X_h
X	XX	XX_h
Y	XY	X_hY

Results of the cross:

XX = healthy female
XX_h = carrier but healthy female
XY = healthy male
X_hY = hemophiliac male

Cross 2: Here's a cross between a carrier female and a male hemophiliac:

$XX_h \times X_hY$:

	X	X_h
X_h	XX_h	XhX_h
Y	XY	X_hY

Results of the cross:

X_hX_h= hemophiliac female (very rare)
XX_h = carrier but healthy female
XY = healthy male
X_hY = hemophiliac male

Codominance

Certain cells contain multiple alleles, which implies that more than two alleles may act in the inheritance of certain traits. For example, a combination of two of three possible alleles determines blood type.

A cell has two alleles, one from its mother and one from its father. The combination of these alleles will result in four possible bloodtypes. The allele for blood type A, I_A, and the allele for blood type B, I_B, are both dominant to the third allele, i. I_A and I_B may appear together to form blood type AB; however, when both are absent, blood type O results. To summarize:

- I_A = dominant gene for producing antigen A on the red blood cell
- I_B = dominant gene for producing antigen B on the red blood cell
- i = recessive gene, does not produce either antigen

And these genes combine in various ways to form the following possible genotypes and blood types (phenotypes):

- I_AI_A or I_Ai = Type A blood
- I_BI_B or I_Bi = Type B blood
- I_AI_B = Type AB blood
- ii = Type O blood

Mutations

Mutations are changes in the genes that are inherited. It is the accumulation of mutations that makes a species evolve. In order to be transmitted to the succeeding generation, mutations must occur in sex cells—eggs and sperm—rather than somatic cells (body cells. Mutations in nonsex cells are called somatic cell mutations and affect only the individual involved, not subsequent generations (for example, some tumors). Most mutations are recessive and deleterious (harmful). Because they are recessive, these mutations can be masked or hidden by the dominant normal genes and are retained in subsequent generations; although they are deleterious, they are not eliminated.

Chromosomal Mutations

These mutations result in changes in chromosome structure or abnormal chromosome duplication. In crossing over, segments of chromosomes switch positions during meiotic synapsis. This process breaks linkage patterns.

Flashback

Remember that in protein transcription, discussed in our chapter on cellular and molecular biology, DNA mutations could occur in two ways:

- Point mutations involve changes in single nucleotide bases in the DNA sequence
- Frameshift mutations involve the insertion or deletion of nucleotides, changing the reading frame of the protein

Nondisjunction, meanwhile, is the failure of some homologous pairs of chromosomes to separate following meiotic synapsis. The result is an extra chromosome or a missing chromosome for a given pair. For example, Down's syndrome is due to an extra chromosome #21 (Trisomy 21). The number of chromosomes in a case of single nondisjunction is $2n + 1$ or $2n - 1$. In Trisomy 21, the individual has 47 chromosomes instead of the usual 46.

The Two Faces of Mutations

Mutations may be beneficial (enabling organisms to develop new adaptations to the changing environment and leading to evolution of a species). Yet they can also damage the organisms in which they occur, sometimes fatally. Mutations are the root cause of diseases like cancer.

Polyploidy ($3n$ or $4n$) involves a failure of meiosis during the formation of the gametes. The resulting gametes are $2n$. Fertilization can then be either $n + 2n = 3n$ or $2n + 2n = 4n$. Polyploidy is always lethal in humans. Finally, *chromosome breakage* might be induced by environmental factors or mutagenic drugs.

Gene Mutations

As discussed in the chapter on cellular and molecular biology, there might be changes in the base sequence of DNA that result in changes in single genes, and these are the most important type of mutation.

Mutagenic Agents

Mutagenic agents induce mutations. For example, *uv* light, X-rays, radioactivity, and some chemicals will cause mutations. Such agents are also typically carcinogenic.

This concludes our chapter on classical genetics. With the knowledge you have gained here, you should be able to look at your own family and study the heritability of certain traits. If you have blue eyes, a recessive trait, you must be homozygous for that trait; each of your parents gave you an allele for blue eyes. If one of your parents has brown eyes, he or she must be a heterozygote, possessing alleles for both blue and brown eyes. Now let's see how you do on some quiz questions.

Classical Genetics Quiz

1. Breeding animals of close genotypes or phenotypes is known as

   (A) inbreeding
   (B) hybridization
   (C) cross-breeding
   (D) selective breeding
   (E) test breeding

2. A process that CANNOT take place in haploid cells is

   (A) mitosis
   (B) meiosis
   (C) cell division
   (D) growth
   (E) digestion

3. A B C D

   If the diagram above represents genes on a chromosome, which genes would have the highest frequency of crossover?

   (A) A and B
   (B) A and D
   (C) B and C
   (D) B and D
   (E) the frequencies are the same for all crossovers

4. Laboratory mice are to be classified based on genes *A*, *B*, and *C*. How many genetically different gametes can be formed by a mouse that is genotypically *AaBbCc*? (Assume that none of these is a lethal gene.)

   (A) 3
   (B) 6
   (C) 8
   (D) 9
   (E) 12

5. The A-B-O human blood groups are inherited through a system of

   (A) multiple alleles
   (B) dihybrid crosses
   (C) recessive alleles
   (D) independent assortments
   (E) spontaneous mutations

6. The gene for red-green color blindness is located on the *X* chromosome. The offspring of a man suffering from red-green color blindness would have which of the following characteristics?

   (A) 50% of the females would be carriers; 100% of the males would be affected.
   (B) 100% of the females would be normal; 50% of the males would be affected.
   (C) 100% of the females would be carriers; 100% of the males would be normal.
   (D) 50% of the females would be affected; 100% of the males would be affected.
   (E) 100% of the females would be normal; 50% of the males would be carriers.

GO ON TO THE NEXT PAGE

7. A mutation in a gene in a somatic cell is deleterious because

(A) it will affect gamete formation
(B) it will be dominant
(C) it may be passed on to subsequent generations
(D) it may lead to a tumor in that tissue
(E) none of the above

8. Which of the following genetic mutations will NEVER affect the protein produced?

(A) point
(B) silent
(C) insertion
(D) frame shift
(E) all of the above

9. Which of the following will NOT affect the frequency of a gene in an ideal population?

(A) environmental selective pressure
(B) mutation
(C) random breeding
(D) nonrandom matings
(E) selective emigration

10. A typical human gamete

(A) contains a haploid number of genes
(B) always contains an *X* or *Y* chromosome
(C) is a result of the meiotic process
(D) has genetic material that has undergone recombination
(E) all of the above

11. Spermatogenesis and oogensis differ in that

(A) spermatogenesis is mitotic while oogenesis is meiotic
(B) oogenesis is mitotic while spermatogenesis is meiotic
(C) spermatogenesis produces gametes while oogenesis does not
(D) spermatogenesis produces four haploid sperm cells while oogenesis produces one egg cell and more than one polar body
(E) spermatogenesis involves unequal division of cytoplasm

12. If a male with blood type A marries a female with blood type B, which of the following types would be impossible for a first generation child?

(A) type B
(B) type A
(C) type O
(D) type AB
(E) all types are possible

GO ON TO THE NEXT PAGE

13. Polar bodies are formed during

(A) male mitosis
(B) female mitosis
(C) male meiosis
(D) female meiosis
(E) two of the above

14. Red is dominant over white in a certain flower. To test whether a red offspring is homozygous or heterozygous in this flower, one would

(A) cross it with a red plant that had a white parents
(B) cross it with a red plant that had two red parents
(C) cross it with a white plant
(D) two of the above will work
(E) none of the above will work

15. Green (*Y*) if dominant over yellow (*y*) in peas, and smooth peas (*W*) are dominant over wrinkled peas (*w*). Which cross must produce all green, smooth peas?

(A) *YyWw* X *YyWw*
(B) *Yyww* X *YYWw*
(C) *YyWW* X *yyWW*
(D) *YyWw* X *YYWW*
(E) none of the above

16. Unequal division of the cytoplasm occurs in

(A) production of sperm cells
(B) production of egg cells
(C) mitosis of an epidermal cell
(D) binary fission in bacteria
(E) none of the above

17. Disjunction is the process whereby

(A) paternal chromosomes separate from maternal chromosomes
(B) homologous pairs of chromosomes recombine
(C) the spindle apparatus is formed from the centrioles
(D) the cell membrane invaginates to form two daughter cells
(E) none of the above

18. Tall is dominant over short in a certain plant. A tall plant was crossed with a short plant, and both tall and short offspring were produced. This demonstrates

(A) the law of segregation
(B) incomplete dominance
(C) linkage
(D) mutation
(E) the law of independent assortment

Answers and Explanations to Classical Genetics Quiz

1. **(A)** Inbreeding occurs when animals that are closely related, i.e., that have close genotypes or phenotypes, are bred to create progeny that are homozygous at all the alleles. (B) Hybridization occurs when one breeds animals that are phylogenetically distinct in order to develop an animal that has characteristics of both parents. (C) Cross-breeding, the crossing of two animals, can be used to determine phenotype when the genotype is known, to determine genotype when the phenotype is known, or to investigate codominance, expressivity, or penetrance. Selective breeding (D) is defined as the creation of certain strains of specific traits through controlled breeding, while test breeding (E) is the breeding of an organism with a homozygous recessive in order to determine whether that organism is homozygous dominant or heterozygous dominant for a given trait.

2. **(B)** A cell that is n (haploid) cannot undergo meiosis to become $\frac{1}{2}n$. (A), (C), and (D) are incorrect because there are a number of organisms that are haploid. These organisms undergo mitosis, divide, and grow. An example of such an organism is a braconid wasp. In the haploid form, these animals are males (n) and females ($2n$), and are formed only when a female mates. Meanwhile, (E) is incorrect because an organism, whether it is diploid or haploid, must be able to digest to maintain life.

3. **(B)** Homologous recombination occurs during metaphase I tetrad formation. The farther apart two genes are, the more likely a homologous recombination will occur between them. Therefore, the genes that are farthest apart are also those most likely to cross over. It is also important to note that the farther away from the centromere genes are, the more likely they are to recombine.

4. **(C)** The gametes that can be formed by the mouse in this case are *ABC, ABc, AbC, Abc, aBC, aBc, abC*, and *abc*.

5. **(A)** Multiple alleles describe more than two allele possibilities. In the ABO blood group, there are three alleles: *A, B*, and *O*. *A* and *B* are codominant over the recessive allele *O*. There are four possible phenotypes, *A, B, AB*, and *O*, and six possible genotypes, *AA, AO, BB, BO, AB*, and *OO*. Meanwhile, dihybrid crosses (B) are crossing organisms heterozygous for two traits, such as *AaBb* x *AaBb*. These crosses will have the characteristic 9: 3: 3: 1 ratio of offspring. (C) refers to recessive alleles that are not expressed unless they are homozygous, while (D) refers to the fact that two genes will be inherited independently of each other if they are not linked physically on the chromosome. Finally, spontaneous mutations (E) are changes in genes that may be traced to spontaneous changes in base sequence.

6. **(C)** A male affected with red-green color blindness would have a genotype of $X_{cb}Y$. If he mated with a normal female, *XX*, all their female offspring would be $X_{cb}X$, receiving one good copy of the *X* chromosome from their mother and the color blindness gene from their father. Since color blindness is a recessive trait, all the offspring would be carriers. All the male offspring of this mating would all be *XY*, receiving one good copy of the *X* chromosome from their mother and the *Y* chromosome from their father.

7. **(D)** Mutations in somatic cells (cells of the body, not germ tissue) affect only the individuals involved. They cannot be passed on to the next generation and will not affect gamete formation. These mutations are typically recessive, although there are

some instances of dominant negative mutations. The major concern about mutations in somatic cells is that they are linked to the development of tumors; these are produced by proteins that have lost their functions due to somatic mutation. Tumors like malignant melanomas, on the other hand, are caused by mutations to the DNA as a result of *uv* irradiation.

8. (B) A silent mutation is a point mutation that either occurs in a noncoding region or does not change the amino acid sequence, due to degeneracy of the genetic code. Therefore, they do not affect the protein produced. Point mutations, meanwhile, occur when a single nucleotide base is substituted for another nucleotide base. A frameshift mutation is either an insertion or deletion of a number of nucleotides. These mutations have serious effects on the protein coded for, since nucleotides are read as series of triplets. The addition or loss of nucleotides (except in multiples of three) will change the reading frame of the mRNA.

9. (C) The Hardy-Weinberg Law states that for a population to have stability in its gene frequencies, it must be large, with no migration, no mutation, and random breeding. Only if these criteria are met will gene frequencies not change. Only (C) is one of these criteria; the other answer choices will all affect gene frequency.

10. (E) During meiosis, the gamete reduces its genetic component from $2n$ to n, resulting in a haploid cell with half the normal chromosome number. When a haploid egg and sperm unite, they form a diploid organism known as a zygote. All ova will contain an *X* chromosome, and all sperm will contain either an *X* or a *Y* chromosome. These gametes are formed during the two reductional divisions of meiosis. During Metaphase I of Meiosis I, tetrads form and sister chromatids undergo a homologous recombination known as crossing over.

11. (D) Spermatogenesis and oogenesis may both be categorized as forms of gametogenesis in that haploid gametes are produced through reductional division (meiosis) of diploid cells. Both processes occur in the gonads. They differ, however, in that in spermatogenesis, the cytoplasm is equally divided during meiosis, and four viable sperm are produced form one diploid cell. In oogenesis, the cytoplasm is divided unequally and only one ovum, containing the bulk of the cytoplasm, is produced, along with two or three inert polar bodies.

12. (E) A and B blood groups are codominant over blood antigen O. Therefore, if a man heterozygous for blood type A (AO) married a woman heterozygous for blood type D (BO), they could have children with the possible genotypes *AO*, *BO*, *OO*, or *AB*. Therefore, all blood types are possible in this mating.

13. (D) Polar bodies may be haploid or monoploid; they are nonfunctional, gametelike cells that are formed during female meiosis. Recall that meiosis is a two-stage process. In the first stage, a diploid cell undergoes a reduction division to form two haploid cells. In the second division, each of those haploid or monoploid cells undergoes a second division to form two haploid or monoploid cells each. So, there are a total of four haploid or monoploid cells formed from each original diploid germ cell. In the case of sperm cells, four functional haploid or monoploid gamete sperm cells are formed. In the case of egg cells, the first meiotic division involves an unequal division of cytoplasm, resulting in the formation of one large cell and one small cell. The large cell will go on to divide again, while the smaller cell is known as the first polar body. This small cell polar body may then divide again to form two other polar bodies. The large cell, meanwhile, undergoes a second meiotic division, again unequally dividing the cytoplasm into one larger resulting cell, the final haploid or monoploid ovum, and another small cell, which again becomes a polar body. Hence it is possible that during a female meiotic division, one large ovum or functional egg cell and three polar bodies may be formed. Polar bodies are not formed in mitosis, which is how all other cell divisions occur in the body; in other words, equal cell divisions do not form sex cells.

14. (C) This question illustrates a test cross. A test cross is performed to determine if a particular dominant individual's phenotype is a homozygous or heterozygous genotype. In this case, there are two possible red genotypes: *RR*, the pure homozygous red, and *Rr*, the hybrid heterozygous red. These two individuals would have the same phenotype. In order to determine the genotype, the unknown red plant would be mated with a recessive, white plant. If the red plant and the white plant produce only red offspring, then it can be assumed that the original red plant is homozygous. If the mating of the unknown red organism and the white organism produces any white offspring at all, then we know that the original unknown red was heterozygous. This is because white offspring can be produced only through the production of one white gamete by each parent.

15. (D) Both green and smooth are dominant phenotypes. The goal in this question is to produce only green smooth peas, so we want only dominant phenotype offspring. Therefore, we must avoid any crossing that may result in the mating or combining of two recessive alleles. In (A), crossing *Yy* and *Yy* could result in approximately one quarter of the offspring turning out yellow. Similarly, in (B), *ww* crossed with *Ww* would produce approximately half of the offspring possessing a wrinkled phenotype. In (C), *Yy* crossed with *yy* is likely to produce offspring which are approximately half yellow. (D) is correct because one of the parents is a double dominant, meaning that all offspring will have the dominant phenotype, regardless of the genotype of the other parent.

16. (B) Unequal cytoplasm division occurs when egg cells are produced during the meiotic process of oogenesis. All meiotic divisions may be divided into two stages. In the first stage of egg production, also known as oogenesis, the precursor diploid cell produces two daughter cells, but one of the daughter cells receives almost the entire amount of cytoplasm, while the other becomes a nonfunctioning polar body. In the second division of oogenesis, the large daughter cell divides again, and once again one of the new daughter cells receives almost all of the cytoplasm, and the other becomes a small nonfunctional polar body. The original first polar body may also divide to form two nonfunctional polar bodies. The final result is a potential four haploid or monoploid cells, but only one of them—the one that received a greater amount of cytoplasm during each meiotic division—becomes a functional egg cell. (A) is incorrect; during spermatogenesis, one diploid precursor cell forms four functional haploid, or monoploid, sperm cells. In this case, both divisions are equal, and all sperm cells are equal in amount of cytoplasm. As for (C), during mitosis of epidermal cells, cytoplasm is distributed equally. Likewise, in binary fission of bacteria, in which bacterial cells are dividing as a means of reproduction, cytoplasmic division will be equal.

17. (A) Disjunction is defined as the separation of maternal and paternal chromosomes during meiosis. Each tetrad is separated into two halves. One of each pair of chromosomes (each containing two chromatids) is pulled to opposite ends of the cell. Note that some of the maternal chromosomes can go to one end and some to the other end of the cell. The distribution of homologous chromosomes between the two resultant nuclei is random.

18. (A) The law of segregation states that when gametes are formed, the two alleles for a particular trait will separate or segregate into the gametes, so that each of the gametes only contains one of the alleles for a given trait. So if tall is dominant over short, and both tall and short offspring were produced, then the tall plant is a hybrid or heterozygous plant. This means that the genotype of the tall plant contains both one tall allele and one short allele. The short plant contains two short alleles. When the gametes are formed for this mating, the two alleles in the tall plant, the tall and short, will segregate into the gametes, forming tall-containing gametes and short-containing gametes. When these meet and fertilize, the short-containing gametes from the other plant, half the offspring produced will be tall because they are the result of the tall gamete's fertilization of a short gamete, and the other half will be the result of the short gamete's fertilization of the other short gamete.

In (B), incomplete dominance, or blending, occurs when two individuals mate and the resulting offspring is a phenotype that appears to be midway between the two phenotypes of the two parents. For example, if a tall and short plant were crossed in this case and they produced an offspring of medium height, the result would be incomplete dominance. Note that in this problem, tall is dominant over short, resulting in a complete dominance. As for (C), linkage refers to genes or alleles that travel with each other on the same chromosome. Hence linked genes cannot segregate into two separate gametes because they are on the same chromosome. Genes or alleles which separate must, by definition, be on separate chromosomes. (D) is not correct either; mutation refers to changes in the DNA sequence of a chromosome, and there is no evidence of mutation occurring in this question. Finally, in (E) the law of independent assortment states that when we are dealing with more than one trait at a time and these traits are not linked, they are carried on different chromosomes, and the inheritance of these traits is not connected.

CHAPTER SIX

ORGANISMAL BIOLOGY

Complex organisms like humans have developed a number of systems to ensure genetic diversity, maintain homeostasis, circulate food, and transport wastes, among other essential bodily functions. Without these systems, humans would never have advanced past the stage of unicellular organisms; we would have developed neither basketball nor CD players. You'll probably see many questions on organismal biology on your SAT II exam. The following chapter will give you the basics of what you'll need to know about topics like reproduction, physiology, and animal behavior.

Reproduction

One of the most essential systems associated with organismal biology is reproduction, the mechanism by which organisms propagate themselves. Without reproduction, no organism would survive past the first generation. Reproduction can be divided into three broad topics: cell division, asexual reproduction, and sexual reproduction.

Cell Division

Cells must replicate if an organism is to grow or repair itself. The process by which they accomplish this is known as cell division. In unicellular organisms, cell division is a means of reproduction, while for multicellular organisms it serves as a method of growth, development, and replacement of worn-out cells. Bacteria replicate themselves through the relatively simple practice of binary fission (which we will discuss in further detail below), while higher organisms must engage in the more complex process of *mitosis.* Without mitosis, which may be defined as the splitting of genetic material in a cell to form new cells, we would never get any taller, and our cuts and scrapes would never heal.

Believe It or Not

It's hard to believe, but true—all the nucleated cells of your body, regardless of structure or function, have *exactly* the same number of chromosomes (including two sex chromosomes). The only exceptions are gametes, which have half the normal number of chromosomes. This means that different cell types have different structures and functions not because their DNA is different, but because what is *expressed* by that DNA is different.

Mitosis

The four stages of the cell cycle are designated as G_1, S, G_2, and M. The first three stages of this cell cycle are interphase stages—that is, they occur between cell divisions. The fourth stage, mitosis, incorporates the actual division of the cell.

A Typical Cell's Day

Typically, cells spend 90 percent of their time in interphase, and only 10 percent in the mitotic phase.

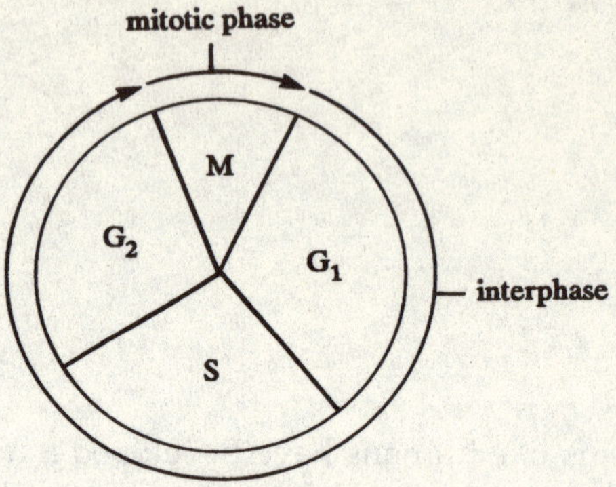

The Cell Cycle

Stage G_1. G_1 is characterized by intense biochemical activity and growth. The cell doubles in size, and new organelles such as mitochondria, ribosomes, and centrioles are produced.

Stage S. This is the stage during which synthesis of DNA takes place. Each chromosome is replicated so that during division, a complete copy of the genome can be distributed to both daughter cells. After replication, the chromosomes consist of two identical sister chromatids held together at a central region called the centromere. The ends of the chromosomes are called telomeres. Each cell is still $2n$, but each chromosome now consists of two chromatids; cells entering G_2 actually contain twice as much DNA as cells in G_1.

Stage G_2. The cell continues to grow in size, as assembly of new organelles and other cell structures continues.

Stage M (Mitosis and Cytokinesis). Mitosis is broken down into four stages: prophase, metaphase, anaphase, and telophase. Upon completion of mitosis, the cell engineers its split into daughter cells via cytokinesis.

Prophase. In this phase, chromosomes condense as centriole pairs separate and move toward opposite poles of the cell and the spindle apparatus forms between them. The nuclear membrane dissolves, allowing spindle fibers to enter the nucleus. During this time, the nucleoli become less distinct or disappear. Kinetochores with attached kinetochore fibers appear at the chromosome centromere.

Metaphase. Centriole pairs are now at opposite poles of the cell. The kinetochore fibers interact with the spindle apparatus to align the chromosome at the metaphase plate (equatorial plate), which is equidistant to the two poles of the spindle fibers.

Anaphase. Centromeres now split so that each chromatid has its own distinct centromere, allowing the sister chromatids to separate. The telomeres are the last part of the chromatids to separate. The sister chromatids are pulled towards the opposite poles of the cell through the shortening of the kinetochore fibers.

Telophase. The spindle apparatus disappears. A nuclear membrane forms around each set of chromosomes, while the nucleoli reappear. The chromosomes uncoil, resuming their interphase form. Each of the two nuclei receives a complete copy of the genome identical to the original genome and to each other.

Cytokinesis. After the cell has divided its DNA in the processes described above, it enters the final stage of mitosis. In cytokinesis, the cytoplasm and all the organelles of the cell divide. This division ultimately results in the formation of two daughter cells from one parent cell.

In a Nutshell

Mitosis proceeds in the following stages:

- *Prophase*: chromosomes condense, spindles form
- *Metaphase*: chromosomes align
- *Anaphase*: sister chromatids separate
- *Telophase*: new nuclear membranes form

Asexual Reproduction

Asexual reproduction is defined as any method of producing new organisms in which fusion of nuclei (fertilization) does not take place. This means that in this type of reproduction, only one parent organism is involved in reproduction. The new organisms produced through asexual reproduction form daughter cells through mitotic cell division. Thus, offspring are genetically identical to their parents, unless mutations occur.

Asexual reproduction serves primarily as a mechanism for perpetuating primitive organisms and plants, especially in times of low population density. It is also utilized, in some form, by all plants. However, in more complex organisms, a new organism must form from undifferentiated cells, necessitating sexual reproducation. This is the reason that among animals, asexual reproduction is more prevalent in invertebrates than vertebrates. Complex animals like humans, meanwhile, are incapable of reproducing asexually.

Asexual Reproduction in Primitive Organisms

Primitive organisms reproduce asexually through the processes of binary fission and budding.

Binary Fission. Binary fission occurs in one-celled organisms such as amoebae, paramecia, algae, and bacteria. In this process, a single DNA molecule attaches to a plasma membrane during replication and duplication, while the cell continues to grow in size. Both the cytoplasm of the organism and its nucleus divide into equal parts; the former process is termed *cytokinesis* and the latter *karyokinesis*. Hence each daughter cell receives a complete copy of the original parent cell's chromosomes.

This type of reproduction occurs at a rapid pace. Undesirable, potentially harmful bacteria cells, for example, can reproduce every 20 minutes under optimal conditions. This is one reason it's a good idea to wash your hands often!

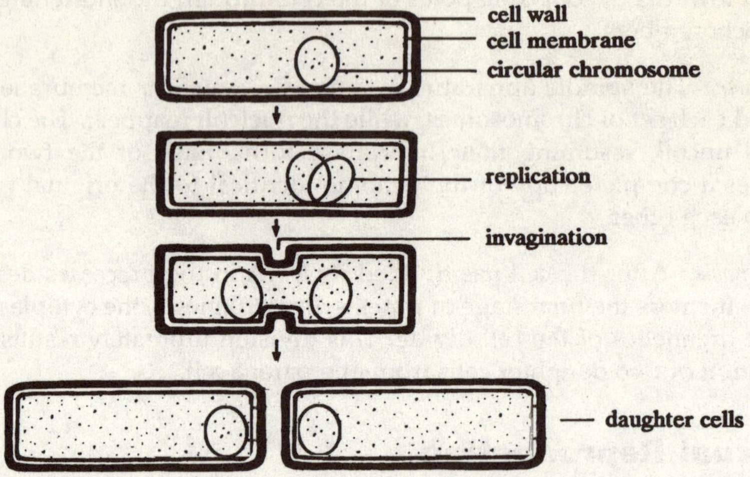

Binary Fission

Budding. Budding is very similar to binary fission, but it differs from it in that it involves an unequal division of cytoplasm (cytokinesis) and equal division of the nucleus (karyokinesis). The parent cell forms a group of many cells termed a bud, which extends from the parent and develops a mouth and tentacles. Eventually, this bud breaks off and grows independently into an adult organism. Although budding is common among unicellular organisms like yeast, it also occurs in some multicellular organisms, such as hydra.

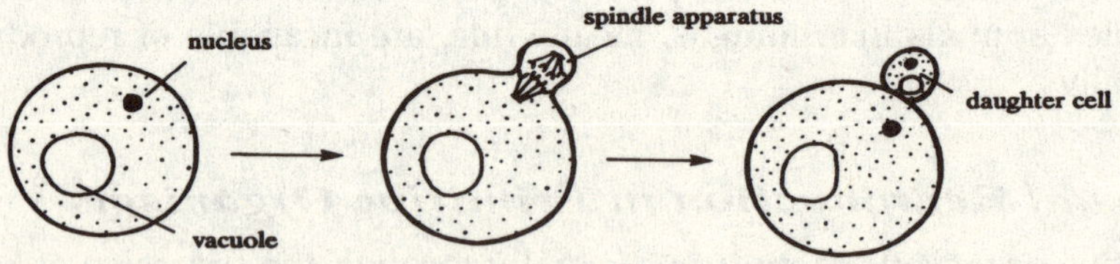

Budding

Don't Mix These Up on Test Day

Binary fission results in equal division of the cytoplasm and symmetrical daughter cells.

Budding involves an unequal division of the cytoplasm and asymmetrical daughter cells.

Asexual Reproduction in Plants

Many plants, meanwhile, rely on spore formation and vegetative propagation to increase their numbers.

Spore Formation. Some plants and bacteria produce specialized cells with hard coverings known as spores. The covering prevents loss of water from the spore contents. The spores are scattered by the air. Under favorable conditions (warmth, food, moisture), these spores will germinate.

Vegetative Propagation. Undifferentiated tissue (meristem) in plants provides a source of cells from which new plants can develop. Vegetative propagation offers a number of advantages to plants, including a lack of genetic variation, speed of reproduction, and production of seedless fruit. It can occur either naturally or artificially.

Natural forms of vegetative propagation include:

- *Bulbs.* These are parts of the root that split to form several new bulbs (an example is the tulip).
- *Tubers.* These modified underground stems have buds, such as the eye of a potato, which develop into new plants.
- *Runners.* Runners are plant stems that run above and along the ground, extending from the main stem. Near the main plant, new plants develop which produce new roots, as well as upright stems at intervals (as in lawn grasses).

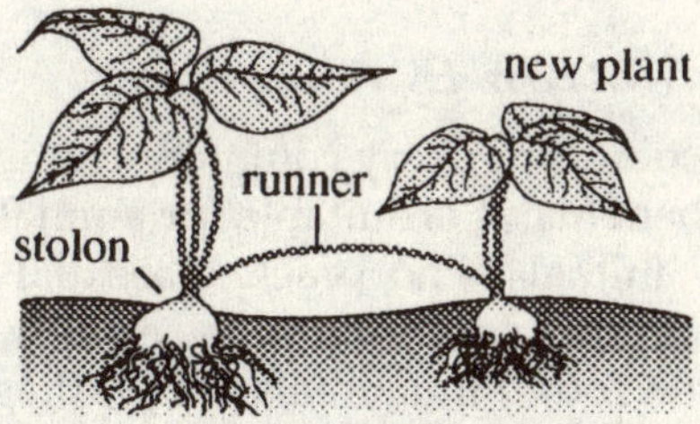

Fig. 4.4 Runner

- *Rhizomes.* Stolons is another term used for these woody, underground stems. They reproduce through new upright stems that appear at intervals, eventually growing into independent plants. The iris is a rhizome.

Meanwhile, artificial forms of vegetative propagation include:

- *Cutting.* When cut, a piece of stem will develop new roots in water or moist ground. Examples include the geranium and the willow. Plant growth hormones like auxins accelerate root formation in cuttings.
- *Layering.* The stems of certain plants, when bent into the ground and covered by soil, will take root. The connection between the main plant and this offshoot can then be cut, resulting in the establishment of a new plant. Blackberry and raspberry bushes reproduce in this manner.
- *Grafting.* Desirable types of plants can be developed and propagated using this method, in which the stem of one plant (scion) is attached to the rooted stem of another closely related plant

Quick Quiz

What kinds of vegetative propagation do the following plants utilize?

1) daffodil
2) strawberry
3) fern
4) yam

Answers:

1) bulb
2) runner
3) rhizome
4) tuber

(stock). One prerequisite for successful grafting is that the *cambium* (the tissue in stem that is not differentiated and allows stems to grow thicker) of the scion must be in contact with the cambium of the stock, since these two masses of undifferentiated cells must grow together to make one. Grafting does not allow for any mixing of hereditary characteristics.

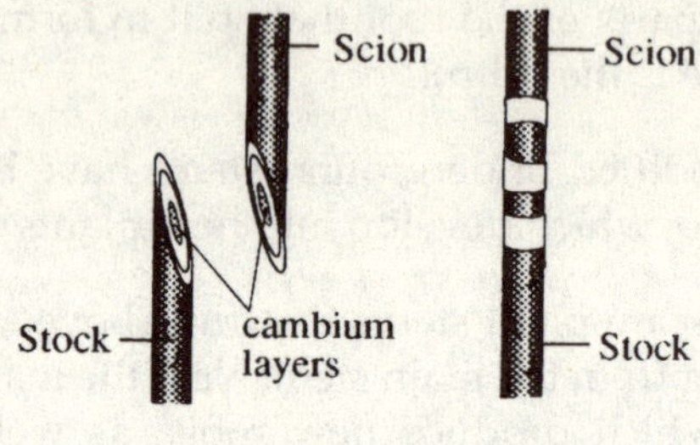

Grafting

The Eggs of the Rotifer

The rotifer, a tiny aquatic animal, has two types of eggs that develop via parthenogenesis. These eggs mature to produce females and degenerate males that cannot feed themselves. These males produce sperm that will fertilize eggs to produce zygotes. These zygotes are so resistant that they can survive even if the pond in which they live dries up.

Asexual Reproduction in Animals

This form of reproduction is not very common in animals, although it does bring with it certain benefits. It is suitable for animal populations that are widely dispersed, as animals who practice asexual reproduction do not need to find another animal to fertilize them sexually. The two major types of asexual reproduction found in animals are parthogenesis and regeneration.

Parthenogenesis. When people think about asexual reproduction, they usually have this process in mind. In parthenogenesis, an egg develops in the absence of fertilization by sperm. With no sperm present, the normal diploid number of chromosomes is not restored in the offspring. Parthenogenic organisms are haploid and are identical to their mothers.

This form of reproduction occurs naturally in bees: Fertilized eggs develop into worker bees and queen bees, while unfertilized eggs form male drone bees. Artificial parthenogenesis can be performed in some animals. The eggs of rabbits and frogs, for example, can be stimulated to develop without fertilization, by giving them an electron shock or a pin prick.

Regeneration. Regeneration, the ability of certain animals to regrow a missing body part, is far less common than parthogenesis. Sometimes parts of an animal grow into a complete animal, resulting in reproduction. For example, the planaria (a flatworm), the earthworm, the lobster, and the sea star can all regenerate limbs or entire organisms. This process is similar in nature to vegetative propagation.

Both regeneration and vegetative propagation are imperfectly understood. One possible explanation is that both are enabled by the presence of undifferentiated (embryonic) cell in certain plants or animals; these unspecialized cells can differentiate as they divide and grow to form new organs. If scientists could determine exactly how these processes occur, revolutionary advances could be made in, for example, medicine; doctors might be able to help victims of accidents grow back limbs or spinal columns.

Sexual Reproduction

Sexual reproduction is the method of reproducation practiced by most complex organisms. Utilizing this method enables such organisms to ensure genetic diversity and variability in their offspring. Sexual reproduction allows for the introduction of new alleles; it also maintains genetic diversity by allowing the organism to avoid *homozygosity* (i.e., the possession of two identical alleles for a given trait). Research has shown that *heterozygous* organisms, which possess two contrasting alleles for a given trait, have what is known as hybrid vigor. This hybrid vigor makes the offspring organism more "fit" in an evolutionary sense than either of its parents.

The major players in sexual reproduction are the sexual organs known as *gametes*. When the male gamete (the *sperm*) and the female gamete (the *egg*) join, a *zygote* is formed that develops into a new organism genetically distinct from both its parents.

Sexual Reproduction in Animals

The range of mechanisms involved in sexual reproduction in animals are detailed below.

Gamete Formation. Specialized organs called *gonads* produce *gametes*, which is the term given to functional sex cells. Male gonads, called *testes*, produce sperm in the seminiferous tubules, while female gonads, called *ovaries*, produce eggs. Organisms like the hydra and the earthworm are hermaphrodites, which implies that they have functional male and female gonads.

Spermatogenesis. Spermatogenesis is the term assigned to the development of sperm. In the testes, one diploid primary male sex cell continuously undergoes meiosis to form four functional haploid sperm. The testes are located outside the body because they must remain 2–4° cooler than the rest of the body to ensure proper development of sperm. The sperm itself has a head that contains DNA and a tail for motility; it contains either an X or a Y chromosome.

There is a specialized sac at the tip of the sperm called the *acrosome*; it is full of enzymes that allow the sperm to break into the egg. Researchers on male birth control have attempted to nullify this enzyme so that it cannot reach the egg.

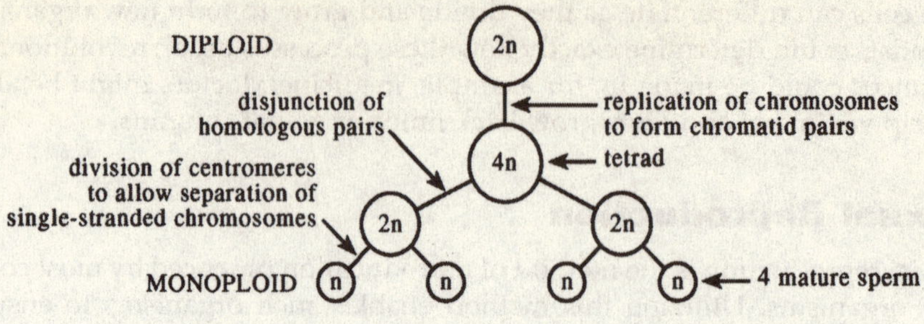

Spermatogenesis

Oogenesis. The egg, meanwhile, develops in a discontinuous process called oogenesis. This process is arrested in Meiosis 2 until a hormonal stimulus occurs in the menstrual cycle that allows egg development to be completed. One diploid primary female sex cell undergoes meiosis in the ovaries to produce one mature egg and three polar bodies. These eggs contain only *X* chromosomes. The mature egg, or ovum, contains most of the cytoplasm, while the polar bodies degenerate.

One salient difference between oogenesis and spermatogenesis is that females are born with all the eggs they will ever have, while males produce fresh sperm daily. This is the reason that genetic anomalies are more common in the eggs of older women; these anomalies have had more time to accumulate.

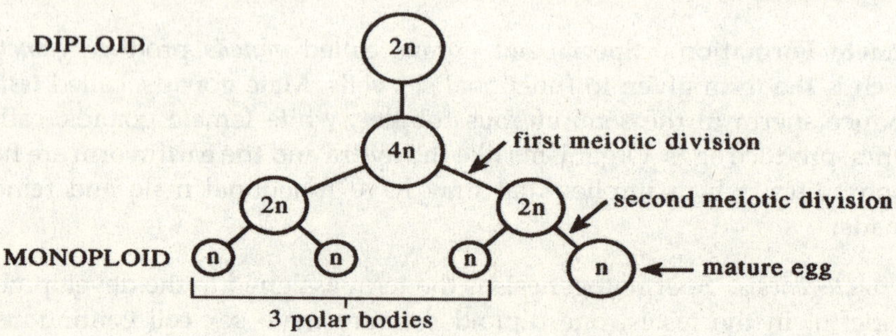

Oogenesis

Don't Mix These Up on Test Day

Spermatogenesis:

- Produces 4 mature sperm
- Produces *X* or *Y* chromosomes
- Is a continuous process
- Does not donate mitochondria to the embryo
- Produces fresh sperm daily

Oogenesis:

- Produces one egg and 2–3 polar bodies
- Produces *X* chromosomes
- Is a discontinuous process
- Donates mitochondria to the embryo
- Produces eggs only once in a female's lifespan; eggs age and have a higher incidence of mutations than sperm

Embryonic Development in Animals

After conception, animals that practice sexual reproduction must be provided with an environment, either internal or external, that allows the cells of their future offspring to differentiate and mature. Different types of embryos require different organic evironments in order to develop successfully.

Zygote Formation. Zygotes, or single diploid cells, form as a result of *fertilization*. In fertilization, the egg nucleus (containing the *haploid* number, or *n* chromosomes) unites with the sperm nucleus (containing *n* chromosomes). This union produces a zygote of the original *diploid* or 2*n* chromosome number. In this way, the normal (2*n*) somatic number of chromosomes is restored, and homologous pairs are formed.

If sperm are present in the oviduct at the time of ovulation, fertilization takes place in the oviduct. If there are two or more eggs released by the ovaries, it is possible that all of them will be fertilized, due to the extremely high number of sperm. The result of multiple fertilizations will be fraternal (dizygotic) twins, which are produced when two separate sperm fertilize two eggs. These twins are related in the same way that any two siblings are. If there is only one fertilized egg, twins may still result. This may occur through separation of identical cells during the early stages of cleavage (for example, the two-, four-, or eight-cell stage) into two or more independent embryos. These develop into identical (monozygotic) twins, triplets, and so forth, since they all came from the same fertilized egg.

When the egg and the sperm join, they trigger a cascade of events that occur as the zygote begins to divide rapidly. These events, which are part of the process of fertilization, may occur either externally or internally.

External Fertilization. External fertilization occurs in vertebrates that reproduce in water (fish and amphibians). Eggs are laid in the water, and sperm fluids are deposited near them. The sperm have flagella, enabling them to swim through the water to the eggs. Since there is no direct passage of sperm from the male to the female, the chances of fertilization are reduced considerably.

Internal Fertilization. Meanwhile, internal fertilization is found in vertebrate land animals (like reptiles, birds, and mammals). The moist passageway of the female reproductive tract from the vagina through the oviducts provides a direct route to the egg for mobile sperm. This direct path increases chances of fertilization success.

Number of Eggs Produced. The number of eggs produced depends upon a number of factors. One of these factors is the type of fertilization employed. Because very few sperm actually reach the egg during external fertilization, this process requires large quantities of eggs to ensure at least some degree of fertilization success. The type of development practiced by the organism is also significant. If development occurs outside the mother's body from the very beginning, many eggs are required to ensure survival of at least some of the offspring. Finally, the less care the parents provide, the more eggs are required to guarantee survival of enough offspring to continue the species.

Development of the Embryo. Cleavage of the embryo starts in the oviduct immediately after fertilization. The developing embryo travels down the oviduct, and, within five to ten days, implants itself in the uterine wall. Initially, the fertilized embryo divides into many undifferentiated cells.

In the earliest stages, mitotic divisions result in one cell producing two cells, which produce four cells, which produce eight cells, and so on. This

How Many Eggs?

Animals that practice external fertilization and external development must produce many eggs and sperm in order to ensure survival of the species. On the other hand, animals that practice internal fertilization and development invest their energy in taking very good care of a smaller number of gametes.

ultimately creates what is known as a *morula*, or a solid ball of cells. This morula continues to divide and form the *blastula*, which is a hollow ball of cells (a single layer thick) produced by the pressure of cells on one another. The central cavity is filled with fluid secreted by the cells, and is referred to as the *blastocoel*. More rapid division of cells at one end of the blastula causes an inpocketing or involution known as the *two-layer gastrula*. Two germ layers, ectoderm and endoderm, are initially present. In a *three-layer gastrula*, mesoderm cells are interposed between the ectoderm and endoderm. This formation and rearrangement of the three germ layers is known as *gastrulation*.

Ecotopic Pregnancy

Sometimes the blastula implants itself outside the uterus, a situation referred to as an ectopic pregnancy.

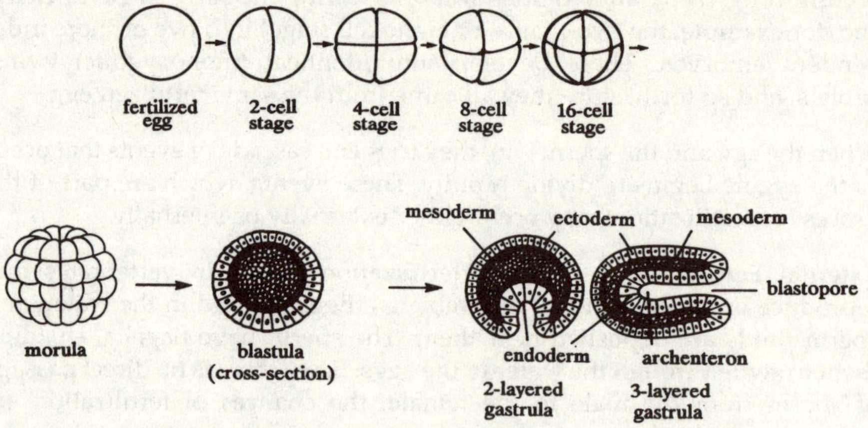

Cleavage of the Egg

Don't Mix These Up on Test Day

The *ectoderm* develops into the skin, the nervous system, and the eyes, hair, and teeth.

The *mesoderm* is associated with the muscles, the skeleton, the circulatory system, the kidney, and the gonads.

The *endoderm* develops into the lining of the digestive and respiratory tracts, and the lining of the bladder, pancreas, and liver.

Differentiation of Embryonic or Germ Layers. In the next stage of embryonic development, the cells of each germ layer begin to differentiate and specialize to form tissues, organs, and organ systems. The *ectoderm* develops into the epidermis of skin, nervous system, and sweat glands. The *endoderm* becomes the lining of digestive and respiratory tracts, parts of the liver and the pancreas, and the bladder lining. Finally, the *mesoderm* develops into the muscles, skeleton, circulatory system, excretory system (except bladder lining), gonads, and the inner layer of skin (dermis).

Site of Embryonic Development of Animals. The embryo may develop either outside or inside the animal. These two forms of development, external and internal development, are described below.

- *External Development.* External development occurs outside the female's body, in water or on land. The eggs of fish and amphibia, for example, are fertilized externally in water. The embryo then develops in water inside the egg, feeding on the yolk in egg. Such embryos are given very little parental care. Thus many embryos must be born to ensure survival of at least some, due to threats from predators, environmental extremes, and other hazards. External development on land occurs in reptiles, birds, and a few mammals, such as the duck-billed platypus. Fertilization in most animals must be internal, though, since sperm must have a watery environment to survive and move.

There are many adaptations for embryonic development within eggs and on land. One of these is a hard shell for protection, which is brittle in birds and leathery in reptiles. Embryonic membranes also help to provide a favorable environment for the developing embryo.

Types of embryonic membranes include the *chorion*, which lines the inside of the egg shell. This moist membrane permits gas exchange through the shell. The *allantois*, a saclike structure developed from the digestive tract, is another embryonic membrane. It carries out functions like respiration and excretion, particularly the exchange of gases with the external environment. The allantois layer has many blood vessels to take in O_2 and give off CO_2, water, salt, and nitrogenous wastes.

A third embryonic membrane, the *amnion*, encloses the amniotic fluid. Amniotic fluid provides a watery environment for the embryo to develop in, and provides protection against shock. Finally, the yolk sac encloses the yolk. Blood vessels in the yolk sac transfer food to the developing embryo.

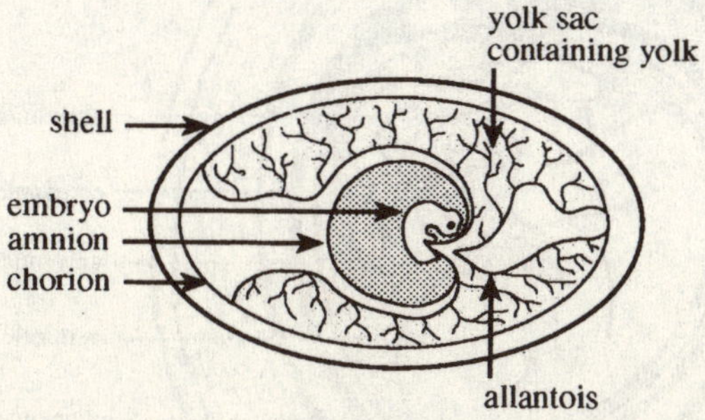

Egg

- *Internal Development.* In animals that develop internally, fertilization and embryo development occur within the mother. This internal development can take a number of different forms, depending on whether or not placenta is utilized in sustaining the embryo. The *placenta* is an intermingling of tissues of embryo and mother. It is the site at which exchange of food, oxygen, waste, and water can take place.

  In nonplacental animals, development occurs inside the mother, but the embryo lacks a placenta. Thus, there is no region of exchange between blood of the mother and the embryo. Eggs must therefore be relatively large, as the yolk must supply the developing embryo's needs. Tropical fish and opossums are exam-

In a Nutshell

There are four types of embryonic membranes:

- Yolk sac
- Allantois
- Chorion
- Amnion

Egg-laying Mammal

The duck-billed platypus, native to Australia, does not develop its young internally as do most other mammals. Instead, it lays a leathery egg. It is classified as a mammal because of its fur and milk glands.

A Tough First Challenge

Marsupials, such as the kangaroo, are nonplacental animals. The newborn kangaroo is very immature. After birth, the underdeveloped infant must climb up its mother's stomach to the pouch, where it is protected and fed milk.

ples of nonplacental animals. They develop inside the oviduct, obtaining food from the yolk of the egg, and are born alive.

In placental animals, on the other hand, there is no direct contact between the bloodstreams of the mother and the embryo. Transport is accomplished by diffusion and active transport between juxtaposed blood vessels of the mother and embryo. The eggs of placental animals are very small, since the embryo requires only a small amount of yolk to be maintained until a placental connection is completed. Humans, for example, have no yolk, but they do have a yolk sac. The umbilical cord attaches the embryo to the placenta; it is composed completely of tissues of embryonic, not maternal, origin. This cord contains the umbilical artery and vein.

As in birds and reptiles, the amnion of placental mammals provides a watery environment to protect the embryo from shock.

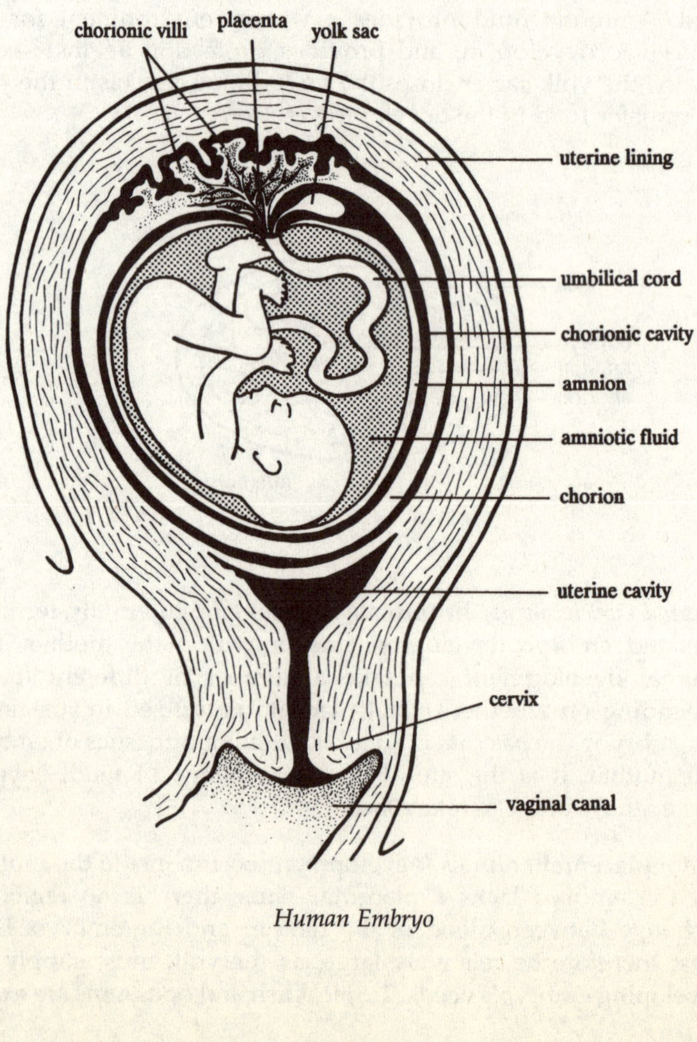

Human Embryo

Postembryonic Development. The development of the embryo to the adult is termed *maturation*. Maturation involves cell multiplication, differentiation, increase in size, and development of a distinctive shape. Maturation can be interrupted, such as in the metamorphosis of arthropods (the pupa stage), or uninterrupted, as in mammals. Differentiation of cells is complete when all organs reach adult form. Further cell division is needed only for repair and replacement of tissues.

Sexual Reproduction in Humans

Like other animals, humans practice sexual reproduction; they also utilize internal fertilization and development, ensuring a high survival rate and continuation of their species. Humans also engage in extensive parental care after birth (sometimes, this care lasts well beyond the college years). However, human sexual reproduction is somewhat more complex than that of most other animals. Read on for a description of the various aspects of human sexual reproduction.

Male Reproductive System. As previously mentioned, the male produces sperm in the testes, located in an outpocketing of the abdominal wall called the scrotum. Ducts, principally the vas deferens, carry sperm to the outside. Glands along the path secrete a liquid (semen) that carries and provides nutrients for the sperm.

As gonads, the testes have a dual function; they produce both sperm and male hormones (such as testosterone). These hormones regulate secondary sexual characteristics of the male, such as facial and pubic hair and deepening voice changes.

Female Reproductive System. Ovaries are paired structures in the lower portion of the abdominal cavity. Through meiosis, ovaries produce eggs in cavities within the ovary called follicles. During ovulation, the eggs leave the follicle and are ejected into the upper end of the oviduct. At birth, all the eggs that a female will ovulate during her lifetime are already present in the ovaries, but these eggs develop and ovulate at a rate of one every 28 days (approximately), starting in puberty.

The ovaries also produce female sex hormones such as estrogen. Like male sex hormones, the female sex hormones regulate the secondary sexual characteristics of the female, including the development of the mammary (milk) glands and wider hip bones (pelvis).

They also play an important role in the menstrual cycle, which involves the interaction of the pituitary gland, ovaries, and uterus. There are four stages in the menstrual cycle: the follicular stage, ovulation, the corpus luteum (luteal) stage, and menstruation.

Menstruation Timeline

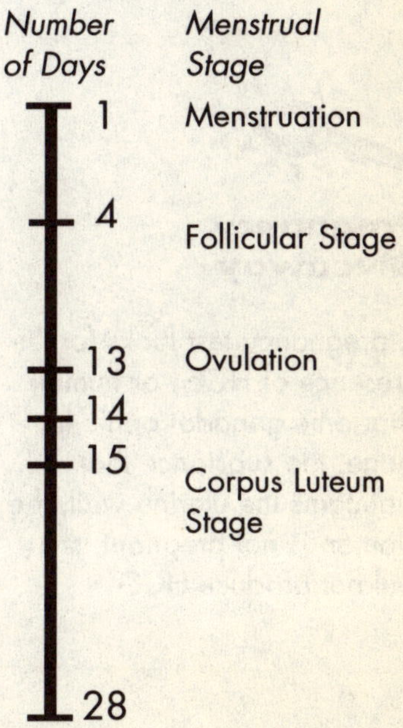

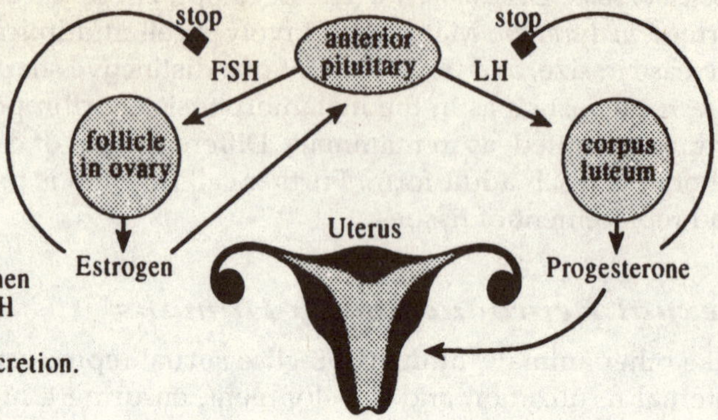

Menstrual Cycle

Follicular Stage. In this stage of menstruation, *FSH* (follicle stimulating hormone) from the anterior pituitary gland stimulates the follicle to develop and produce estrogen. Estrogen promotes thickening of the uterine lining. This stage lasts approximately nine to ten days.

Ovulation. Here the mature egg is released from the follicle as estrogen instructs the pituitary to inhibit FSH (through a feedback mechanism) and, at the same time, to release LH (lutenizing hormone).

Corpus Luteum Stage. Although corpus luteum means "white body" in latin, it is actually a yellow organ within the ovary. During the corpus luteum or luteal stage, it forms within the ruptured follicle. Lutenizing hormone from the pituitary stimulates the corpus luteum to produce another hormone, progesterone, which stimulates vascularization (growth of blood vessels) and lining formation of the uterus in preparation for implantation of the fertilized egg. This stage lasts 12 to 15 days.

Menstruation. If no fertilization or implantation has occurred, the corpus luteum atrophies, reducing progesterone levels. Without progesterone, the thickened, spongy uterine wall that had been prepared for implantation breaks down. The degenerating tissue, blood, and unfertilized egg are passed out as menstrual flow. This stage lasts approximately four days, bringing the total to 28 days for the entire cycle.

If fertilization occurs, the developing placenta produces HCG (human chorionic gonadotrophic), which acts to maintain the corpus luteum. Progesterone further ensures that the uterine wall is thickened so that embryonic development can occur and pregnancy can continue.

Let's now turn to sexual reproduction in the plant kingdom.

Pregnancy Giveaway

A pregnancy test looks for the presence of HCG, or human chorionic gonadotrophic hormone, the substance that maintains the uterine wall. If a woman is not pregnant, she will not produce HCG.

Sexual Reproduction in Plants

Most plants are able to reproduce both sexually and asexually; some do both in the course of their life cycles, while others do one or the other. In the life cycles of mosses, ferns, and other vascular plants, there are two kinds of individuals associated with different stages of the life cycles: the *diploid* and the *haploid*.

Diploid and Haploid Generations. In the diploid or sporophyte generation, the asexual stage of a plant's life cycle, diploid nuclei divide meiotically to form monoploid spores (not gametes) and the spores germinate to produce the haploid or gametophyte generation.

The haploid or gametophyte generation, a more evolved, sexual stage, is concerned with the production of male and female gametes. Union of the gametes at fertilization restores the diploid sporophyte generation. Since there are two distinct generations, one reproducing sexually and the other asexually, this cycle is sometimes referred to as the alternation of generations. The relative lengths of the two stages vary with the plant type.

In general, the evolutionary trend has been toward a reduction of the gametophyte generation, and increasing importance of the sporophyte generation. This trend can be attributed to the fact that plants are often *polyploid*, meaning that they have a chromosome number ranging from 4*n* to 16*n*. Hence plants inherently possess more genetic diversity than animals, and have no need for the increased genetic variation that sexual reproduction offers.

How do these generations express themselves in common plants? In moss, the gametophyte is the green plant that you see growing on the north side of trees. The sporophyte variety is smaller, nongreen (nonphotosynthetic), and short-lived. It is attached to the top of the gametophyte, and is dependent upon it for its food supply. Spores from the sporophyte germinate directly into gametophytes.

In ferns, on the other hand, the reverse pattern may be observed, with the sporophyte of the species dominant. The gametophyte is a heart-shaped leaf the size of a dime. Fertilization produces a zygote from which the commonly seen green fern sporophyte develops. The sporophyte fern's leaves (the fronds) develop spores beneath the surface of the leaf. These spores germinate to form the next generation of gametophyte.

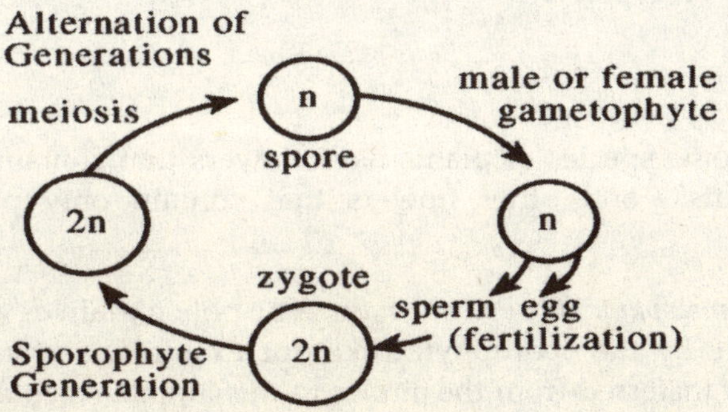

Alternation of Generations

Don't Mix These Up on Test Day

You will be expected to know that in sexual reproduction in animals, the haploid cell (gametophyte) is always unicellular, while in the alternation of generations in plants, the haploid gametophyte is multicellular.

The Birds, the Bees, and the Flowers

We enjoy sexual reproduction in flowering plants when we admire blooming trees or eat fruits and vegetables. Flowering is the elaborate mechanism that plants have evolved to attract pollinators such as birds and bees, guaranteeing for themselves the genetic diversity that sexual reproduction provides.

Sexual Reproduction in Flowering Plants (Angiosperms). In flowering plants or angiosperms, the evolutionary trend mentioned above continues; the gametophyte consists of only a few cells and survives for a very short time. The woody plant that is seen (for example, a rose) is the sporophyte stage of the species.

The Flower. This reproductive organ of the plant consists of male and female organs. The flower's male organ is known as the *stamen.* It consists of a thin, stalklike filament with a sac at the top (analogous to the scrotum). This structure is called the anther, and produces monoploid spores. The monoploid spores develop into pollen grains. The monoploid nuclei within the spores will become the sperm nuclei, which fertilize the ovum.

Meanwhile, the flower's female organ is termed a *pistil.* It consists of three parts: the *stigma*, the *style*, and the *ovary*. The stigma is the sticky top part of the flower, protruding beyond the flower, which catches the pollen. The tubelike structure connecting the stigma to the ovary at the base of the pistil is known as the style; this organ permits the sperm to reach the ovules. And the ovary, the enlarged base of the pistil, contains one or more ovules. Each ovule contains the monoploid egg nucleus.

Petals are specialized leaves that surround and protect the pistil. They attract insects with their characteristic colors and odors. This attraction is essential for cross-pollination—that is, the transfer of pollen from the anther of one flower to the stigma of another (introducing genetic variability).

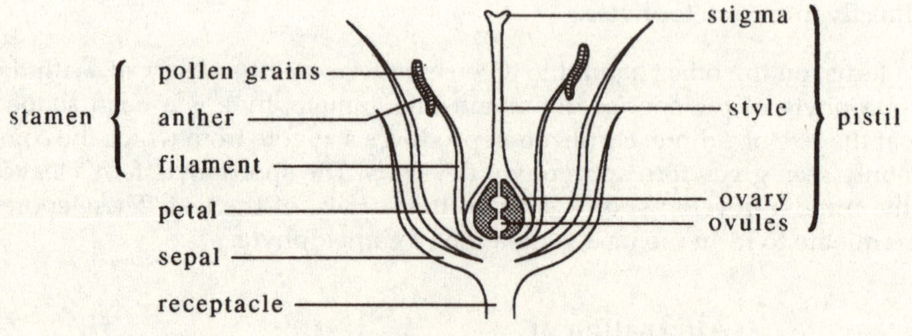

Parts of the Flower

Note that some species of plants have flowers that contain only stamens ("male plants") and other flowers that contain only pistils ("female plants").

The Male Gametophyte (the Pollen Grain). The pollen grain develops from the spores made by the sporophyte (like, for example, a rose bush). Pollen grains are transferred from the anther to the stigma. Agents of cross-pollination include insects, wind, and water. The flower's reproductive organ is brightly colored and fragrant in order to attract insects and birds, which help to spread these male gametophytes. When the pollen grain reaches the stigma (pollination), it releases enzymes that enable it to absorb and utilize food and water from the stigma and to germinate a *pollen tube*. The pollen tube is the remains of the evolutionary gametophyte. The pollen's enzymes proceed to digest a path down the pistil to the ovary. Contained within the pollen tube is the tube nucleus and two sperm nuclei; all are monoploid.

Female Gametophyte. The female gametophyte develops in the ovule from one of four spores. This embryo sac contains nuclei, including the two polar (endosperm) nuclei and an egg nucleus.

Fertilization. The gametes involved in this cycle of reproduction are nuclei, not complete cells. The sperm nuclei of the male gametophyte (pollen tube) enters the female gametophyte (embryo sac), and a double fertilization occurs. One sperm nucleus fuses with the egg nucleus to form the diploid zygote, which develops into the embryo. The other sperm nucleus fuses with the two polar bodies to form the endosperm (triploid or $3n$). The endosperm provides food for the embryonic plant. In dicotyledonous plants, the endosperm is absorbed by the seed leaves (cotyledons).

1 sperm nuclei + 1 egg nuclei = zygote = embryo
1 sperm nucleus + 2 polar nuclei = $2n$ endosperm

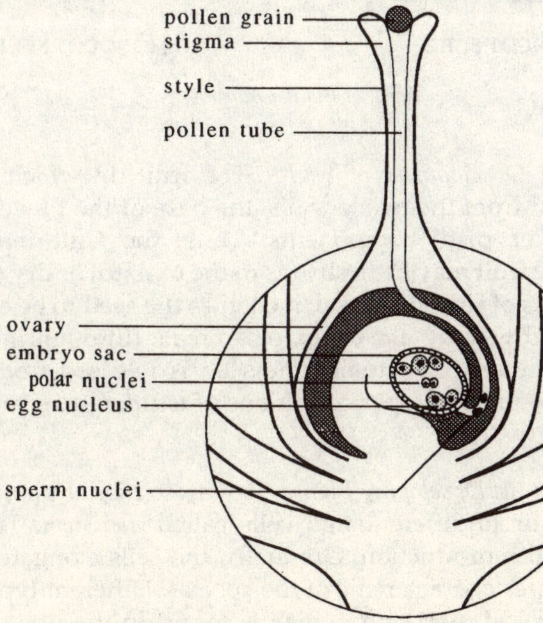

Fertilization in Angiosperms

Don't Mix These Up on Test Day

The *endosperm* is a nutrient-rich structure formed when a sperm nucleus fuses with two polar nuclei. This fusion provides nourishment for the developing *embryo*.

Development of the Plant Embryo-Seed Formation. The zygote produced in the sequence above divides mitotically to form the cells of the embryo. This embryo consists of the following parts, each with its own function:

- The *epicotyl* develops into leaves and the upper part of the stem.
- The *cotyledons* or seed leaves store food for the developing embryo.
- The *hypocotyl* develops into the lower stem and root.
- The *endosperm* grows and feeds the embryo. In dicots, the cotyledon absorbs the endosperm.
- The *seed coat* develops from the outer covering of the ovule. The embryo and its seed coat together make up the seed. Thus, the seed is a ripened ovule.

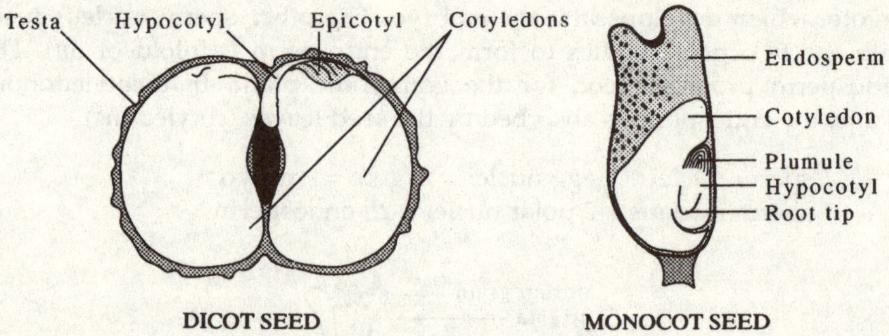

Types of Plant Embryos

Fruit Ovaries

Fruits may be thought of as ripened ovaries. They nourish their seed embryos until these can germinate under the right conditions.

Seed Dispersal and Development of Fruit. The fruit, in which most seeds develop, is formed from the ovary walls, the base of the flower, and other consolidated flower pistil components. Thus, the fruit represents the ripened ovary. The fruit may be fleshy (as in the tomato) or dry (as in a nut). It serves as a means of seed dispersal; it enables the seed to be carried more frequently or effectively by air, water, or animals (through ingestion and subsequent elimination). Eventually, the seed is released from the ovary, and will germinate under proper conditions of temperature, moisture, and oxygen.

Growth of Tissues in the Developing Plant. Growth in higher plants is restricted to embryonic or undifferentiated cells called *meristem*. These tissues undergo active cell reproduction. Gradually, the cells elongate and differentiate into cell types characteristic of the species. Different types of meristem include the apical meristem, which is found in the tips of roots and stems. These cells provide for growth in length, which occurs only at the root and stem tips.

Meanwhile, the lateral meristem or *cambium* is located between the xylem and phloem. This tissue permits growth in diameter. It is not an active tissue in monocots (grasses) or herbaceous dicots (alfalfa). Instead, it is predominant in woody dicots (like oak).

Physiology

In order to survive, all living organisms must perform a variety of vital functions. We have seen the various ways in which organisms perform one of these functions, reproduction. It is now time to turn to some of the ways in which organisms perform other necessary functions.

Physiology is the branch of biology that is devoted to these functions of living organisms and their parts. Physiologists study topics like nutrition, circulation, respiration, homeostasis, and the endocrine and nervous systems.

Nutrition

Animals are *heterotrophic*, which means that they are unable to synthesize their own nutrients. Instead they must obtain these nutrients from food, which provides the raw material for energy, repair, and growth of tissues. The process of digestion encompasses the events that take place after food material is ingested or eaten; in other words, it is the preparation of nutrients for use by the cells of the body. Digestion, therefore, involves mechanical and chemical breakdown of foodstuff and absorption of digested nutrients.

Unicellular organisms capture food primarily via *phagocytosis*—that is, by engulfing it. Food vacuoles form immediately after ingestion is completed. Multicellular organisms rely upon a wide variety methods of food capture and ingestion.

Digestion may be defined as the transformation of large, insoluble molecules into small, soluble molecules. Digestion can be *intracellular*, occurring through the action of intracellular enzymes. It can also be *extracellular*; in this process, enzymatic secretions in a cavity break down nutrients into simpler compounds, which are in their turn absorbed by cells lining the gut to undergo further processing.

Digestion may also be broken down into *mechanical* and *chemical* aspects. In many organisms, mechanical digestion, or the physical breakdown of large particles of food into small particles, occurs through cutting and grinding in the mouth and churning in the digestive tract. The molecular composition of these food particles is unchanged; however, the surface area of the substrates on which the enzymes act is increased.

Meanwhile, chemical breakdown of molecules in digestion is often accomplished through enzymatic hydrolysis. The smaller digested nutrients (glucose, amino acids, fatty acids, and glycerol) pass through the semipermeable plasma membrane of the gut cells to be metabolized or transported to other parts of the body.

Let's take a look at how different organisms ingest and digest their food.

Don't Mix These Up on Test Day

Mechanical digestion involves chewing in the mouth, grinding in the gizzard, and churning in the digestive tract.

Chemical digestion involves enzymes that break down carbohydrates (e.g., amylase), proteins (e.g., pepsin), or lipids (e.g., lipase).

They Grab Their Food Too

Many lower organisms, such as paramecia or hydra, sweep food into their oral grooves or mouths with their cilia or tentacles.

Ingestion and Digestion in Protozoa

These simple organisms utilize intracellular digestion. In *amoebae*, pseudopods surround and engulf food (in phagocytosis) and enclose it in food vacuoles. Lysosomes (containing digestive enzymes) fuse with the food vacuole and release their digestive enzymes, which act upon the nutrients. The resulting simpler molecules then diffuse into the cytoplasm. The unusable end products are eliminated from the vacuoles.

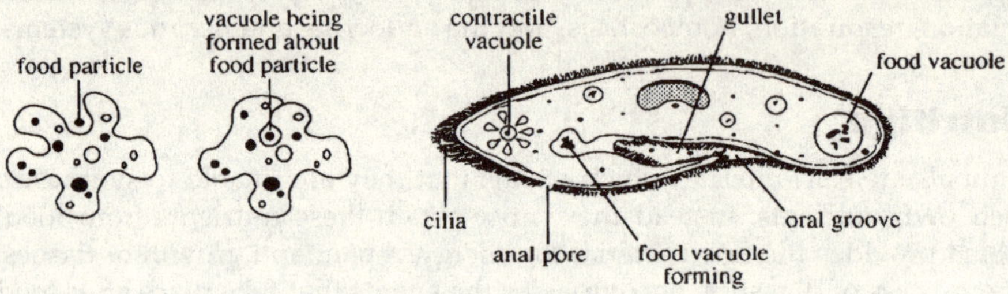

Protozoan Digestive System

In the *paramecium*, cilia sweep food into the oral groove and cytopharynx. A food vacuole forms around food at the lower end of the cytopharynx. Eventually, the vacuole breaks off into the cytoplasm and progresses toward the anterior end of the cell. Enzymes are secreted into the vacuole and the products diffuse into the cytoplasm. Solid wastes are expelled at the anal pore.

Hydra (phylum Coelenterata) employ both intracellular and extracellular digestion. Tentacles bring food to the mouth (ingestion) and release the particles into a cuplike sac. The endodermal cells lining this gastrovascular cavity secrete enzymes into the cavity. Thus, digestion principally occurs outside the cells (extracellularly). However, once the food is reduce to small fragments, the gastrodermal cells engulf the nutrients and digestion is completed intracellularly. Undigested food is expelled through the mouth. Every cell is exposed to the external environment, thereby facilitating intracellular digestion.

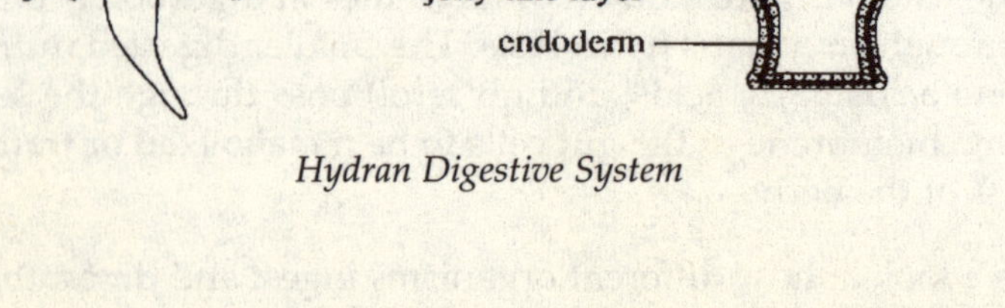

Hydran Digestive System

Ingestion and Digestion in Annelida

Since the earthworm's body is many cells thick, only the outside skin layer contacts the external environment. For this reason, this species requires an advanced digestive system and circulatory system. Like higher animals, earthworms have a complete one-way, two-opening digestive tract. This enables specialization of different parts of the tract for mechanical and chemical digestive processes and absorption of the food that has been ingested by the body tissue. These parts include the mouth, pharynx, esophagus, crop (to store the food), gizzard (to grind the food), intestine (which contains a large dorsal fold that provides increased surface area for digestion and absorption), and anus (where undigested food is released).

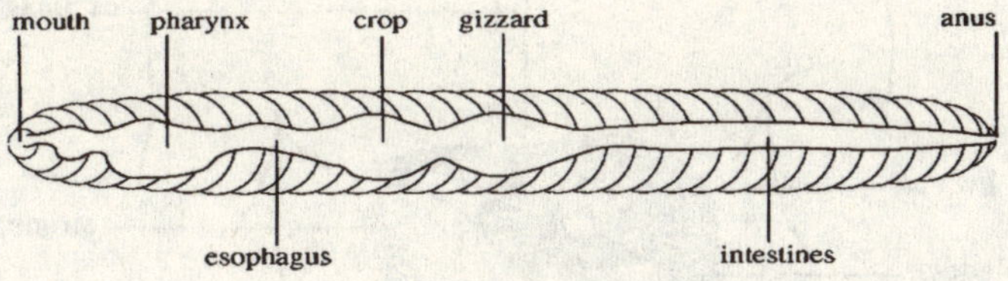

Digestive System of Annelida

Crops, Stomachs, and Gizzards

Crops and stomachs store food, while **g**izzards **g**rind it.

Ingestion and Digestion in Arthropoda

Insects have the same basic system as the earthworm, except that they utilize jaws for chewing and salivary glands for better digestion.

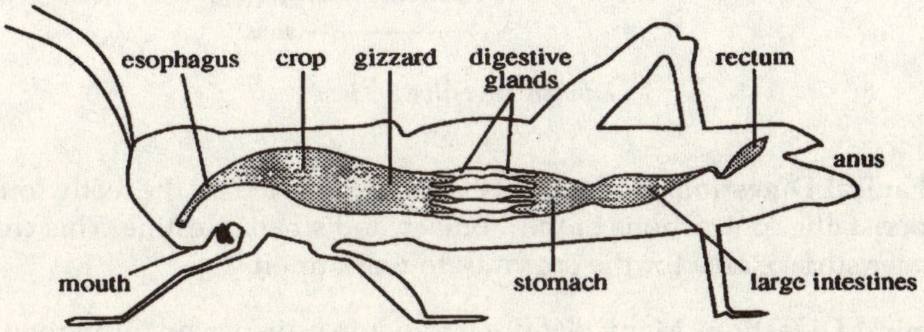

Digestive System of Arthropoda

Ingestion and Digestion in Humans

The human digestive system consists of the *alimentary canal* and the associated glands that pour secretions into this canal. The alimentary canal consists of the oral cavity, pharynx, esophagus, stomach, small intestine, large intestine, and rectum. Many glands line this canal, such as the gastric gland in the wall of the stomach and intestinal glands in the small intestine. Other glands, like the pancreas and liver, are outside the canal proper, and pour their secretions into the canal via ducts.

Quick Quiz

Where does digestion of the following foods begin?

1) carbohydrates
2) proteins
3) lipids

Answers:

1) in the mouth (through the action of salivary amylase)
2) in the stomach (through the action of pepsin)
3) in the small intestine (through the action of bile and lipase)

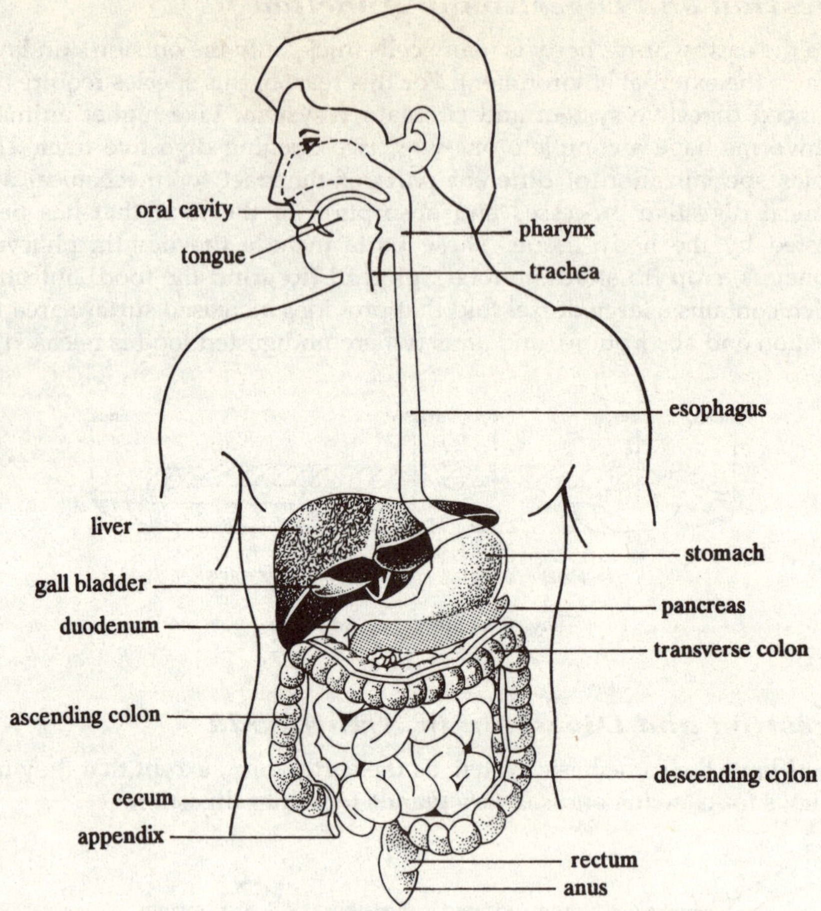

Human Digestive System

Mechanical Digestion. Food is crushed and liquefied by the teeth, tongue, and peristaltic contractions of the stomach and small intestine. This creates a greater surface area for the enzymes to work upon.

Chemical Digestion. Many glandular structures, dispersed throughout the body, are involved in this process. In the *mouth*, the salivary glands produce saliva, which both lubricates food and begins the process of starch digestion. Saliva contains salivary amylase (ptyalin), which digests starch to maltose (a disaccharide). As food leaves the mouth, the *esophagus* conducts it to the stomach via the cardiac sphincter by means of peristaltic waves of smooth muscle contraction.

In the *stomach*, gastric glands produce hydrochloric acid and the enzyme pepsin. The acidity of the stomach provides the low pH environment necessary for the optimum enzymatic activity of pepsin. In addition, the acidity destroys ingested microorganisms, sterilizing the stomach. Chyme (partially digested food in the stomach) enters the duodenum through the pyloric sphincter.

The *liver* is the next link in the digestive process. It produces bile, which is essential to the digestion of fats. Bile is stored in the gall bladder prior to its release into the small intestine. It contains no enzymes; instead, it induces emulsification of fats—that is, the breakdown of large fat globules into smaller droplets. Emulsification of fats exposes a greater surface area to the action of pancreatic lipase.

This pancreatic lipase is produced by the *pancreas*, which is also responsible for manufacturing amylase (for starch digestion), trypsin, and chymotrypsin (for protein digestion). Unlike pepsin, these enzymes have a pH optimum in the alkaline range. The necessary alkaline environment is created by the release of large quantities of bicarbonate ion (HCO_3^-) by the pancreas along with the digestive enzymes. This bicarbonate neutralizes the acidity of the chyme released into the duodenum from the stomach.

As food enters the *small intestine*, digestion truly begins. Glands in the wall of the intestine produce lipase (for fat digestion), aminopeptidases (for polypeptide digestion), and disaccharidases (for digestion of maltose, lactose, and sucrose). The pancreatic and intestinal enzymes in the small intestine are responsible for the bulk of digestion in the gastrointestinal tract. In addition, most of the absorption of the digested nutrients occurs here. Meanwhile, the *large intestine* is mainly devoted to activities like water and Vitamin K absorption, and the *rectum* acts as a transient storage place for feces prior to their elimination through the anus.

Sites of Absorption

Amino acids and monosaccharides are absorbed into the bloodstream by the microvilli of the small intestine.

Lipid molecules, however, are picked up by the lymphatic system via lacteals.

Adaptations for Absorption

It is not enough for organisms to simply digest food. They must also be able to transport this food to their cells in a form in which it can be utilized for energy. Absorption is the process by which water and other nutrients pass through membranes to cells. As we ascend the evolutionary scale of organisms, we can observe that adaptations for absorption become increasingly complex.

The *protozoa* is one of the simpler organisms under discussion in this book. In the protozoa, digested food in the food vacuole passes by simple diffusion into cell cytoplasm. The *hydra* (also known as the coelenterata) absorbs extracellularly digested nutrients into its body cells; some nondigested material is also taken up into the cells and digested intracellularly.

The earthworm (*annelida*) and grasshopper (*arthropoda*) absorb nutrients in similar ways. Soluble food passes, by diffusion, through the walls of these creatures' small intestines into blood. The infolding (typholosole) of the digestive canal increases the absorptive surface.

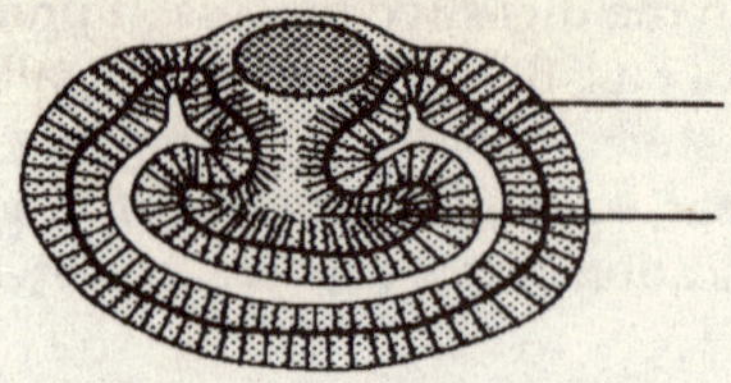

Absorption in Annelida

As for *humans,* most absorption of nutrients occurs in the small intestine, which is adapted for absorption in the following ways. Perhaps the most important absorbing agents located here are the *villi.* These microscopic outpocketings of the lining of the surface of the small intestine significantly increase its absorptive surface. Villi contain *capillaries* and *lacteals* (projections of the lymphatic system). They are also covered with microvilli, "hairs" that further increase surface area and aid in absorption.

Amino acids, small fatty acids, and glucose pass through the villi walls into the capillary system. Large fatty acids and glycerol pass into the lacteals and are then reconverted into fats (fatty acids + glycerol). Note that some nutrients, such as glucose and amino acids, are actively absorbed (that is, requiring energy), while others are passively absorbed. More than 20 feet of intestine provide a large surface area for absorption.

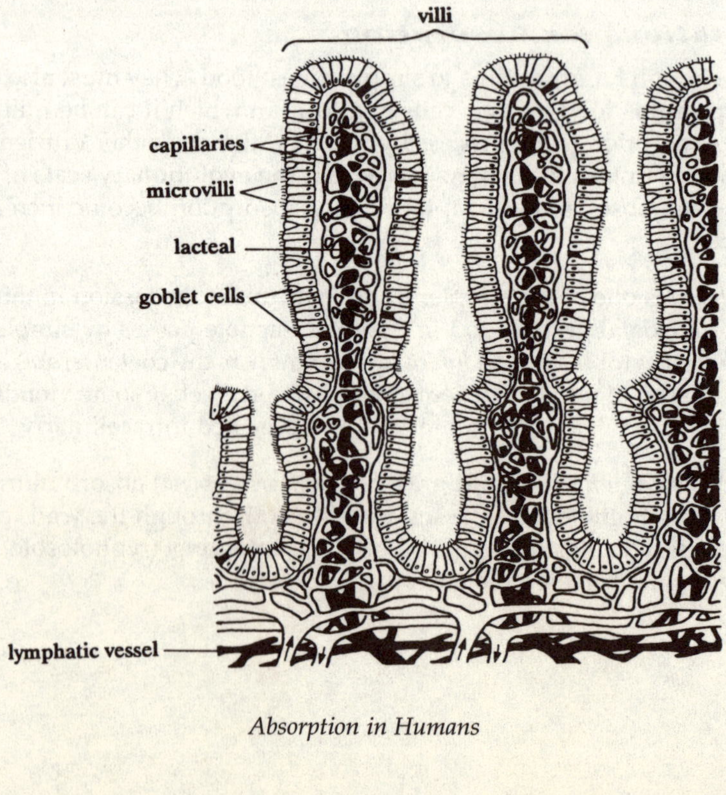

Absorption in Humans

Circulation

As we saw earlier, organisms make nutrients available to cells through the process of absorption. But these nutrients, along with gases and wastes, must also be transported throughout these organisms' bodies in order for them to survive. The process by which this transportation is accomplished is known as circulation. While smaller organisms may achieve this by allowing contact between themselves and the external environment, more complex organisms require circulatory systems.

Circulation in Protozoa and Coelenterates

Circulation in these organisms occurs through simple diffusion and streaming of cytoplasm within the cell (cyclosis). In *hydra,* for example, water simply circulates into and out of the body cavity, because all cells are in direct contact with the external or internal environment.

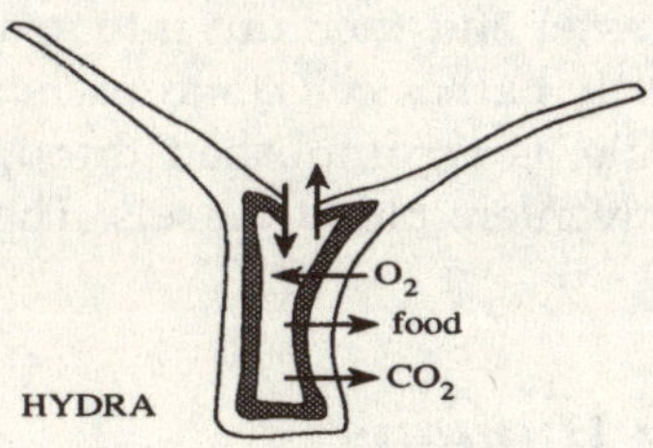

Circulation in Hydra

Circulation in Phylum Annelida

Just as in all higher forms of life, most cells in *annelids* (earthworms) are not in direct contact with the external environment. An internal closed circulatory system indirectly brings materials (food, water, and oxygen) from the external environment to cells. Blood travels toward the head (anterior) through dorsal blood vessels. Five aortic arches or "hearts" force blood down to the ventral vessel, which carries blood to the posterior and up to complete the circuit (see figure).

Annelids have no red blood cells, unlike higher animals; rather, a hemoglobin-like pigment is dissolved in their blood. Nourishment in the form of food and gases is diffused into the cells from the capillaries.

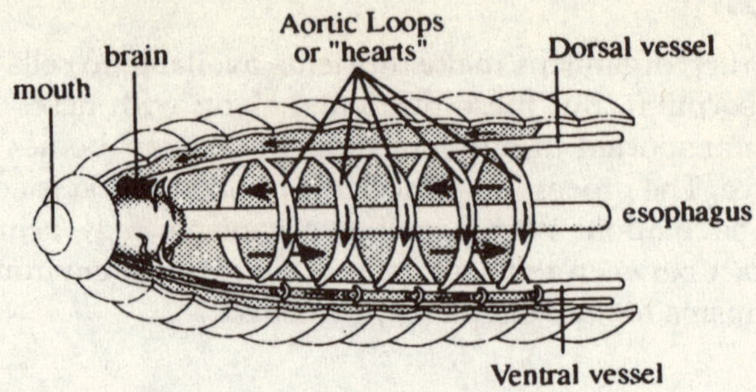

Circulation in Annelida

Circulation in Phylum Arthropoda

Arthopods (grasshoppers) utilize an open circulatory system: Blood flows through a dorsal vessel and then out into spaces called sinuses. In these sinuses, blood directly bathes cells, and exchange of food takes place (air exchange, meanwhile, is accomplished through a tracheal system of air tubes). Blood then reenters blood vessels. The heart consists of a simple beating tube.

Circulation in Humans

Given our complexity, our circulatory system is an elaborate maze of arteries, veins, and capillaries. This is supplemented by the *lymphatic system*, which transports lymph and recaptures fluid from the extracellular spaces. Let's take a look at the primary participants in human circulation.

Heart. The human heart is a four-chambered pump, with two collecting chambers called atria and two pumping chambers called ventricles. This organ is central to the smooth functioning of the circulatory system, as it is the heart that keeps blood, the circulatory fluid, moving throughout the body.

Arteries. The arteries carry blood from the heart to the tissue of the body. They repeatedly branch, eventually supplying blood to the tissues via the capillaries. Arteries are thick-walled, muscular, and elastic. They conduct blood at high pressure and have a pulse (periodic surges of blood from the heart). Arterial blood is oxygenated (except for the pulmonary artery, which goes to the lungs to renew the oxygen supply).

Veins. Veins, on the other hand, carry blood back to the heart from the capillaries. Veins are relatively thin-walled, conduct at low pressure (because they are at some distance from the pumping heart), and contain many valves to prevent backflow. Veins have no pulse; they usually carry dark red, deoxygenated blood (except for the pulmonary vein, which carries recently oxygenated blood from the lungs).

Don't Mix These Up on Test Day

Arteries:

- Are thick-walled
- Are usually oxygenated
- Conduct blood at high pressure
- Have a pulse
- Have no valves

Veins:

- Are thin-walled
- Are usually deoxygenated
- Conduct blood at low pressure
- Have no pulse
- Have valves to prevent backflow

Capillaries. Capillaries are very thin-walled vessels. In fact, their walls are only one endothelial cell thick. Capillaries permit exchange of materials between the blood and the body's cells. Intercellular fluid (ICF), containing plasma with digested nutrients, enzymes, and hormones, seeps from capillaries through diffusion and bathes the cells. This fluid then returns to the blood via lymph tubes.

Lymph Vessels. These vessels are a separate system of tubes semi-independent of the blood system. This system carries ICF (at this stage known as lymph) from one lymph node to the next at very low pressure. The lymph nodes are responsible for filtering lymph to rid it of foreign particles. The system ultimately returns lymph to the blood system via the largest lymph vessel, the *thoracic duct*, which empties lymph back into circulation shortly before it enters the heart.

The Other Circulatory System

The *lymphatic system* is a semi-independent network of vessels that joins the circulatory system at the thoracic duct. It transports lymph, not blood.

Circulatory Fluid. Circulation would not be possible without circulatory fluid. In humans, this circulatory fluid is blood. One of blood's most vital components is *plasma*, the liquid part of blood that contains dissolved nutrient wastes, proteins, hormones, and fibrinogen. Several vital solid materials are suspended in plasma and transported throughout the body, such as:

Red Blood Cells. These are the most common cells found in blood, and their primary function is to transport oxygen. After they are formed in the bone marrow, red blood cells (erythrocytes) lose their nuclei and become bi-concave discs. They live for about four months; at the end of this period, they are destroyed in the spleen. Red blood cells contain hemoglobin (the red pigment containing iron), which unites with oxygen to form oxyhemoglobin; it is in this form that oxygen is carried in the blood.

Red blood cells also lose *mitochondria*, which renders them incapable of performing aerobic respiration. If they were able to carry on this form of respiration, they would themselves use up the oxygen that they were assigned to carry to the tissues of the body. Instead, they produce energy in the form of ATP without using oxygen, through a process termed glycolysis (see the section on respiration for more details of the process).

White Blood Cells. Meanwhile, white blood cells are generally used for protective and defensive functions in the body. White blood cells include *phagocytes*, which engulf bacteria with amoeboid motion, and various types of *lymphocytes*, which are involved in the immune response.

Phagocytes consist of neutrophils, which are the first cells to arrive at a site of inflammation to eat bacteria and other foreign particles. They are the primary component of pus. Macrophages and monocytes are also phagocyte cells that engulf and present foreign components, such as bacteria and viruses, to specific arms of the immune system.

Unhealed Belly Button

If your neutrophils were faulty, your belly button would never have healed after the umbilical cord between you and your mother was cut.

Quick Quiz

What do *B cells* do?

(A) They produce antibodies.

(B) They lyse infected cells.

(C) They produce proteins that stimulate killer T and B cells to mature.

(*Answer* = (A))

Meanwhile, lymphocytes are composed of *B cells* and *T cells*. B cells produce antibodies in response to pathogens, and are also responsible for most allergic reactions, such as hay fever. Killer T cells "kill," or lyse, virally infected cells. Helper T cells, on the other hand, secrete proteins that stimulate killer T or B cells to mature and differentiate.

Platelets. These small cells are actually cell fragments. At the site of a bleeding injury, platelets liberate an enzyme, thromboplastin, which helps to form blood clots. Platelets, along with white blood cells, are responsible for the protective functions of blood.

Functions of Blood. The components of blood each carry out a number of vital bodily functions, including the transport of food, oxygen, and wastes to and from the tissues of the body. They are also responsible for protective mechanisms like clotting, phagocytosis, and immunologic reactions.

- *Clotting.* Platelets in an open wound release the enzyme *thromboplastin*, which initiates a series of reactions that ultimately lead to the formation of a fibrin clot. Thromboplastin, with the aid of calcium and vitamin K as cofactors, leads, in several steps, to the conversion of the inactive plasma prothrombin to its active form, thrombin. Thrombin in its turn converts fibrinogen (dissolved in plasma) into the fibrinous protein called fibrin. Threads of fibrin trap red blood cells to form clots. As the blood clots, serum is the liquid left over. Thus serum is essentially plasma, minus fibrinogen and other clotting factors.

- *Immunologic Reactions.* If foreign proteins (called *antigens*) enter the blood, specialized white blood cells in the blood, lymphatics, and tissues react defensively by manufacturing specific antibodies against the antigens. Red blood cells manufacture two prominent types of antigens, antigen A (associated with blood type A) and antigen B (blood type B). In any given individual, one, both, or neither antigen may be present. The same pattern appears in every red blood cell.

  The plasma of every individual also contains antibodies for the antigens that are not present in the individual's red blood cells (if an individual were to produce antibodies against his or her own red cells, they would agglutinate and the blood would clump). Type A individuals have anti-B antibody, and type B individuals have anti-A antibody. Type O individuals, who have neither A nor B antigens, have both anti-A and anti-B antibodies. Type AB individuals have neither type of antibody. These relationships are depicted in the graph below.

	Blood Type	Antigen on RBC	Antibodies Found in Plasma
	A	A	anti-B
	B	B	anti-A
Universal Recipient	AB	A, B	none
Universal Donor	O	none	anti-A, anti-B

Internal Transport Systems in Plants

Transport in plants, as in animals, encompasses both absorption and circulation. In plants, circulation is called translocation, and mainly involves transporting water and carbohydrates.

Stem. The stem is the principle part of the plant involved in transport. Vascular (conducting) bundles run up and down the stem. The vascular bundle at the center of the stem contains xylem, phloem, and cambium cells.

Xylem. This is a conducting system composed of xylem cells. These cells are thick-walled and, usually, hollow. They are located on the inside of the vascular bundle (towards the center of the stem). They carry water and minerals up the plant, and their thick walls give the plant its rigid support. Older xylem cells at the innermost layer die, forming the heartwood used for lumber. The outer layer of xylem is alive and is called the sapwood. Two types of xylem cells have been identified: vessel cells and tracheids.

The rise of water up the xylem is caused by transpiration pull (as water evaporates from the leaves of plants, a vacuum is created), capillary action (the rise of any liquid in a thin tube because of the surface tension of the liquid), and root pressure exerted by water entering the root hairs.

Phloem. Phloem cells, on the other hand, are thin-walled and are found on the outside of the vascular bundle. They usually transport nutrients (especially carbohydrates produced in the leaves) down the stem. These living cells include the sieve tube cells and companion cells. If a tree is girdled by completely removing a strip of bark around the trunk, the phloem tube connections are severed and the tree dies.

Cambium. Cambium cells (two layers thick) are the actively dividing, undifferentiated cells that upon differentiation give rise to xylem and phloem. The cambium lies between the xylem and phloem cell layers. As it divides, the cells near the phloem differentiate into phloem cells, and those near the xylem differentiate into xylem cells.

Mnemonic

Remember:

XYLEM UP and
PHLOEM DOWN

and

XYLEM = WATER and
PHLOEM = **F**OOD

Don't Mix These Up on Test Day

Cambium is undifferentiated tissue that can develop into either xylem or phloem.

Xylem cells are found in the center of the vascular bundle of the plant, and are responsible for transporting water and minerals up the stem.

Phloem cells are found outside the vascular bundle of the plant, and transport nutrients down the stem.

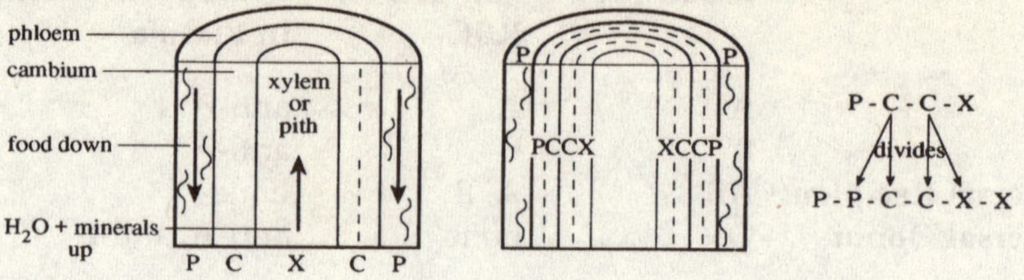

Parts of the Plant

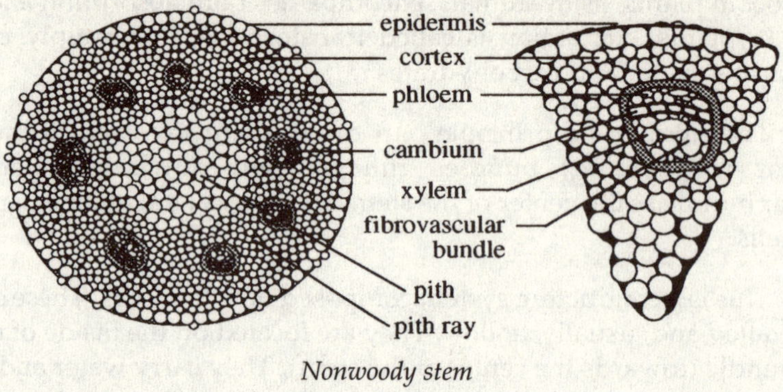

Nonwoody stem

The gross structure of a nonwoody stem is as follows (proceeding from the outside inwards): epidermis (outer bark) cortex, phloem, cambium, xylem, and finally pith (tissue concerned with storage of nutrients and plant support). The phloem, cambium, and xylem layers are known as the fibrovascular bundle.

Root. Like the stem, the root has an epidermis cortex, phloem cells, xylem cells, and cambium cells. The epidermis contains the root hair cells. Root hairs are specialized cells of the root epidermis with thin-walled projections. They provide increased surface area for absorption of water and minerals from the soil through diffusion and active transport. The main functions of the root are absorption, which is accomplished through these root hairs, and anchorage of the plant in the ground. Some roots additionally function in the storage of food (such as the roots of turnips and carrots).

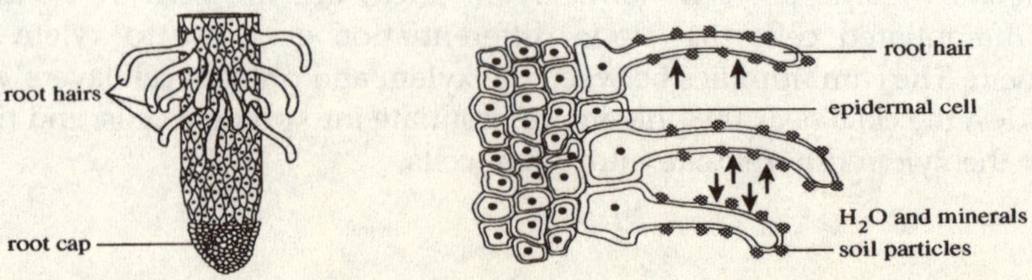

Root

Regions of Growth in the Plant. *Meristem* is a term used for actively dividing, undifferentiated, and, therefore, embryonic cells. Cambium, lying between the phloem and xylem, is a type of meristem. It provides for lateral growth (that is, growth in diameter) of the stem by adding to the phloem or xylem.

Meristem is also located at the tips of roots and stems and produces growth in length in these areas of the plant. After actively dividing, the new cells elongate in what is referred to as the zone of elongation and, finally, differentiate into one of the many types of specialized cells of the plants.

Respiration

Another vital function that living organisms must maintain is that of gas exchange, or respiration. In order to ensure adequate levels of oxygen for survival, organisms must excrete CO_2 while reoxygenating their tissues. If your carbon dioxide level rises too high, for example, you might hyperventilate. Even the most expert pearl divers in the Mediterranean cannot live without oxygen for more than a few minutes. Ordinary mortals can last for a much shorter time before suffering brain damage.

Gas exchange is accomplished via a variety of efficient ways, which range from simple diffusion to complex systems of respiration. We describe the adaptations employed by a range of organisms below.

A Simple Life

Note that in lower organisms, circulation occurs mainly through the process of simple diffusion. This implies that the processes of digestion and respiration are practically identical in such organisms.

Respiration in Unicellular and Simple Multicellular Organisms (Protozoa and Hydra)

Since every cell of these types of primitive organism is in contact with the external environment (in this case, water), respiratory gases can be easily exchanged between the cell and the outside by direct diffusion of these gases through the cell membrane.

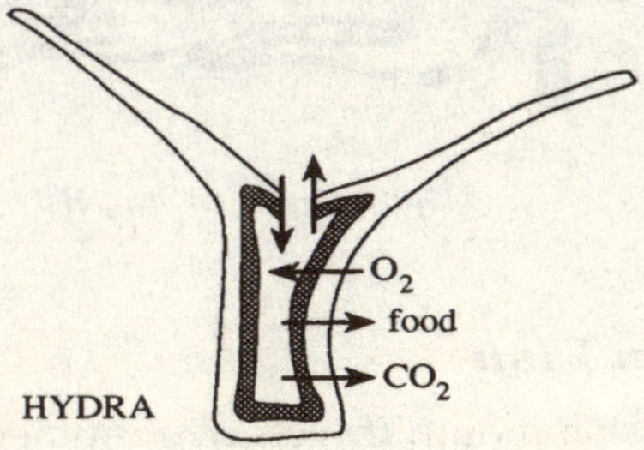

Respiration in Hydra

Respiration in Annelida

Mucus secreted by cells at the external surface of the annelida's body provide a moist surface for gaseous exchange from the air to the blood through diffusion. The annelida's circulatory system then brings O_2 to the cells and waste products such as CO_2 back to the skin, excreting them into the outside environment.

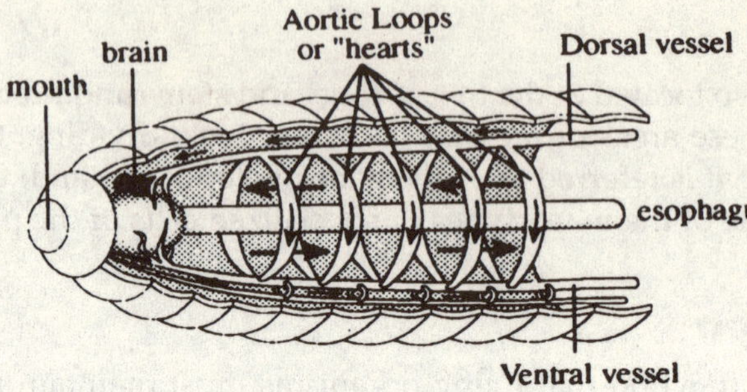

Respiration in Annelida

Respiration in Arthropoda

This organism's respiratory system consists of a series of respiratory tubules called tracheae. These tubules open to the outside in the form of pairs of openings called *spiracles*. Inside the body, the tracheae subdivide into smaller and smaller branches, enabling them to achieve close contact with most cells. In this way, this system permits the direct intake, distribution, and removal of respiratory gases between the air and the body cells. As with invertebrates, no oxygen carrier is needed. Since a blood system does not intervene in the transport of gases to the body's tissues, this system is very efficient and rapid, enabling most arthropods to produce large amounts of energy relative to their weights.

Ephemeral Arthropoda

Mayflies belong to the order Ephemerata, which means shortlived. The adult flies emerge from the pupae, mate, and die within 24 hours. These insects do not eat during their lifetimes; in fact, they do not even possess functioning mouths.

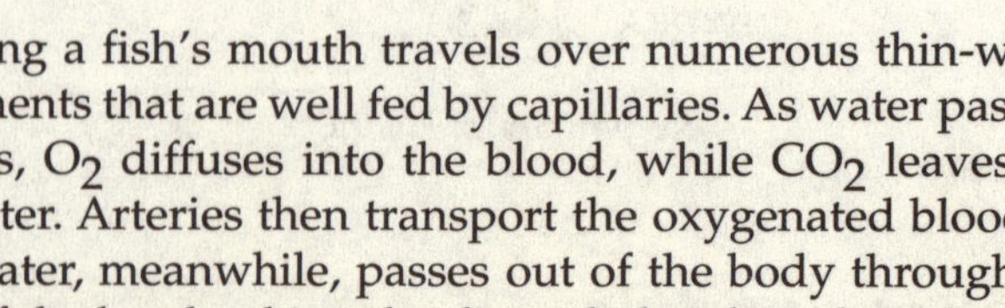

Respiration in Arthropoda

Respiration in Fish

Water entering a fish's mouth travels over numerous thin-walled, threadlike gill filaments that are well fed by capillaries. As water passes over these gill filaments, O_2 diffuses into the blood, while CO_2 leaves the blood to enter the water. Arteries then transport the oxygenated blood through the body. The water, meanwhile, passes out of the body through openings on either side of the head, taking the discarded carbon dioxide with it.

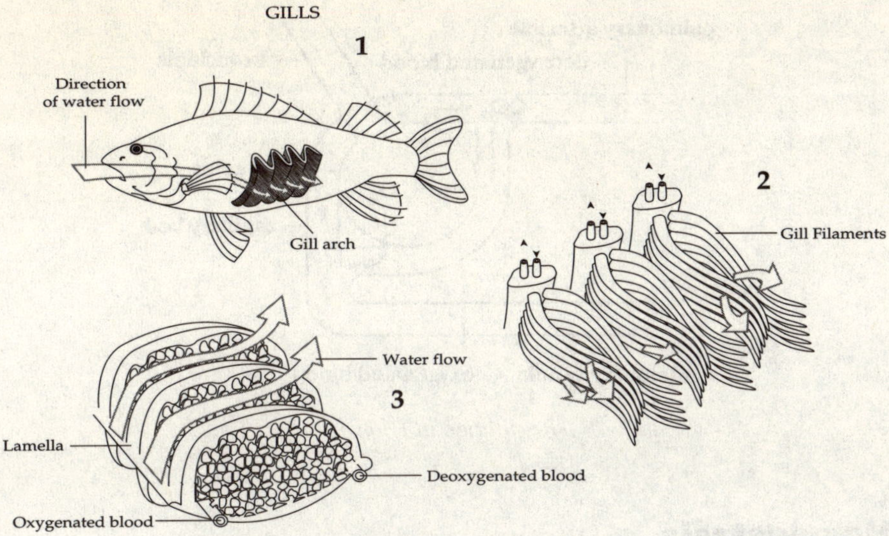

Respiration in Fish

Countercurrent Exchange

Respiration in fish is very efficient, due to a process called countercurrent exchange. As blood flows through the capillaries, it becomes loaded with oxygen. Concurrently, it encounters water flowing through the fish which has an even higher percentage of oxygen, having just entered the gills (see the figure opposite).

Respiration in Humans

As complex organisms, humans have developed a complex system of respiration in order to transport oxygen to their cells and to rid their bodies of waste products like carbon dioxide. Human respiration may be carried out both externally and internally. External respiration is the term given to the system of respiratory tubes which brings gases to and from the blood. Thus *external respiration* incorporates breathing and gas exchange between the alveoli (air sacs) and the blood. *Internal respiration,* on the other hand, refers to the exchange of gases between the blood and cell, as well as to the intracellular processes of respiration.

The air passages involved in respiration consist of the nose, pharynx, larynx, trachea, bronchi, bronchioles, and the alveoli. The major functions of respiration are outlined in the following paragraphs.

Breathing. Chest muscles move the ribs up and out, as the diaphragm moves down; this creates both a larger chest cavity and a vacuum that forces air into the respiratory passages (in a process known as inhalation). The reverse process decreases the size of the chest cavity and forces air out of the lungs (exhalation). The breathing rate is controlled by a part of the brain, the medulla oblongata, that monitors carbon dioxide content in the blood. Excess CO_2 in the blood stimulates the medulla to send messages to the rib muscles and the diaphragm to increase the frequency of respiration.

Alveolar Exchange of Gases. The *alveoli,* located in the lungs, have thin, moist walls and are surrounded by thin-walled capillaries. Oxygen passes from the alveolar air into the blood by diffusion through the alveolar and capillary walls. CO_2 and H_2O pass out in the same manner. Note that all exchanges at the alveoli involve passive diffusion.

Don't Smoke

Emphysema is a disease characterized by the destruction of the alveolar walls. This results in reduced elasticity of the lungs, making exhalation difficult. Most cases can be traced to excessive cigarette smoking.

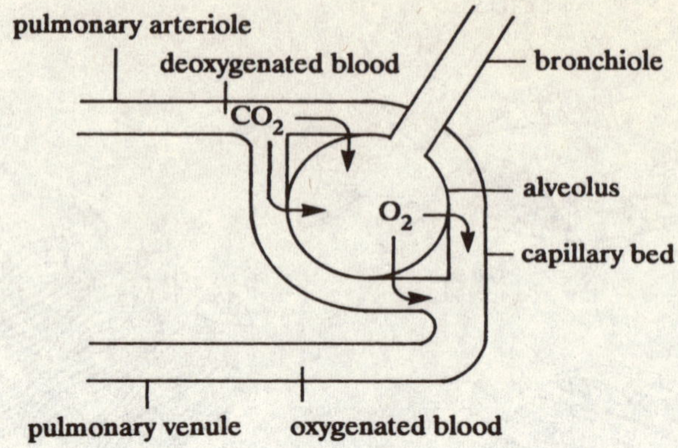

Respiration in Humans

Homeostasis

Homeostasis is the process by which a stable internal environment within an organism is maintained. Some important homeostatic mechanisms include the maintenance of a constant internal body temperature (thermoregulation), the removal of metabolic waste products (excretion), the maintenance of water and solute balance (osmoregulation), and the regulation of blood glucose levels.

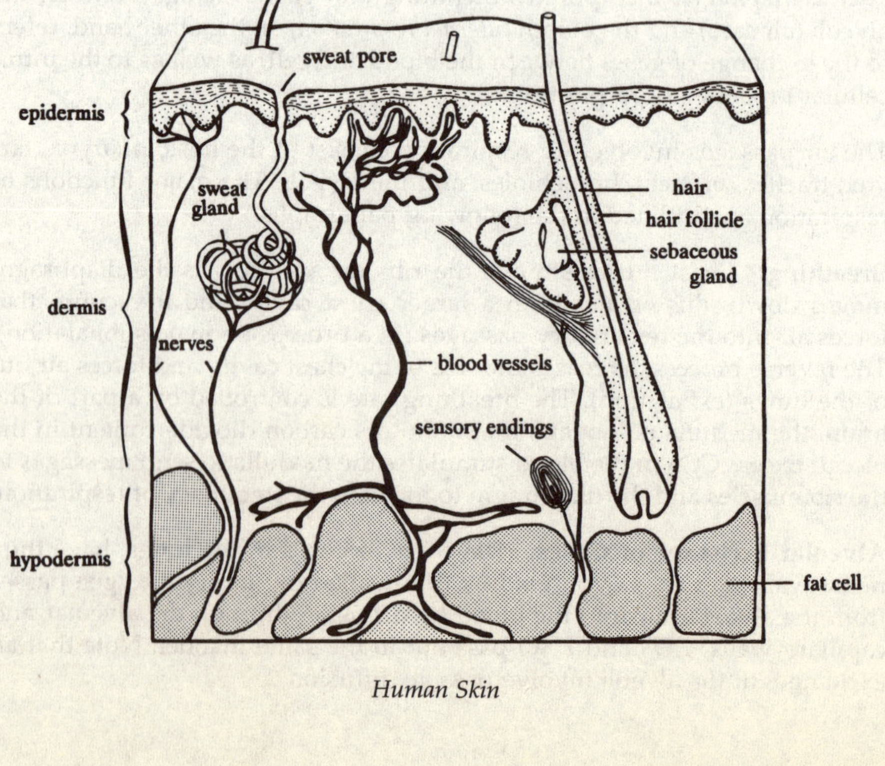

Human Skin

Thermoregulation

In humans, the skin protects the body from microbial invasion and from environmental stresses, such as dry weather and wind. Specialized epidermal cells called *melanocytes* synthesize the pigment *melanin*, which protects the body from ultraviolet light. The skin is a receptor of stimuli, such as pressure and temperature. The skin is also an excretory organ (removing excess water and salts from the body) and a thermoregulatory organ (helping control both the conservation and release of heat).

Extra Sensitive

People who are albinos cannot synthesize the pigment-melanin. This autosomal recessive disease results in an exquisite sensitivity to the sun.

Sweat glands secrete a mixture of water, dissolved salts, and urea via sweat pores. As sweat evaporates, the skin is cooled. Thus, sweating has both an excretory and a thermoregulatory function. Sweating is under autonomic (involuntary) nervous control.

Subcutaneous fat in the hypodermis insulates the body. Hair entraps and retains warm air at the skin's surface. Hormones such as epinephrine can increase the metabolic rate, thereby increasing heat production. In addition, muscles can generate heat by contracting rapidly (shivering). Heat loss can be inhibited through the constriction of blood vessels (vasoconstriction) in the dermis. Likewise, dilation of these same blood vessels (vasodilation) dissipates heat.

Don't Mix These Up on Test Day

Ectoderms warm their bodies by absorbing heat from their surroundings.

Endoderms derive their body heat from their own metabolism.

Alternative mechanisms are used by some mammals to regulate their body temperature. For example, *panting* is a cooling mechanism that evaporates water from the respiratory passages. Most mammals have a layer of fur; fur traps and conserves heat. Some mammals exhibit varying states of *torpor* in the winter months in order to conserve energy; their metabolism, heart rate, and respiration rate greatly decrease during these months. *Hibernation* is a type of intense or extreme torpor during which the animal remains dormant over a period of weeks or months with body temperature maintained below normal. Animals with a constant body temperature are referred to as *homeotherms (endotherms)*.

Excretion

Excretion is the term given to the removal of metabolic wastes produced in the body. (Note that it is to be distinguished from elimination, which is the removal of indigestible materials.) Sources of metabolic waste include:

Waste	Metabolic Activity Producing the Waste
carbon dioxide	aerobic respiration
water	aerobic respiration, dehydration synthesis
nitrogenous wastes (urea, ammonia, uric acid)	deamination of amino acids
mineral salts	all metabolic processes

Our select group of organisms has developed various types of adaptations for excretion:

Protozoa and Colenterates. Remember that all of these organisms' cells are in contact with the external, aqueous environment. Water-soluble wastes such as carbon dioxide and the highly toxic ammonia can therefore exit via simple diffusion through the cell membrane. Some freshwater protozoa, such as the paramecium, possess a contractile vacuole, an organelle specialized for water excretion by active transport. Excess water, which continually diffuses into the hyperosmotic cell from the hypo-osmotic environment (in this case, fresh water), is collected and periodically pumped out of the cell. This permits the cell to maintain its volume and pressure.

Annelida. Carbon dioxide excretion occurs directly through the moist skin of this organism. Two pairs of *nephridia* (small nephrons that enter each cell) in each body segment excrete water, mineral salts, and nitrogenous wastes in the form of urea.

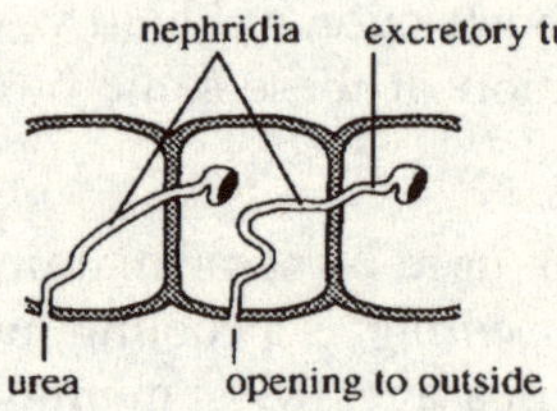

Excretion in Annelida

Arthropoda. As previously mentioned in our discussion of respiration, carbon dioxide is released from the tissues of this organism into adjacent tubelike tracheae that communicate with the external air, and these tracheae lead to openings to the outside called spiracles through which the carbon dioxide is expelled. Nitrogenous wastes, meanwhile, are excreted in the form of solid uric acid crystals. The use of solid nitrogenous wastes is an adaptation that allows the arthropoda to conserve water. Mineral salts and uric acid accumulate in the *Malphigian tubules*; they are then transported to the intestine to be expelled along with solid wastes of digestion.

Human Excretion. The principle organs of excretion in humans are the lungs, skin, liver, and kidneys.

Lungs. In the lungs, carbon dioxide and water vapor diffuse from the blood into the alveoli of the lungs, and are continually exhaled in the process outlined earlier in this book. Additional functions of the lungs are described in more detail under the "Respiration" heading above.

Lungs play a relatively small role in excretion. The major organs of the human excretory system are depicted in the diagram below.

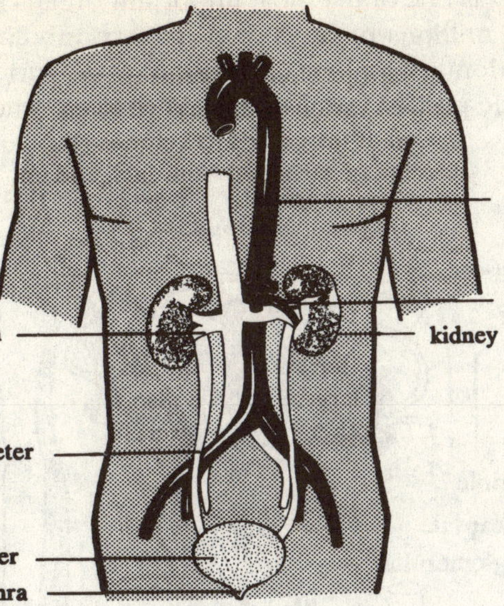

Human Excretory System

Liver. The liver helps regulate blood glucose levels and produces urea. Glucose and other monosaccharides absorbed during digestion are delivered to the liver via the *hepatic portal vein*. Glucose-rich blood is processed by the liver, which converts excess glucose to glycogen for storage. If the blood has a low glucose concentration, the liver converts glycogen into glucose and releases it into the blood, restoring blood glucose levels to normal. In addition, the liver synthesizes glucose from noncarbohydrate precursors via the process of gluconeogenesis. Glycogen metabolism is under both hormonal and nervous control.

The liver is also responsible for the processing of nitrogenous wastes. Excess amino acids are absorbed in the small intestine and transported to the liver via the hepatic portal vein. There, the amino acids undergo a process called *deamination*, in which the amino group is removed from the amino acid and converted into ammonia, a highly toxic compound. In a complex biochemical process, the liver combines ammonia with carbon dioxide to form urea, a relatively nontoxic compound, which is released into the blood and eventually excreted by the kidneys.

Busy Liver

The liver is also responsible for:

- Detoxification of toxins
- Storage of iron and vitamin B_{12}
- Destruction of old erythrocytes
- Synthesis of bile
- Synthesis of various blood proteins
- Defense against various antigens
- Beta-oxidation of fatty acids to ketones
- Interconversion of carbohydrates, fats, and amino acids

Functions of the Kidney

- Filtering of blood
- Reabsorption of amino acids, glucose, and salts
- Secretion of urea, uric acid, and other wastes

Kidney. The kidney is one of the most important organs involved in excretion. It consists of millions of functional urinary tubules called *nephrons*. There are three anatomical regions of the kidney: the cortex (outer portion), the medulla (middle portion), and the pelvis (inner portion).

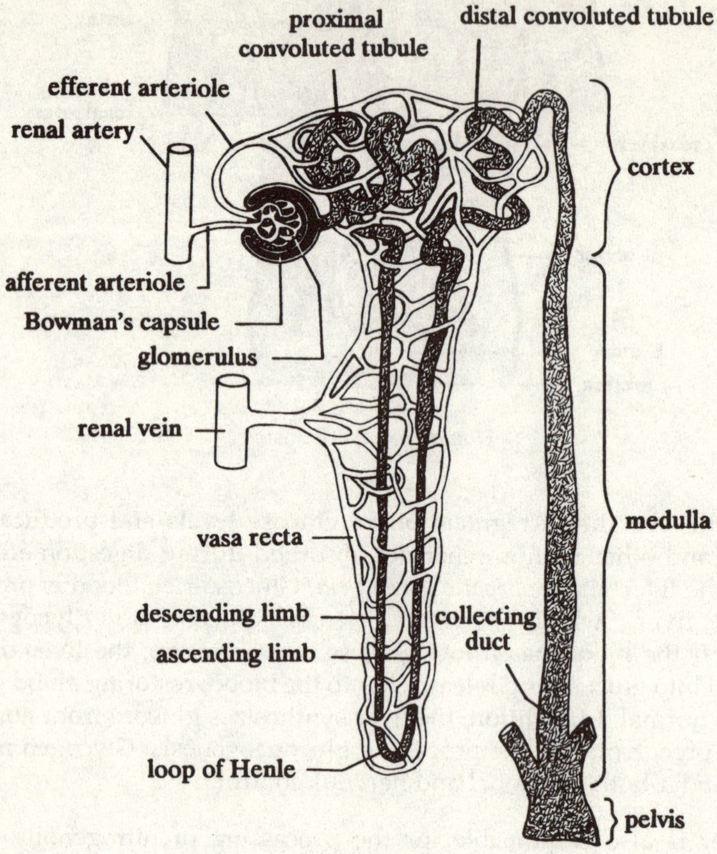

Nephron

Within each nephron may be found a *Bowman's capsule*, containing the *glomerulus*, in which blood is filtered. The filtrate flows first into the proximal convoluted tubule, then into the *Loop of Henle* and the distal convoluted tubule, and, finally, into collecting tubules, as depicted in the diagram above. Generally, the nephrons are arranged so that the Bowman's capsule, the glomerulus, and the proximal and distal convoluted tubules are located in the cortex, while the loops of Henle extend into the medulla. The collecting tubules are found in the medulla; these empty urine into the ureter.

The kidney's primary function, the formation of urine, encompasses the processes of filtration, reabsorption, and secretion. During *filtration*, blood passes through the capillary tuft of the glomerulus and is filtered as it passes through the capillary walls. Only blood cells and protein remain in the blood, while water, salts, urea, glucose, and amino acids are filtered into the capsular space. Filtration is passive, driven by the hydrostatic pressure of the blood.

Meanwhile, *reabsorption* of amino acids, glucose, and salts occurs predominantly in the proximal convoluted tubules. These are actively transported across the membrane and create an osmotic gradient that drives some simultaneous diffusion of water out of the tubules.

Visual Aid

Picture the *glomerulus* as a kind of sieve or collander. Small molecules, such as amino acids and glucose, can pass through (to be subsequently reabsorbed), while large molecules like proteins and blood cells cannot pass through.

Nephrons are characterized by *selective permeability*. The walls of the proximal tubule and the descending limb of the loop of Henle are permeable to water. Meanwhile, the walls of the lower ascending limb of the loop of Henle are impermeable to water; they are in fact permeable only to salt. In the presence of ADH, the walls of the collecting cut are permeable to water and urea, but only slightly permeable to salt.

Filtrate enters Bowman's capsule and flows into the proximal convoluted tubule, where virtually all glucose, amino acids, and other important organic molecules are reabsorbed via active transport. In addition, 60–70 percent of the Na^+ in the filtrate is reabsorbed (by both active and passive mechanisms); water and Cl^- passively follow. The filtrate then flows down the descending limb into the renal medulla, where there is an increasing ionic concentration in the interstitial fluid, causing more water to diffuse out of the nephron.

Next, the filtrate flows through the ascending limb, which is impermeable to water, and then into the distal convoluted tubule. It continues through the collecting duct, where water reabsorption is under hormonal (ADH) control. The remaining filtrate, *urine*, is hypertonic to the blood and highly concentrated in urea and other solutes.

By the time urine exits the nephron, most of the water in it has been reabsorbed. The remaining fluid, composed of urea, uric acid, and other wastes, leaves the collecting tubule and exits the kidney via the *ureter*, a duct leading to the bladder. Urine is stored there until it is excreted from the body through the *urethra*.

In a healthy individual, the nephron reabsorbs all of the glucose entering it, producing glucose-free urine. The urine of a diabetic, however, is not glucose-free. The high blood glucose concentration in a diabetic overwhelms the nephron's active transport system, leading to the excretion of glucose in the urine.

Osmoregulation

Osmoregulation may be defined as the maintenance of blood volume levels and osmolarity (in other words, how much blood there is in the body, and how concentrated that blood is). This process is under the control of hormonal regulation.

Hormonal regulation plays a key role in urine formation. Two hormones that regulate water reabsorption are *aldosterone* and *ADH*. Aldosterone, a mineral cortoid that is produced by the adrenal cortex, stimulates both the reabsorption of Na^+ from the distal convoluted tubule and the secretion of K^+. Na^+ reabsorption increases water reabsorption, leading to a rise in blood volume, and a corresponding rise in blood pressure. Aldosterone secretion is regulated by the renin-angiotensin system.

In a person suffering from Addison's disease, aldosterone is produced insufficiently or not at all. This disease causes overexcretion of urine with a high Na^+ concentration, which results in a considerable drop in blood pressure.

Meanwhile, *ADH*, also known as vasopressin, is formed in the hypothalamus and stored in the posterior pituitary. As an antidiuretic, it causes increased water reabsorption. It acts directly on the collecting duct, increasing its permeability to water.

The amount of ADH produced is dependent on plasma osmolarity. A high solute concentration in the blood causes increased ADH secretion, while a low solute concentration reduces ADH secretion. Alcohol and caffeine inhibit ADH secretion, causing excess excretion of dilute urine and dehydration.

Don't Mix These Up on Test Day

ADH (vasopressin):

- Is produced by the hypothalamus and secreted by the posterior pituitary
- Makes the collecting duct more permeable to water
- Leads to a rise in blood volume

Aldosterone:

- Is produced by the adrenal cortex
- Stimulates the reabsorption of Na^+ and the secretion of K^+
- Leads to a rise in blood volume and blood pressure

Endocrine System

The endocrine system acts as a means of internal communication for organisms, coordinating the activities of the organ systems. Compared to the nervous system, this system acts fairly slowly. Endocrine glands synthesize and secrete chemical substances called *hormones* directly into the circulatory system.

Hormones are circulated all over the body, and can affect almost every tissue in it. Without these chemical messengers, your heart would not beat quickly when you were frightened, and you wouldn't sweat while explaining to your dad why there are three new dents in the car.

The Changing Role of the Thymus

The thymus plays a crucial role during the embryonic development of the immune system. However, it begins to atrophy after birth and is very small by the time humans reach puberty.

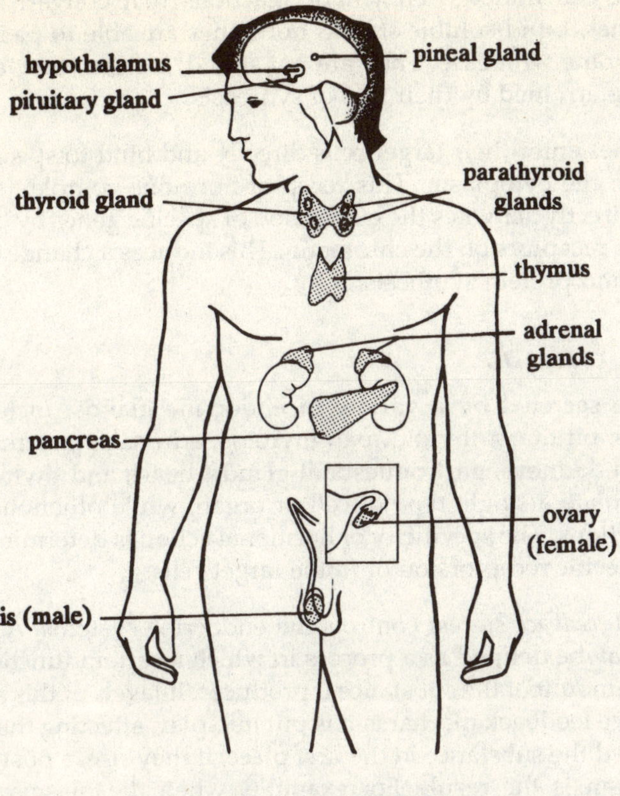

Human Endocrine System

Mechanisms of Action

Hormones are classified on the basis of their chemical structure into two major groups, peptide hormones and steroid hormones. They affect the activities of their target cells via extracellular receptors (located on the cell surface) or intracellular receptors (inside the cell).

Peptide Hormones. The extracellular peptide hormones range from simple short peptides, such as ADH, to complex polypeptides like insulin. Synthesis of peptide hormones begins with the synthesis of a large polypeptide. The polypeptide is then cleaved into smaller protein units and transported to the Golgi apparatus, where it is further modified into the active hormone. The hormone is packaged into secretory vesicles and stored until it is released by the cell via exocytosis.

Peptide hormones act as first messengers. Their binding to specific receptors on the surface of their target cells triggers a series of enzymatic reactions within each cell.

Steroid Hormones. Steroid hormones, such as estrogen and aldosterone, belong to a class of intracellular, lipid-derived molecules with a characteristic ring structure. They are produced by the testes, ovaries, placenta, and adrenal cortex. As steroid molecules are synthesized, precursors already

Don't Mix These Up on Test Day

Peptide hormones, such as insulin and ADH, are proteins, have membrane receptors, act as second messengers, modify existing proteins, and have a quick onset time.

Steroid hormones, such as testosterone and estrogen, are lipid-derived, have intracellular receptors, act as first messengers, affect transcription directly, and have a longer onset time.

present in the cell undergo enzymatic reactions that convert them into active hormones. Lipid-soluble steroid hormones are able to pass through the cell membrane with ease. They are not stored; instead, they are secreted, at a rate determined by their rate of synthesis.

These hormones enter their target cells directly and bind to specific receptor proteins in the cytoplasm. This receptor-hormone complex enters the nucleus and directly activates the expression of specific genes by binding to corresponding receptors on the chromatin. This induces a change in mRNA transcription and protein synthesis.

Endocrine Glands

Hormones are secreted by a variety of endocrine glands, including the hypothalamus, pituitary, thyroid, parathyroids, adrenals, pancreas, testes, ovaries, pineal, kidneys, gastrointestinal glands, heart, and thymus. Some hormones regulate a single type of cell or organ, while others have more widespread actions. The specificity of hormonal action is determined by the presence of specific receptors on or in the target cells.

A complicated *feedback system* controls the endocrine systems. A feedback mechanism may be defined as a process in which a certain function is regulated by the amount of the substance it produces. If levels of this substance drop, a negative feedback mechanism is put into play, affecting the function which produced the substance in the first place; if they rise, a positive feedback mechanism is the result. For example, when the plasma levels of adrenal cortical hormones drop, hypothalamic cells give out ACTH-releasing factor into the portal system via a negative feedback mechanism.

Hypothalamus and Pituitary Gland. The *hypothalamus*, a section of the posterior forebrain, is located above the pituitary gland and is intimately associated with it via a portal system of capillaries (two capillary beds). When the hypothalamus is stimulated (by feedback from endocrine glands or by neurons innervating it), it releases hormonelike substances called releasing factors into the anterior-pituitary-hypothalamic portal circulatory system. In their turn, these releasing factors stimulate cells of the anterior pituitary to secrete the specific hormone coded for by its specific releasing factor.

The pituitary gland is a small bilobed gland lying at the base of the brain. The two lobes, anterior and posterior, function as independent glands. The anterior pituitary secretes the following hormones:

- *Growth hormone* fosters growth in a variety of body tissues.
- *Thyroid stimulating hormone (TSH)* stimulates the thyroid gland to secrete its own hormone, thyroxine.
- *Adrenocorticotrophic hormone (ACTH)* stimulates the adrenal cortex to secrete its corticoids.

- *Prolactin* is responsible for milk production by the female mammary glands.
- *Follicle-stimulating hormone (FSH)* spurs maturation of seminiferous tubules in males and causes maturation of ovaries in females. It also encourages growth of follicles in the ovaries.
- Finally, *luteinizing hormone (LH)* induces interstitial cells of the testes to mature by beginning to secrete the male sex hormone testosterone. In females, a surge of LH stimulates ovulation of the primary oocyte from the follicle. LH then induces changes in the follicular cells and converts the old follicle into a yellowish mass of cells rich in blood vessels. This new structure is the corpus luteum, which subsequently secretes progesterone and estrogen.

Thyroid Gland. The thyroid hormone, thyroxine, is an amino acid that contains four atoms of iodine. It accelerates oxidative metabolism throughout the body. An abnormal deficiency of thyroxine causes goiter, decreased heart rate, lethargy, obesity, and decreased mental alertness. In contrast, hyperthyroidism (too much thyroxine) is characterized by profuse perspiration, high body temperature, increased basal metabolic rate, high blood pressure, loss of weight, and irritability.

Parathyroid Glands. The parathyroid glands are small pea-like organs located on the posterior surface of the thyroid. They function by secreting parathyroid hormone, which regulates the calcium and phosphate balance between the blood and other tissues.

Pancreas. The pancreas is a multifunctional organ. It has both an exocrine and an endocrine function. The *exocrine function* secretes hormones through ducts into whatever part of the body that needs them. The *endocrine function*, on the other hand, secretes hormones directly into the blood stream.

The endocrine function is centered in the islets of Langerhans, which contain alpha and beta cells that secrete glucagon and insulin respectively. Insulin stimulates the muscles to remove glucose from the blood. It is also responsible for spurring both muscles and liver to convert glucose to glycogen, the storage form of glucose. The islets of Langerhans also secrete glucagon, which responds to low concentrations of blood glucose by stimulating the breakdown of glycogen into glucose.

Adrenal Glands. The adrenal glands are situated on top of the kidneys and consist of the *adrenal cortex* and the *adrenal medulla*.

Adrenal cortex. In response to stress, ACTH stimulates the adrenal cortex to synthesize and secrete the steroid hormones, which are collectively known as corticosteroids. The corticosteroids, derived from cholesterol, include glucocorticoids, mineralocorticoids, and cortical sex hormones.

Mnemonic: Flat Pig

The anterior pituitary gland secretes the following hormones:

FSH
LH
ACTH
TSH

Prolactin
I(gnore)
Growth

Don't Mix These Up on Test Day

In the pancreas, the *exocrine function* secretes digestive enzymes and bicarbonate.

The *endocrine function* of the pancreas, meanwhile, secretes insulin and glucagon.

Glucocorticoids, such as cortisol and cortisone, are involved in glucose regulation and protein metabolism. Glucocorticoids raise blood glucose levels by promoting gluconeogenesis and decrease protein synthesis. They also reduce the body's immunological and inflammatory responses. Cortisol secretion is governed by a negative feedback mechanism.

Mineralocorticoids, particularly aldosterone, regulate plasma levels of sodium and potassium, and consequently, the total extracellular water volume. Aldosterone causes active reabsorption of sodium and passive reabsorption of water in the nephron, a topic we dealt with in more detail earlier in this book.

Fight or Flight

Hormones like epinephrine and norepinephrine are responsible for the body's physical reactions to stress. They temporarily increase heart beat and blood supply to the active organs in what is known as "the fight or flight response."

The adrenal cortex also secretes small quantities of androgens (male sex hormones) in both males and females. Since, in males, most of the androgens are produced by testes, the physiologic effect of the adrenal androgens is quite small. In females, however, overproduction of the adrenal androgens may have masculinizing effects, such as excessive facial hair.

Adrenal Medulla. The secretory cells of the adrenal medulla can be viewed as specialized sympathetic, postganglionic nerve cells that secrete hormones into the circulatory system. This organ produces *epinephrine* (adrenaline) and *norepinephrine* (noradrenaline), both of which belong to a class of amino acid-derived compoinds called catecholamines.

Epinephrine increases the conversion of glycogen to glucose in liver and muscle tissue, causing a rise in blood glucose levels and an increase in the basal metabolic rate. Both epinephrine and norepinephrine increase the rate and strength of the heartbeat, and dilate and constrict blood vessels. These in turn increase the blood supply to skeletal muscle, the heart, and the brain, while decreasing the blood supply to the kidneys, skin, and digestive tract. These effects are known as the *"fight or flight response,"* and are elicited by sympathetic nervous stimulation in response to stress. Both of these hormones are also neurotransmitters.

Control of Growth in Plants

The regulation of growth patterns is largely accomplished by plant hormones, which are almost exclusively devoted to this function. These hormones are produced by actively growing parts of the plant, such as the meristematic tissues in the apical region (apical meristem) of shoots and roots. They are also produced in young, growing leaves and developing seeds. Some of these hormones and their specific functions are discussed below.

Auxins. This is an important class of plant hormone associated with several growth patterns, including the following types.

Phototropism of Shoots. Auxins are responsible for phototropism, the tendency of the shoots of plants to bend toward light sources (particularly the sun). When light strikes the tip of a plant from one side, the auxin supply on that side is reduced. Thus the illuminated side of the plant grows more slowly than the shaded side. This asymmetrical growth in the cells of the stem causes the plant to bend toward the slower growing light side; thus the plant turns toward the light. Indoleacetic acid is one of the auxins associated with phototropism.

Geotropism. This is the term given to the growth of portions of plants towards or away from gravity. With negative geotropism, shoots tend to grow upward, away from the force of gravity. If the plant is turned on its side (horizontally), the shoots will eventually turn upward again. Gravity itself increases the concentration of auxin on the lower side of the horizontally placed plant, while the concentration on the upper side decreases. This unequal distribution of auxins stimulates cells on the lower side to elongate faster than cells on the upper side. Thus the shoots turn upward until they grow vertically once again.

With positive geotropism, roots, unlike shoots, grow toward the pull of gravity. In a horizontally placed stem, however, the effect on the root cells is the opposite. Those exposed to a higher concentration of auxin (the lower side) are inhibited from growing, while the cells on the upper side continue to grow. In consequence, the root turns downward.

Inhibition of Lateral Buds. Auxins produced in the terminal bud of a plant's growing tip move downward in the shoot and inhibit development of lateral buds. Auxins also initiate the formation of lateral roots, while they inhibit root elongation.

Gibberellins. This second class of hormones stimulates rapid stem elongation, particularly in plants that normally do not grow tall (for example, dwarf plants). Gibberellins also inhibit the formation of new roots, and stimulate the production of new phloem cells by the cambium (where the auxins stimulate the production of new xylem cells). Finally, these hormones terminate the dormancy of seeds and buds, and induce some biennial plants to flower during their first year of growth.

Kinins. Kinins make up another general class of hormone compounds that promote cell division. Kinetin is an important type of cytokinin, and is involved in general plant growth, breaking seed dormancy, and expanding leaves. The action of kinetin is enhanced when auxin is present. The ratio of kinetin to auxin is of particular importance in the determination of the timing of the differentiation of new cells.

Ethylene. Ethylene stimulates the ripening of fruit and induces aging. Its functioning is slowed by inhibitors, which block cell division as part of a number of important control mechanisms. Inhibitors are particularly important to the maintenance of dormancy in the lateral buds and seeds of plants during autumn and winter. They break down gradually with time (and, in some cases, are destroyed by the cold), so that buds and seeds can become active in the next growing season.

Sun Worshippers

When sunflowers bend to follow the sun, they are exhibiting phototropism.

Treating the Fruit You Eat

Ethylene and one of its most important inhibitors, abscissic acid, are commonly used to treat fruits for human consumption. Bananas, for example, are picked unripe, then injected with ethylene so that the ripening process will continue after the fruit is off the tree. As bananas are transported to markets and sold, they are treated with abscissic acid to prevent rotting and extend their shelf life.

Antiauxins. These compounds regulate the activity of auxins. For example, indoleacetic acid oxidase regulates the concentration of indoleacetic acid. An increase in the concentration of indoleacetic acid increases the amount of indoleacetic acid oxidase produced.

Nervous System

The nervous system enables organisms to receive and respond to stimuli from their external and internal environments. Your brain and spinal column control your breathing, your movement, and your ability to feel, see, touch, hear, and taste, as well as your ability to understand complex emotions. Without these vital body parts, you wouldn't be capable of taking the SAT II at all.

It is just as vital, however, that these organs undergo some form of regulation, or you would be inundated with distracting stimuli. For example, if you were incapable of regulating noise stimuli, what is background noise now would become intolerably loud, and if your sense of touch were not controlled, you would become acutely aware of the fact that you were wearing clothes.

Those Who Feel Too Much

The inability to regulate all the stimuli that are encountered in the course of everyday life can lead to a variety of learning disabilities. Those who are incapable of shutting out some of the distracting stimuli around them cannot concentrate on what they are supposed to be learning.

Regulation

The environment provides information in the form of stimuli to the organism, which in turn responds through movement, secretion, and growth. *Regulation*, one of the most basic functions of the nervous system, is accomplished in most animals through the interaction of a nervous system and a system of the specialized secretions we have already discussed, hormones.

All of the knowledge we obtain from our environment comes through sensory receptors—eyes for light, ears for sound, noses for smell, tongues for taste, and skin for tactile stimuli. These receptors need a mechanism by which they can inform the rest of the body of the specific information they have processed. *Nerves* serve this important function.

A *stimulus* is a change in the environment that produces a response in a nerve. In multicellular animals, the stimulus initiates an electrochemical impulse that moves rapidly through the specialized nerve cells called neurons. This impulse elicits the secretion of chemical *neurotransmitters*, which travel across a narrow junctional space, the synapse, stimulating the dendrites of the adjacent neuron, where another impulse may be initiated.

Functional Units of the Nervous System

Let's now turn to the functional units of the nervous system. The most basic of these functional units is the *neuron*.

Neurons. There are many different kinds of neurons in the vertebrate nervous system. Neurons that carry information about the external or internal environment to the brain or spinal cord are called *afferent neurons.* Neurons that carry commands from the brain or spinal cord to various parts of the body (e.g., muscles or glands) are called *efferent neurons*. Some neurons (termed *interneurons*) participate only in local circuits; their cell bodies and their nerve terminals are in the same location.

Nerves are essentially bundles of axons covered with connective tissue. A nerve may carry only sensory fibers (a *sensory nerve*), only motor fibers (a *motor nerve*), or a mixture of the two (a *mixed nerve*). Neuronal cell bodies often cluster together; such clusters are called *ganglia* in the periphery, and *nuclei* in the central nervous system.

The neuron's primary function is to convert stimuli into electrochemical signals that are conducted through the nervous system. This functional unit is an elongated cell consisting of *dendrites*, a *cell body*, and an *axon*. Dendrites are cytoplasmic extensions that receive information and transmit it toward the cell body. The cell body (soma) contains the nucleus and controls the metabolic activity of the neuron. Meanwhile, the axon hillock connects the cell body to the axon (nerve fiber), which is a long cellular process that transmits impulses away from the cell body.

Most mammalian axons are ensheathed by an insulating substance known as *myelin*, which allows axons to conduct impulses faster. Myelin is produced by cells known as glial cells. (Oligodendrocytes produce myelin in the central nervous system, and Schwann cells produce myelin in the peripheral nervous system.) The gaps between segments of myelin are called nodes of Ranvier.

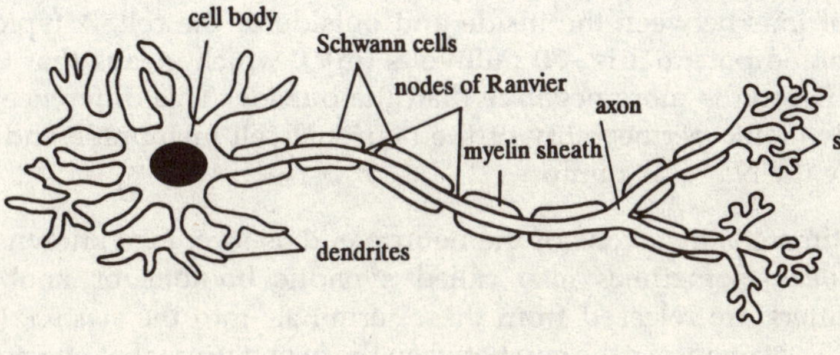

Neuron

Neurons are specialized to receive signals from sensory receptors or from other neurons in the body and to transfer this information along the length of the axon. These impulses, known as *action potentials*, travel the length of the axon and invade the nerve terminal, thereby causing the release of neurotransmitter into the synapse (which will be discussed in more detail below).

If the cell becomes sufficiently excited or depolarized (i.e., the inside of the cell becomes less negative), an action potential is generated. The action potential is often described as an all-or-none response. This means that

Malfunctioning Myelin

When the body mounts a reaction against its own myelin, it begins to impede the myelination of nerves. This results in weakness, lack of balance, vision problems, and/or incontinence. This condition is termed multiple sclerosis (MS).

Watch What You Eat

The puffer fish is considered a delicacy in Japan, but it is not the kind of food you want to bolt down indiscriminately. This fish contains a poison called tetrodotoxin (TTX). TTX blocks the voltage-gated Na^+ channels of the body, which serves to block neuronal transmissions. Death follows soon after.

whenever the threshold membrane potential is reached, an action potential with a consistent size and duration is produced.

Neuronal information is coded by the frequency and number of action potentials rather than the size of the action potential. In other words, the harder you hit your thumb with a hammer, the more action potentials will travel up your pain fibers, but the size and duration of each individual action potential will remain the same.

If there is an adequate stimulus, the action potential will first be initiated at the axon hillock. Although axons can theoretically propagate action potentials bidirectionally, information transfer will occur only in one direction: from dendrite to synaptic terminal. (This is because synapses operate only in one direction and because refractory periods make the backward travel of action potentials impossible.)

Different axons can propagate action potentials at different speeds. The greater the diameter of the axon and the more heavily it is myelinated, the faster the impulses will travel. Many invertebrates, for example, have giant axons that speed conduction; examples include the squid. Myelin increases conduction velocity by insulating segments of the axon, so that the membrane is permeable to ions only in the nodes of Ranvier. In this way, the action potential "jumps" from node to node; this process is called *saltatory conduction*.

When a neuron is at rest, the potential difference between the extracellular space and the intracellular space is called the *resting potential*. Even at rest, a neuron is polarized. This potential difference is the result of an unequal distribution of ions between the inside and outside of the cell. A typical resting membrane potential is –70 millivolts (mV), which means that the inside of the neuron is more negative than the outside. This difference is due to selective ionic permeability of the neuronal cell membrane and is maintained by the Na^+/K^+ pump.

Synapse. Ultimately, the axons of the neuron end as swellings known as *synaptic terminals* (sometimes also called synaptic boutons or knobs). Neurotransmitters are released from these terminals into the *synapse* (or *synaptic cleft*). The synapse is the gap between the axon terminal of one neuron (called the *presynaptic neuron* because it is before the synapse) and the dendrites of another neuron (*postsynaptic neuron*). Neurons may also communicate with postsynaptic cells other than neurons, such as cells in muscles or glands; these are called *effector cells*.

The vast majority of synapses in the human are *chemical synapses*. In these synapses, the nerve terminal contains thousands of membrane-bound vesicles full of chemical messengers known as *neurotransmitters*. When the action potential arrives at the nerve terminal and depolarizes it, the synaptic vesicles fuse with the presynaptic membrane and release neurotransmitter into the synapse. The neurotransmitter diffuses across the synapse and acts on receptor proteins embedded in the postsynaptic membrane.

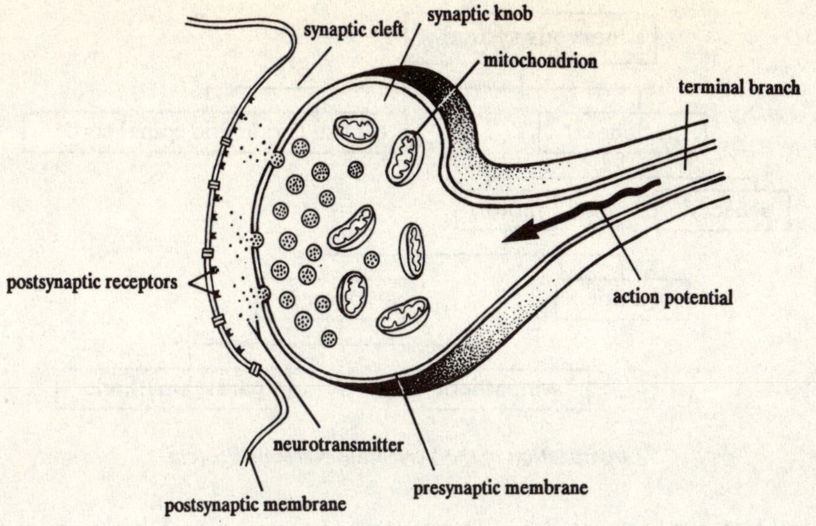

Synapse

Depending on the nature of the receptor, the neurotransmitter may have an excitatory or an inhibitory effect on the postsynaptic cell. The neurotransmitter may be removed from the synapse in a variety of ways. It may be taken back up into the nerve terminal (via a protein known as an uptake carrier) where it may be reused or degraded. It may also be degraded by enzymes located in the synapse (e.g., acetylcholinesterase inactivates the neurotransmitter acetylcholine). Alternatively, it may simply diffuse out of the synapse.

Drugs and Nerves

Many drugs modify processes that occur in the synapse. Cocaine, for example, blocks neuronal uptake carriers, prolonging the action of neurotransmitters. Nerve gases, meanwhile, are potent acetylcholinesterase inhibitors that cause rapid death by preventing the action of skeletal muscles (most importantly, the diaphragm), which leads to respiratory arrest.

Organization of the Nervous System

As organisms evolved and became more complex, their nervous systems must also undergo corresponding increases in complexity. Simple organisms can only respond to simple stimuli, while complex organisms like humans can discern subtle variations of a particular stimuli, such as a particular shade of color.

Invertebrates. *Protozoa* have no organized nervous system. They can only respond to stimuli like touch, heat, light, and chemicals. *Coelenterates*, on the other hand, have a nervous system consisting of a network of cells, the *nerve net*, located between the inner and outer layers of the cells of its body. *Annelids* possess a primitive central nervous system consisting of a solid ventral nerve cord and an anterior "brain" of fused ganglia. Finally, *arthropods* have a better-developed, more refined nervous system than annelids. They have more specialized sense organs, including simple or complex eyes and a tympanum for detecting sound.

Vertebrates. The vertebrate nervous system itself is divided into three major subsystems, the central nervous system, the peripheral nervous system, and the autonomic nervous system.

Insensitive

Lower organisms have very primitive nervous systems and can only respond to strong stimuli like touch, heat, light, and chemicals. Higher organisms like vertebrates, on the other hand, may respond to a wide range of stimuli.

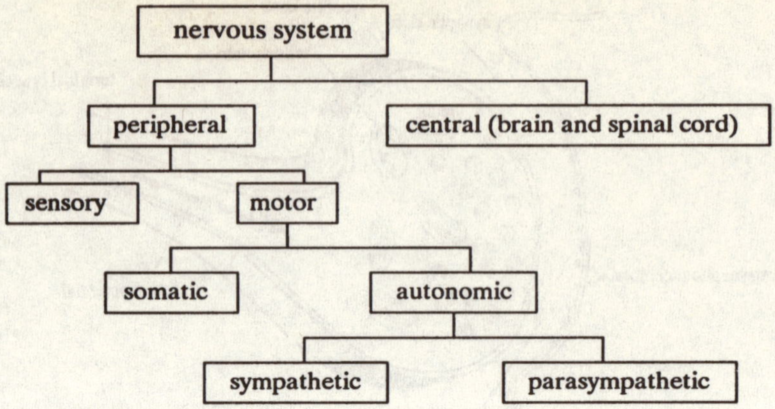

Organization of the Vertebrate Nervous System

Central Nervous System (CNS). This system is composed of the brain and the spinal cord. The *brain* is a mass of neurons at the upper end of the spinal cord. It consists of an outer portion containing neuronal cell bodies (gray matter) and an inner portion containing axons (white matter). The human brain is composed of the following regions:

- *Cerebral cortex.* This controls all voluntary motor activity by initiating the responses of motor neurons present within the spinal cord. It also controls higher functions, such as memory and creative thought.

- *Olfactory lobe.* This serves as the center for reception and integration of olfactory input.

- *Thalamus.* Nervous impulses and sensory information are relayed and integrated en route to and from the cerebral cortex by this region.

- *Hypothalamus.* Such visceral functions as hunger, thirst, pain, temperature regulation, and water balance are controlled by this center.

- *Cerebellum.* Muscle activity is coordinated and modulated here.

- *Pons.* This serves as the relay center for cerebral cortical fibers en route to the cerebellum.

- *Medulla oblongata.* This influential region controls vital functions like breathing, heart rate, and gastrointestinal activity. It has receptors for carbon dioxide; when carbon dioxide levels become too high, the medulla oblongata forces you to breathe. This is why when you hold your breath until carbon dioxide levels rise so high in your blood that you pass out, you will breathe involuntarily to bring an influx of oxygen into your body.

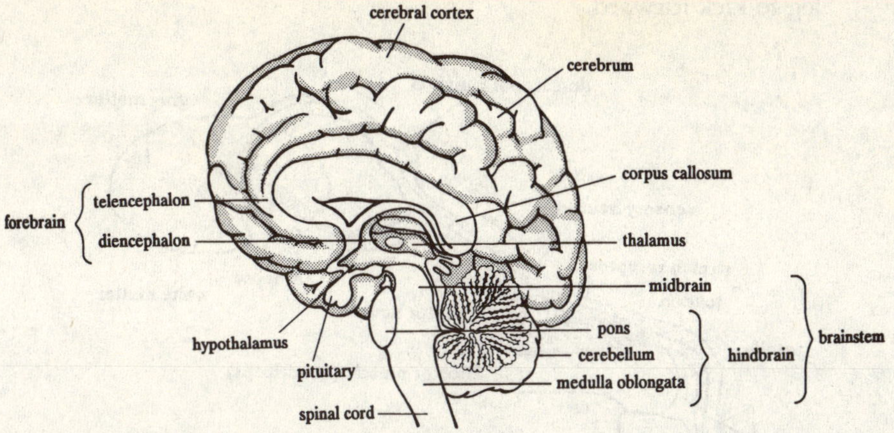

Human Brain

The *spinal cord*, meanwhile, is a hollow, elongated extension from the brain that serves as a center for many reflex actions that do not involve the brain, such as the knee-jerk reflex. The spinal cord consists of two parts. The *dorsal horn* is the entrance point for sensory nerve fibers whose cell bodies are contained within the dorsal root ganglion. The *ventral horn*, on the other hand, contains the cell bodies of motor neurons. Fibers from the cerebral cortex synapse on the ventral horn motor neurons, thereby initiating muscular contractions.

The Peripheral Nervous System (PNS). The peripheral nervous system consists of 12 pairs of cranial nerves, which primarily innervate the head and shoulders, and 31 pairs of spinal nerves, which innervate the rest of the body. Cranial nerves exit from the brainstem and spinal nerves exit from the spinal cord. The PNS has two primary divisions: the somatic and the autonomic nervous systems, each of which has both motor and sensory components.

- *The somatic nervous system (SNS).* This system innervates skeletal muscles and is responsible for voluntary movement. Motor neurons release the neurotransmitter acetylcholine (ACh) onto ACh receptors located on skeletal muscle. This causes depolarization of the skeletal muscle, leading to muscle contraction. In addition to voluntary movement, the somatic nervous system is also important for reflex action. There are both monosynaptic and polysynaptic reflexes.

  Monosynaptic reflex pathways have only one synapse between the sensory neuron and the motor neuron. The classic example is the *knee-jerk reflex.* When the tendon covering the patella (kneecap) is hit, stretch receptors sense this and action potentials are sent up the sensory neuron and into the spinal cord. The sensory neuron synapses with a motor neuron in the spinal cord, which, in turn, stimulates the quadriceps muscle to contract, causing the lower

Don't Mix These Up on Test Day

In the spinal cord, the *dorsal horn* serves as the entrance point for sensory nerve fibers or afferent neurons.

The *ventral horn*, meanwhile, contains the cell bodies of motor or efferent neurons.

leg to kick forward.

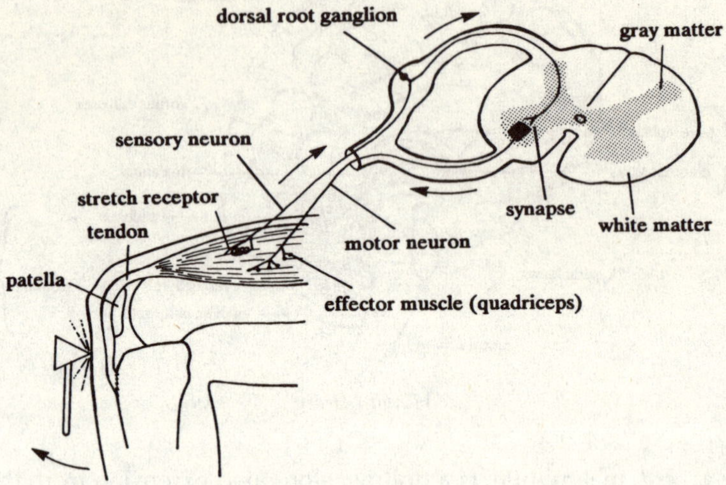

Reflex Arc for Knee Jerk

In *polysynaptic reflexes*, sensory neurons synapse with more than one neuron. A classic example of this is the *withdrawal reflex*. When a person steps on a nail, the injured leg withdraws in pain, while the other leg extends to retain balance.

The Autonomic Nervous System (ANS). The autonomic nervous system is neither structurally nor functionally isolated from the CNS or the PNS. Its function is to regulate the involuntary functions of the body. The ANS innervates the heart and blood vessels, the gastrointestinal tract, urogenital organs, structures involved in respiration, and the intrinsic muscles of the eye. It is made up of the sympathetic nervous system and the parasympathetic nervous sytem.

- *The sympathetic nervous system.* This utilizes norepinephrine as its primary neurotransmitter. It is responsible for activating the body for emergency situations and actions (the *fight-or-flight response*), including strengthening of heart contractions, increases in the heart rate, dilation of the pupils, bronchodilation, and vasoconstriction of vessels feeding the digestive tract.

- *The parasympathetic nervous system.* Here, acetylcholine serves as the primary neurotransmitter. One of this system's main functions is to deactivate or slow down the activities of muscles and glands (the *rest-and-digest response*). These activities include pupillary constriction, slowing down of the heart rate, bronchoconstriction, and vasodilation of vessels feeding the digestive tract. The principal nerve of the parasympathetic system is the vagus nervous.

Most of the organs innervated by the ANS receive both sympa-

Don't Mix These Up on Test Day

The *sympathetic nervous system*:

- Is associated with the fight-or-flight response
- Increases the heart rate
- Increases the breathing rate
- Lowers the digestive rate
- Causes pupil dilation

The *parasympathetic nervous system*:

- Is associated with the rest-and-digest response
- Lowers the heart rate
- Does not affect the breathing rate
- Increases the digestive rate
- Does not cause pupil dilation

thetic and parasympathetic fibers, the two systems being antagonistic to one another.

Sensory Systems of the Nervous System

All complicated nervous systems are made more useful through input mechanisms that we know as our senses. Sight, hearing, balance, taste, smell, and touch provide an influx of data for the nervous system to assimilate. They also make life more enjoyable.

Sight. The eye detects light energy and transmits information about intensity, color, and shape to the brain. The transparent cornea at the front of the eye bends and focuses light rays. These rays then travel through an opening called the pupil, whose diameter is controlled by the pigmented, muscular iris. The iris responds to the intensity of light in the surroundings (light makes the pupil constrict). The light continues through the lens, which is suspended behind the pupil. This lens focuses the image onto the retina, which contains photoreceptors that transduce light into action potentials.

There are two types of photoreceptors: *cones* and *rods*. Cones respond to high-intensity illumination and are sensitive to color, while rods detect low-intensity illumination and are important in night vision. The photoreceptor cells synapse onto bipolar cells, which in their turn synapse onto ganglion cells. Axons of these cells bundle to form the optic nerves, which conduct visual information to the brain.

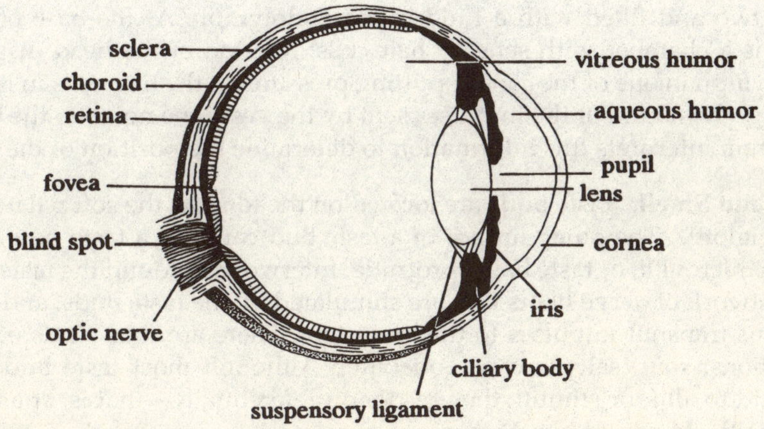

Human Eye

Hearing and Balance. The ear transduces sound energy into impulses that are perceived by the brain as sound. It is also responsible for maintaining equilibrium (balance) in the body. Sound waves pass through three regions as they enter the ear. First, they enter the outer ear, which consists of the auricle (pinna) and the auditory canal. Located at the end of the auditory canal is the tympanic membrane (eardrum) of the middle ear,

Mnemonics

These two mnemonics will help you to get a handle on the following concepts:

With the exceptions of sexual activity and digestion,

The **s**ympathetic nervous system **s**peeds things up.

The parasympathetic nervous system slows things down.

and:

Cones respond to **c**olor.

Rods respond to black and white.

Don't Mix These Up on Test Day

Cones are photoreceptors that respond to high intensity illumination and color.

Rods, on the other hand, respond to low intensity illumination (they are important in night vision).

which vibrates at the same frequency as the incoming sound. Next, three bones, or ossicles (malleus, incus, and stapes), amplify the stimulus, and transmit it through the oval window, which leads to the fluid-filled inner ear.

This inner ear consists of the cochlea and semicircular canals. The *cochlea* contains the organ of Corti, which has specialized sensory cells called hair cells. Vibration of the ossicles exerts pressure on the fluid in the cochlea, stimulating the hair cells to transduce the pressure into action potentials, which travel via the auditory nerve to the brain for processing.

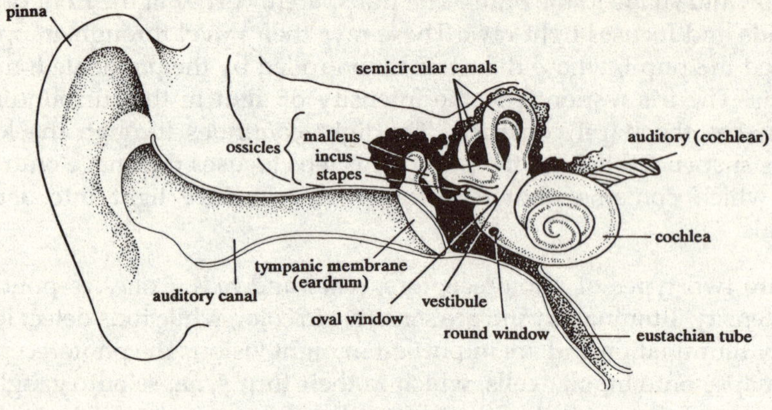

Human Ear

Each of the three semicircular canals in the inner ear is perpendicular to the other two and filled with a fluid called endolymph. At the base of each canal is a chamber with sensory hair cells; rotation of the head displaces endolymph in one of the canals, putting pressure on the hair cells in it. This changes the nature of the impulses sent by the vestibule nerve to the brain. The brain interprets this information to determine the position of the head.

Taste and Smell. Taste buds are located on the tongue, the soft palate, and the epiglottis. The outer surface of a taste bud contains a taste pore, from which microvilli, or taste hairs, protrude. Interwoven around the taste buds is a network of nerve fibers that are stimulated by the taste buds, and these neurons transmit impulses to the brainstem. There are four kinds of taste sensations: sour, salty, sweet, and bitter. Although most taste buds will respond to all four stimuli, they respond preferentially—that is, at a lower threshold—to one or two of them.

Olfactory receptors are found in the olfactory membrane, which lies in the upper part of the nostrils over a total area of about 5 cm^2. The receptors are specialized neurons from which olfactory hairs project. When odorous substances enter the nasal cavity, they bind to receptors in the cilia, depolarizing the olfactory receptors. Axons from the olfactory receptors join to form the olfactory nerves, which project direction to the olfactory bulbs in the base of the brain.

It's Too Spicy!

We lose taste buds as we age. This is why young children seem to enjoy bland food such as macaroni and cheese, while adults tend to be more interested in spicier food.

Motor Systems of the Nervous System

Physical support and locomotion are the functions of animal skeletal systems. Neither the skeleton nor its complementary muscles are regarded as parts of the nervous system; we discuss them here because they are the components of the organism that enable it to carry out responses to stimuli (that is, the effectors innervated by the nervous system).

The muscular system generates force, and the skeletal system is the mass against which the muscles contract to produce motion. Read on for explanations of the characteristic motor systems of members of the select group of organisms we have been returning to throughout this book.

Cilia and Flagella. Ciliates and flagellates are unicellular organisms that do not have discrete skeletal-muscular systems. *Protozoans* and primitive algae move by the beating of cilia or flagella. *Amoebae,* meanwhile, use cell extensions called pseudopodia for locomotion; the advancing cell membrane extends, and the cytoplasm liquefies and flows into the pseudopods.

The cilia and flagella of all eukaryotic cells possess the same basic structure. Each contains a cylindrical stalk that consists of eleven microtubules, nine pairs arranged in a peripheral circle with two single microtubules in the center of the stalk. Movement is affected by means of the power stroke, a thrusting movement in which the microtubule cylinders slide past each other. Return of the cilium or flagellum to its original position is called the recovery stroke.

Hydrostatic Skeletons. The muscles within the body wall of advanced flatworms such as planaria are arranged in two antagonistic layers, longitudinal and circular. As the muscles contract against the resistance of the incompressible fluid within the animal's tissues, this fluid functions as a hydrostatic skeleton. Contraction of the circular layer of muscles causes the incompressible interstitial fluid to flow longitudinally, lengthening the animal. Conversely, contraction of the longitudinal layer of muscles shortens the animal.

The same type of hydrostatic skeleton assists in the locomotion of *annelids.* Each segment of this animal can expand or contract independently. Annelids advance principally through the action of muscles on a hydrostatic skeleton, as well as through bristles in the lower part of each segment. These bristles, called *setae,* anchor the earthworm temporarily in earth while muscles push the earthworm ahead.

Exoskeleton. This is the hard skeleton that covers all the muscles and organs of some invertebrates. Exoskeletons, found principally in *arthropods,* are composed of chitin. In all cases, the exoskeleton is composed of noncellular material secreted by the epidermis. Although it serves the additional function of protection, an exoskeleton imposes limitations on growth. Thus periodic molting and deposition of a new skeleton are necessary to permit body growth.

Effectors of the Nervous System

Motor systems allow organisms to respond to stimuli from their nervous systems.

Quick Quiz

Match the numbered organism with the correct lettered motor system below.

1. earthworm
2. primitive algae
3. bear

(A) endoskeleton
(B) hydrostatic skeleton
(C) cilia and flagella

Answers:
1. = (B)
2. = (C)
3. = (A)

Build and Destroy

Bone tissue is made up of *osteoblasts*, which build bone, and *osteoclasts*, which destroy bone. The reason that the body builds bone is obvious; the reason it destroys it is less obvious. Bone acts as a calcium reservoir for the entire body. When the body needs calcium, bone is destroyed in order to provide it.

Mnemonic

Remember:

Osteo**b**lasts **b**uild **b**one!

Endoskeleton. The endoskeleton serves as a framework within all *vertebrate* organisms. In this framework, muscles are attached to bones, permitting movement. The endoskeleton also provides protection, since bones surround delicate vital organs. For example, the rib cage protects the thoracic organs (heart and lungs), while the skull and vertebral column protect the brain and spinal cord.

Cartilage, although firm, is not as hard or as brittle as bone. It makes up the skeletons of lower vertebrates, such as sharks and rays. In higher animals, cartilage is the principle component of embryonic skeletons, and is replaced during development by the aptly termed replacement bone. Because cartilage has no vessels or nerves, it takes longer to heal than bone. The long bones of the arms and legs (the tibia, the humerus, and the femur) are examples of this class of bone.

Bone makes up the skeleton of mature higher vertebrates, including humans. Bone arises through the replacement of cartilage or through direct ossification. Bone produced through the latter process is called *dermal* bone; the bones of the skull are examples of this. In *replacement* bone, on the other hand, osteoblasts replace the cartilage that has already formed. As a result, a hollow cavity is formed within each bone; this cavity is subsequently filled with *bone marrow* (the site of formation of blood cells).

While the division between dermal and replacement bone is based on embryologic origin, the division between *spongy* and *compact* bone is based on function and internal structure. Spongy bone is located in the central portions of bone. It consists of a network of hard spicules separated by marrow-filled spaces. The low density and ability to withstand lateral stress that are characteristic of bone may be attributed to this type of spongy bone.

Meanwhile, compact bone, located on the outer surfaces and articular surfaces, is responsible for the hardness of bone and its ability to withstand longitudinal stress. It consists of cylindrical units called Haversian systems. Each unit consists of cells embedded in a matrix of inorganic material (calcium phosphate) which directly leads to the hardness of bone. These cells radiate around a central capillary within a Haversian canal.

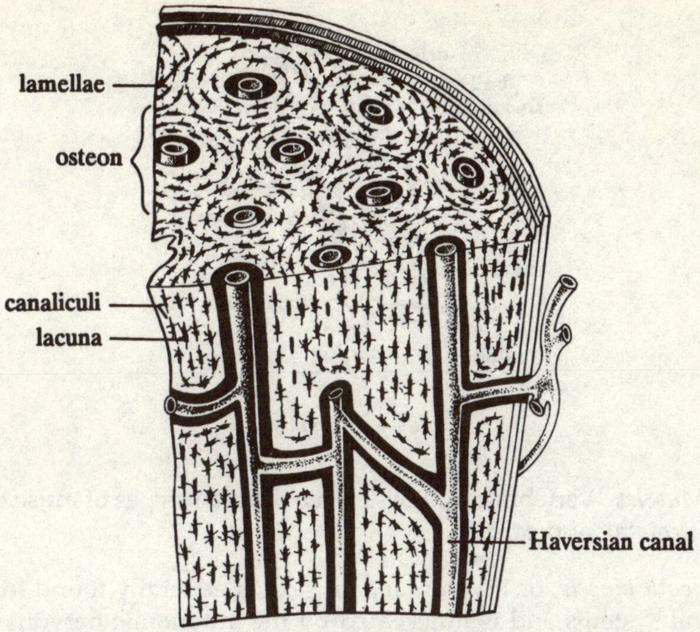

Microscopic Bone Structure

Bones are held together by sutures or immovable joints, such as those in the skull, and movable joints like the hip joint. In the latter type, *ligaments* serve as bone-to-bone connectors, while *tendons* attach skeletal muscle to bones and bend the skeleton at the movable joints.

In the vertebrate skeleton, the *axial skeleton* is the midline basic framework of the body, consisting of the skull, vertebral column, and the rib cage. The *appendicular skeleton*, on the other hand, includes the bones of the appendages and the pectoral and pelvic girdles.

Muscle System. The muscle system serves as the effector of the nervous system. Muscles contract to implement actions after they receive nervous stimuli. For example, your arm muscles will automatically contract if you touch a hot stove.

Attachments. A muscle originates at a point of attachment to the stationary bone. The insertion of a muscle is the portion attached to the bone that moves during contraction.

Muscle Movement. An extensor extends or straightens the bones at a joint—as in, for example, straightening out a limb. A flexor bends a joint to an acute angle, as in bending a limb.

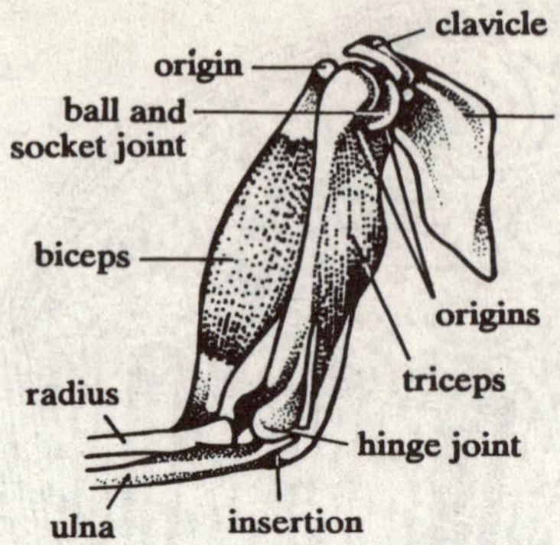

Muscle Movement

Don't Mix These Up on Test Day

Smooth muscle is involuntary muscle.

Skeletel muscles are voluntary muscles.

Cardiac muscle is the tissue that makes up the heart.

Types of Muscles. Vertebrates possess three different types of muscle tissues: smooth, skeletal, and cardiac.

- *Smooth muscle*, or involuntary muscle, is generally found in visceral systems and is innervated by the autonomic nervous system. Each muscle fiber consists of a single cell with one centrally located nucleus. Smooth muscle is nonstriated. Examples are the muscles found in the walls of the arteries and veins, the walls of the digestive tract, the bladder, and uterus.

Smooth Muscle

- *Skeletal muscles*, or voluntary muscles, produce intentional physical movement. Each fiber is multinucleated (that is, it has more than one nucleus in each cell) and is crossed by alternating light and dark bands called *striations.* Skeletal muscles also have sarcomeres, or contractile units. The somatic nervous system innervates skeletal muscle.

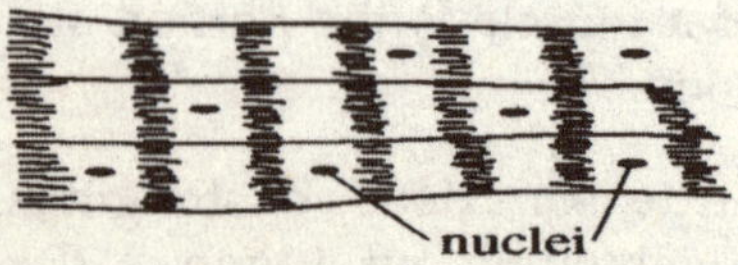

Skeletal Muscle

- *Cardiac muscle* is the tissue that makes up the heart. It has characteristics of both skeletal and smooth muscle—that is, it is both striated and mononucleate. It is controlled by the autonomic nervous system, which implies that it does not need as much nervous stimulation as other muscle types. Instead, the nervous system only modulates its inherent heart beat.

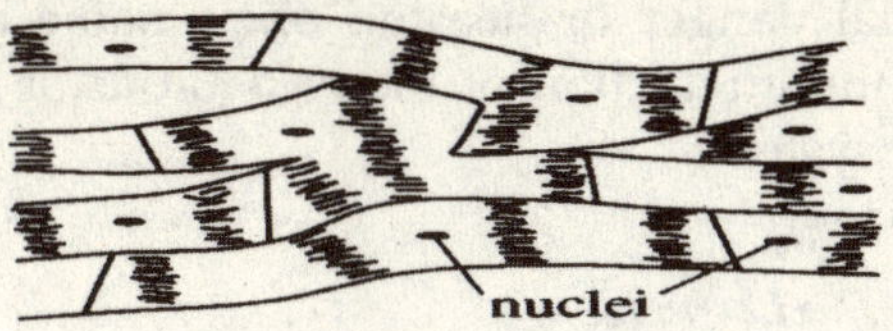

Cardiac Muscle

With that, we've covered the basics of the physiology you might be called upon to know on Test Day. The last topic associated with organismal biology that we need to discuss is that of animal behavior.

Animal Behavior

The ways in which organisms respond to the environments around them are a significant facet of the study of organisms and organismal biology. Remember the old "nature versus nurture" controversy: While our genetic makeup determines a great deal about the kind of people we become, our environment and how we respond to it is also very instrumental in this process, and the same is true of all organisms.

Patterns of Animal Behavior

Animal behavior is characterized by a number of different patterns. Some of these patterns are more strongly associated with the nervous system than others, as we shall see. Simple organisms are capable only of simple, automatic responses to their environment, while more complex organisms are characterized by an increased reliance on mental processes like learning.

Simple Reflexes

Reflexes are simple, automatic responses to simple stimuli. They can be defined functionally as reliable occurrences of particular behavioral responses following a given environmental stimulus. A simple reflex is controlled at the spinal cord level of a vertebrate, involving a pathway from the receptor (afferent nerve) to the efferent or motor nerve. The efferent nerve innervates the effector, a muscle or gland. Thus a simple reflex involves a two-neuron pathway. Although reflex behavior is important in the behavioral response of lower animals, it is relatively less important in the behavioral repertoire of higher forms of life such as vertebrates.

Don't Mix These Up on Test Day

An example of a *simple reflex* is the knee-jerk reflex.

An example of a *complex reflex* is the startle response.

Fixed-action patterns, meanwhile, are the characteristic movements of a herd or flock of animals.

Complex Reflexes

More complex reflex patterns involve neural integration at a higher level, such as the brainstem and even the cerebrum. An example is the "startle response," the alerting response of an animal to a significant stimulus (for example, a potential danger or hearing one's name called). The startle response involves the interaction of many neurons, a system termed the reticular activating system.

Fixed-Action Patterns

Fixed-action patterns are complex, coordinated, innate behavioral responses to specific patterns of stimulation in the environment. The stimulus that elicits the behavior is referred to as the releaser. Fixed-action patterns are not learned; they are innate. For this reason, they are relatively unlikely to be modified by learning. An animal has a repertoire of fixed-action patterns and only a limited ability for developing new ones. The particular stimuli that trigger a fixed-action pattern are more readily modified, provided certain cues or elements of the stimuli are maintained.

An example of a fixed-action pattern is the retrieval-and-maintenance response of many female birds to an egg of their species. Certain kinds of stimuli are more effective than others in triggering a fixed-action pattern. Hence an egg with the characteristics of that species will be more effective in triggering the response than one which only crudely resembles the natural egg. Another type of fixed-action pattern is the characteristic movements made by animals that herd or flock together, such as the swimming actions of fish and the flying actions of locusts.

Even Plants Need to Sleep Sometimes

The first evidence noted for the existence of 24-hour circadian rhythms was the sleep movements of bean plants.

Behavioral Cycles

Daily cycles of behavior are called *circadian rhythms*. Animals who exemplify these behavior cycles lose their exact 24-hour periodicity if they are isolated from the natural phases of light and dark. Cyclical behavior, however, will continue with approximate day-to-day phasing. The cycle is thus initiated intrinsically, but modified by external factors.

Daily cycles of eating, maintained by most animals, provide a good example of cycles characterized by both internal and external control. The internal controls are the natural bodily rhythms of eating and satiation. External modulators include the elements of the environment that occur in familiar cyclic patterns, such as dinner bells and clocks.

Sleep and wakefulness are the most obvious examples of cyclical behavior. These behavior patterns have been associated with particular patterns of brain waves.

Environmental Rhythms

In many situations, patterns of behavior are established and maintained by periodic environmental stimuli. A human example of this is the response to traffic light signals. Just as environmental stimuli influence many naturally occurring biological rhythms, biological factors influence behavior governed by periodic environmental stimuli.

Learning

Learned behavior involves adaptive responses to the environment. Learning is a complex phenomenon that occurs, to some extent, in all animals. In lower animals, instinctual or innate behaviors are the predominant determinants of behavior patterns, and learning plays a relatively minor role in the modification of these predetermined behaviors.

Higher animals, on the other hand, learn the major share of their repertoire of responses to the environment. The capacity for learning adaptive responses is closely correlated with the degree of neurologic development of the organism—that is, the capacity of its nervous system, particularly its cerebral cortex, for flexibility.

Limits of Behavioral Change

There are limits, however, to what an organism is capable of learning. These limits can be either neurologic (the organism simply doesn't have the brain power) or chronologic (learning must occur during a narrow window during the organism's development in order to be successful).

Critical Period

These are specific time periods during an animal's early development when it is physiologically capable of developing specific behavioral patterns. If the proper environmental pattern is not present during this critical period, the behavioral pattern will not develop correctly. In some animals there is also a visual critical period; if light is not present during this period, visual effectors will not develop properly.

Imprinting

Imprinting is a process in which environmental patterns or objects presented to a developing organism during a brief critical period in early life become accepted permanently as an element of its behavioral environment. To put it another way, these patterns are "stamped in" and included in an animal's behavioral response.

A duckling, for example, passes through a critical period in which it learns that the first large moving object it sees is its mother. In the natural environment, its mother usually *is* the first thing a duckling sees. However, if a large, moving object other than its mother is the first thing it sees, the duckling will follow it, as Konrad Lorenz discovered when he was pursued by newborn ducklings that assumed that he was their mother.

Quick Quiz

Pavlov conducted experiments in which he rang a bell every time he fed his dogs. He discovered that eventually, the dogs salivated every time the bell was rung, even when no food was present. This is an example of

(A) imprinting

(B) a circadian rhythm

(C) a learned response

(Answer = (C))

Round and Waggle Dancing

Honey bees have developed a complex system of communication that involves a form of dancing. A "round dance" (rapid sideways movement in tight circles) indicates that food is near. A "waggle dance" (a half-circle swing in one direction, followed by a straight run and another half circle swing in the other direction) indicates that food is further away in a specified direction.

Intraspecific Interactions

Just as an organism communicates within itself via nervous and endocrine systems, it also requires methods to communicate with other members of its species. These methods include behavioral displays, pecking order, territoriality, and responses to chemicals.

Behavioral Displays

A display may be defined as an innate behavior that has evolved as a signal for communication between members of the same species. According to this definition, a song, a call, or an intentional change in an animal's physical characteristics is considered a display.

Reproductive displays are specific behaviors found in all animals, including humans. Many animals have evolved a variety of complex actions that function as signals in preparation for mating. Agonistic displays are intended to appease the observer, such as a dog's display of appeasement when it wags its tail. Meanwhile, antagonistic displays are intended to imply hostility and to instil fear, as when a dog directs its face straight at its opponent and raises its body.

Other displays include various dancing procedures exhibited by honeybees, especially the scout honeybee, which is able to convey information to workers in the hive concerning the quality, direction, and distance from the hive of food sources. Displays utilizing auditory, visual, chemical, and tactile elements are often used as a means of communication.

Pecking Order

Frequently, the relationships among members of the same species living as a contained social group become stable for a period of time. When food, mates, or territory are disputed, a dominant member of the species will prevail over a subordinate one. This social hierarchy is often referred to as the pecking order. This established hierarchy minimizes violent intraspecific aggressions by defining the stable relationships among members of the group; subordinate members dare not seriously challenge the dominant individuals.

Territoriality

Members of most land-dwelling species defend a limited area or territory from intrusion by other members of the same species. These territories are typically occupied by a male or a male-female pair. The territory is frequently used for mating, nesting, and feeding.

Territoriality serves the adaptive function of distributing members of the species so that the environmental resources are not depleted in a small region. Furthermore, intraspecific competition is reduced. Although there is frequently a minimum size for a species' territory, that size varies with population size and density. The larger the population, and the scarcer the resources the available to it, the smaller the territories are likely to be.

Responses to Chemicals

The olfactory sense is immensely important as a means of communication. Many animals secrete substances called pheromones that influence the behavior of other members of the same species. One type of pheromone, the *releaser pheromone*, triggers a reversible behavioral change in the recipient. For example, female silkworms secrete a very powerful attracting pheromone that will attract a male from a distance of two miles or more. In addition to their sex-attracting purposes, releaser pheromones are secreted as alarm and toxic defense substances.

Another type of pheromone, the *primer pheromone*, produces long-term behavioral and physiological alterations in recipient animals. For example, pheromones from male mice may affect the estrous cycles of females. They have also been shown to limit sexual reproduction in areas of high population density.

Follow the Leader

The trailing behavior of ants is based on scent. Scout ants release scent trails to guide the other ants to food sources.

This chapter aimed to provide you with a broad coverage of the different systems that keep organisms going. You should now feel comfortable with the basics of organismal biology, including concepts like the mechanics of sexual reproduction, nutrition, circulation, respiration, and the endocrine system. Find out just how comfortable you have become with these concepts by seeing how you do on the following quiz questions. Good luck!

Organismal Biology Quiz

1. Which enzyme breaks down starch to disaccharides?

(A) amylase
(B) gastrin
(C) secretin
(D) pepsin
(E) maltase

2. Of the choices below, which statement about the blastula stage of embryonic development is false?

(A) It consists of a solid ball of cells.
(B) It contains a fluid-filled center called the blastocoel.
(C) It is the stage of development that precedes the gastrula.
(D) It is a more advanced stage than a morula.
(E) It is a less advanced stage than a neurula.

3. Which of the following associations of germinal tissues and developed tissue is INCORRECT?

(A) mesoderm: heart
(B) ectoderm: nervous sytem
(C) endoderm: intestinal tract
(D) mesoderm: pancreas
(E) ectoderm: epidermis of skin

4. To ensure survival of their species, animals which do not care for their young must

(A) have the ability to live in water and on land
(B) lay eggs
(C) produce many offspring
(D) have protective coloring
(E) have internal fertilization

5. Basal metabolism disorders are most likely caused most directly by

(A) impairment of the pituitary
(B) impairment of the gonads
(C) impairment of the thyroid
(D) impairment of the thymus
(E) impairment of the parathyroid

6. In humans, the site of fertilization is the

(A) ovary
(B) fallopian tube
(C) uterus
(D) cervix
(E) vagina

GO ON TO THE NEXT PAGE

7. Which of the following statements about muscle tissue is true?

(A) In the muscle fiber, actin is the thick filament.
(B) The sarcoplasmic reticulum regulates the level of Ca^{2+} within a muscle cell.
(C) Unlike a nerve cell, a muscle cell does not possess an excitable membrane.
(D) In a muscle fiber, myosin is the thin filament.
(E) Contraction of a muscle fiber can occur in the absence of Ca^{2+}.

8. Which of the following statements is NOT true of the digestive system?

(A) Digestive enzymes from the pancreas are released via a duct into the duodenum.
(B) Peristalsis is a wave of smooth muscle contraction that proceeds along the digestive tract.
(C) In the small intestine, villi absorb nutrients into both the lymphatic and circulatory systems.
(D) The low pH of the stomach is essential in order for carbohydrate digestive enzymes to function.
(E) The release of bile from the gall bladder is triggered by the hormone cholecystokinin.

9. Which of the following statements about blood is false?

(A) Mature red blood cells are not nucleated.
(B) Blood platelets are involved in the clotting process.
(C) The adult spleen is a site of red blood cell development.
(D) White blood cells are capable of phagocytosing foreign matter.
(E) New red blood cells are constantly developing in the bone marrow.

10. Which of the following is a normal pathway of blood flow?

(A) right ventricle to aorta
(B) pulmonary veins to left atrium
(C) inferior vena cava to left atrium
(D) pulmonary veins to left ventricle
(E) left ventricle to pulmonary artery

11. Which of the following associations of brain structure and function is false?

(A) hypothalamus: appetite
(B) cerebellum: motor coordination
(C) cerebral cortex: higher intellectual function
(D) reticular activating system: sensory processing
(E) medulla: basic emotional drives

GO ON TO THE NEXT PAGE

12. Which statement about the respiratory system is NOT true?

(A) Ciliated nasal membranes warm, moisten, and filter inspired air.
(B) Contraction of the diaphragm enlarges the thoracic cavity.
(C) When the thoracic cavity enlarges, the pressure of air within the lungs falls.
(D) When the pressure of air within the lungs is less than the atmospheric pressure, air will flow out of the lungs.
(E) The respiratory process consists of inspiratory and expiratory acts following one another.

13. Which of the following statements about hormones is NOT true?

(A) They are transported by the circulatory system.
(B) They bind to receptors on target cells.
(C) They must be present in large quantities to have an effect.
(D) They are secreted by endocrine glands.
(E) They can affect organs of the body that are far removed from their site of synthesis.

14. Which of the following is an INCORRECT pairing of an endocrine gland and hormone secretion?

(A) posterior pituitary: LH
(B) adrenal cortex: aldosterone
(C) anterior pituitary: TSH
(D) adrenal medulla: epinephrine
(E) hypothalamus: FSH releasing factor

15. Which of the following statements about acetylcholine is NOT true?

(A) ACH is released at the neuromuscular junction.
(B) ACH binds to specific receptors on the postsynaptic membrane.
(C) In a synaptic cleft, there are enzymes that degrade ACH.
(D) ACH diffuses through the presynaptic membrane after its synthesis.
(E) A synapse that is subjected to many action potentials may be depleted of ACH granules.

16. The sympathetic nervous system causes which of the following?

(A) constriction of the pupil
(B) decreased heart rate
(C) increased gastric section
(D) reduction of adrenaline secretion
(E) increased respiration

GO ON TO THE NEXT PAGE

17. In a reflex arc

(A) the sensory neuron synapses directly with the motor neuron
(B) sensory and motor neurons can synapse outside of the spinal cord
(C) sensory neurons synapse in the brain
(D) the motor response occurs without synaptic delay
(E) a minimum of three neurons must participate

18. Hyperthyroidism is always associated with

(A) low blood pressure
(B) severely diminished mental activity
(C) high metabolic rate
(D) low body temperature
(E) decreased heart rate

19. Geotropism is

(A) the growth of parts of plants towards or away from gravity
(B) the tendency of plants to bend towards light
(C) the tendency of plants to have branched roots
(D) the formation of new phloem cells
(E) none of the above

20. Gibberellins

(A) inhibit lateral buds
(B) stimulate fruit ripening
(C) determine timing and differentiation of new cells
(D) stimulate rapid stem elongation
(E) all of the above

21. Rhizomes are

(A) underground stems with buds
(B) stems running above and along the ground
(C) spores
(D) the result of sexual reproduction
(E) woody, underground stems

22. After Konrad Lorenz swam in a pond with newly hatched ducklings separated from their mother, they followed him around as if he were their mother. This is an example of

(A) discrimination
(B) response to pheromones
(C) imprinting
(D) instrumental conditioning
(E) none of the above

Answers and Explanations to Organismal Biology Quiz

1. (A) Amylase is the enzyme that breaks down complex carbohydrates into disaccharides. Amylase is found in both salivary and pancreatic form. Gastrin (B) is a hormone secreted by the stomach wall of mammals when food makes contact with the wall; it stimulates other parts of the wall to secrete gastric juice. Secretin (C), secreted by the small intestine, stimulates the pancreas to secrete pancreatic juice, which contains bicarbonate ions, which buffer the chyme. Pepsin (D) breaks down proteins into amino acids, while (E) maltase breaks down maltose into glucose.

2. (A) The blastula is not a solid ball of cells, as this answer asserts; it is the hollow ball stage of embryonic development. It develops from the morula, a solid, "mulberrylike" cluster of cells. The center of the blastula is termed the blastocoel. This inpockets during gastrulation to form the gastrula. At this stage, the mesoderm, the archenteron, and the blastopore are formed. During the next stage, the neurula, the ectoderm forms the nerve cord. Therefore, all of the remaining answers accurately represent the various stages of embryonic development.

3. (D) The ectoderm develops into the skin, the nervous system, and the eyes, while the mesoderm develops into the musculoskeletal system and the internal organs. The endoderm develops into the digestive tract and its associated organs such as the pancreas and liver, the respiratory tract and the bladder lining. The pancreas would therefore develop from the endoderm, and not from the mesoderm as (D) states.

4. (C) There are many types of fertilization, development, and care of offspring in nature. At one end of the spectrum would be internal fertilization, internal development, and lots of care. Organisms that practice internal fertilization, such as elephants, produce few offspring, but a large percentage of the offspring produced reach adulthood. At the other end of the spectrum would be external fertilization, external development, and no care. Organisms at this end of the spectrum, such as many species of fish, must produce large numbers of both eggs and sperm, as relatively few eggs and sperm will interact to produce a zygote. Lacking protection, these zygotes are also susceptible to predation. Therefore, millions of eggs and sperm must be released in order to perpetuate the species. Whether or not a species lives in water or on land (A) does not affect its chances of survival at all, and while protective coloring (D) might be useful, it has no relation to the type of care a species' young receive.

5. (C) The thyroid gland controls the basal metabolic rate through release of thyroxin. It is stimulated by TSH released by the anterior pituitary gland. (A) is too general an answer; impairment of the pituitary would result in a myriad of disorders. The anterior and posterior pituitary glands secrete a large number of hormones, such as ACTH, LH, FSH, ADH, and GH, which control many bodily functions. Meanwhile, (B) impairment of the gonads would result in a loss of FSH and LH from the anterior pituitary. (D) The thymus is involved in the development, maturation, and education of T cells, and the parathyroid hormone (E) stimulates calcium resorption from bone, increasing plasma Ca^{2+} levels as a result.

6. (B) The eggs are released from the ovaries and picked up by the fallopian tubes. In the case of fertilization, a haploid sperm will swim to the egg and fertilize it through the use of a special granule, termed the acrosome, that allows the sperm to penetrate the cell membrane of the egg. The fertilized egg will then continue down the fallopian tube until it reaches the uterus, where it will implant. If it implants in the fallopian tube, an ectopic pregnancy results. (C) Fertilization in the uterus can occur, but it is rare and will sometimes be associated with complications like cervical implantation. (D) and (E) would both disallow implantation of the fertilized egg into the uterus. The vagina is the birth canal; the cervix is the connection between the uterus and the birth canal that expands during delivery.

7. **(B)** Muscle cells are composed of myofibrils of thick (myosin) and thin (actin) filaments. Contraction occurs through the sliding of these filaments over each other through reversible bridges. ACH released from the motor neuron excites the membrane of the muscle cell and stimulates the sarcoplasmic reticulum to release Ca^{2+} into the cytoplasm. This increase in calcium attaches to the actin filaments and causes troponin and tropomyosin to expose their mysoin binding sites. As this occurs, muscle filaments slide over each other and the muscle contracts, which makes (B) the only correct answer.

8. **(D)** A low stomach pH is essential in order for enzymes that break down proteins into amino acids to function, not enzymes that break down carbohydrates, as this answer suggests. The other choices are all true. In (A), amylases, lipases, and bicarbonate are released through the pancreatic duct, while (B) peristalsis does propel food and waste through the system. (C) Glucose and amino acids are picked up by the blood, while fats are picked up by lacteals (special vessels that connect with the lymphatic system). Finally, cholecystokinin (E) is a hormone released by the wall of the small intestine. It travels through the blood to the gall bladder to release bile into the duodenum, emulsifying fats and decreasing the pH of the chyme.

9. **(C)** In the adult, all hematopoiesis (red blood cell development) occurs in the bone marrow. In the fetus, hematopoiesis occurs in the fetal liver and spleen. The spleen is not involved in hematopoiesis; it acts as a reservoir for red blood cells and filters blood. (A) Mature red blood cells, meanwhile, are not nucleated, in order to create more space for hemoglobin. Blood platelets (B) are crucial to the clotting of blood, and certain white blood cells (D), such as macrophages and neutrophils, engulf foreign matter in a process known as phagocytosis. (E) New blood cells are continually being produced in the bone marrow to replace the cells after their life span of 120 days has ended.

10. **(B)** Blood traveling from the left ventricle flows into the aorta, then goes to all areas of the body except the lungs. For example, blood in the brachiocephalic artery travels to the head and the shoulders, while blood in the renal artery travels to the kidney to be filtered. Blood flows from these arteries into arterioles, then into capillaries, where food, waste, and energy will be exchanged. Next, the blood continues into venules and collects in veins, which then transport this blood to the superior and inferior vena cavas. This process is known as systemic circulation. In pulmonary circulation, the blood enters the right atrium and flows into the right ventricle. It is then transported to the lungs via the pulmonary artery, where capillary beds surround the alveoli so that gas exchange can occur. At this point, the pulmonary veins bring the blood back to the left atrium to start the process all over again.

11. **(E)** The medulla monitors blood carbon dioxide levels and pH and adjusts breathing, temperature, and heart rate. It is also the center for reflex activities such as coughing, sneezing, and swallowing, and is not associated with emotional drives. The other answer choices are true: The hypothalamus (A) is the center that controls thirst, hunger, sleep, blood pressure, and water balance; (B) the cerebellum controls muscle coordination and tone and maintains posture; (C) the cerebral cortex is the center for vision, hearing, smell, voluntary movement, and memory; and (D) the reticular activating system receives and sorts sensory input.

12. **(D)** There is low pressure inside the thoracic cavity due to expansion of the thoracic volume when the diaphragm contracts (as in (B) and (C)). When this pressure drops, air rushes in, and (A) the ciliated membrane warms, moistens, and filters the inspired air. Air then travels through the bronchi, into the bronchioles, and finally into the alveoli, where diffusion occurs to oxygenate the blood and release CO_2 carried back from the tissue. (D) is incorrect because when the pressure of air within the lungs is less than the atmospheric pressure, air actually rushes into the lungs rather than flowing out of them.

13. (C) Hormones are capable of being effective at picomolar, or very small, concentrations. (A) They are also transported by the circulatory system after being secreted by the endocrine (ductless) glands (D), and they may travel far from their site of synthesis to have an affect on their target organ (E). As for (B), cells will respond to hormones only if they have the appropriate receptor on their cell surfaces (in the case of peptide hormones) or inside their cytoplasm (in the case of steroid hormones).

14. (A) The posterior pituitary secretes ADH and oxytocin, not LH. (B) The adrenal cortex does secrete aldosterone, a mineral cortisone that causes increased absorption of Na^+, while (C) TSH, secreted by the anterior pituitary, stimulates thyroid production of thyroxin, which raises the basal metabolic rate. (D) Epinephrine increases heart rate, blood pressure, and the flow of blood to skeletal muscle. Finally, (E) the hypothalamus-secreted FSH releasing factor stimulates the anterior pituitary.

15. (D) ACH is released at the neuromuscular junction when vesicles containing this neurotransmitter merge with the membrane. ACH diffuses across the synapse and binds to specific receptors on the postsynaptic membrane. An enzyme known as acetylcholinesterase degrades ACH very quickly after release in order to prevent constant stimulation of the synapse.

16. (E) The sympathetic nervous system is known as the "fight or flight" response. Stimulation of this branch of the autonomic nervous system is characterized by an increase in heart rate, dilation of the pupils, an increase in respiration and bronchial dilation, and more blood flow to the skeletal muscles and away from the digestive organs. An increase in the amount of adrenaline secretion also occurs. (A)–(D) are all characteristics of the anatagonistic branch of the autonomic nervous system, the parasympathetic nervous system, which is known as the "rest and digest" response.

17. (A) A reflex arc is a quick sensation response coupled to a motor response for quickness or protection. Examples of this are sneezing, coughing, blinking, or moving your hand away from a hot stove. The simplest version of this type of arc would be one sensory neuron linked to one motor neuron. As for (B) and (C), the synapse between the sensory and motor neuron is located in the spinal cord. There is a brief synaptic delay between the motor neuron and the corresponding muscle (D), while this reaction involves a minimum of two neurons (E) (although other neurons might be present, such as neurons that feel pain, they are not considered part of the arc).

18. (C) Hyperthyroidism results in an excess of thyroxin and other thyroid hormones. These hormones increase the basal metabolic rate and blood pressure. Patients with hyperthyroidism are often characterized by sensitivity to heat and nervousness. The other choices are all characteristics of hypothyroidism.

19. (A) Geotropism is the growth of portions of plants towards or away from gravity. Negative geotropism occurs as the shoot grows away from the force of gravity, and positive geotropism describes the growth of roots toward the pull of gravity. Both occur as a result of differential concentrations of auxins (plant hormones associated with growth patterns) on upper and lower sides of the plant. The tendency of plants to bend toward light is termed phototropism (B); the formation of new phloem cells is stimulated by gibberellins, another class of plant hormones (D).

20. (D) Gibberellins are a class of plant hormones that stimulate rapid stem elongation, inhibit the formation of new roots, stimulate the production of new phloem cells by the cambium, and terminate the dormancy of seeds and buds. Auxins, not gibberellins, inhibit the growth of lateral buds (A), while ethylene stimulates fruit ripening (B).

21. (E) Rhizomes are a type of natural form of vegetative propagation and are characterized by woody, underground stems. At intervals, new, upright stems appear. Examples of plants that utilize rhizomes are ferns and irises. Other forms of natural vegetative propagation include bulbs (such as tulips), tubers (underground stems with buds such as potatoes), and runners (stems running above and along the ground, such as strawberries). All of the above forms exemplify asexual reproduction in plants.

22. (C) Imprinting is a process in which environmental patterns or objects presented to a developing organism during a brief "critical period" in early life become accepted as permanent elements of their behavioral environment. A duckling, for example, passes through a critical period in which it learns that the first large moving object it sees is its mother. Discrimination (A) involves the ability of the learning organism to differentially respond to slightly different stimuli, while (B) pheromones are substances secreted by many animals that influence the behavior of other members of the same species. Finally, instrumental conditioning (D) involves conditioning responses to stimuli through reward or reinforcement.

CHAPTER SEVEN

EVOLUTION AND DIVERSITY

On your SAT II exam, you'll probably be faced with quite a few questions dealing with the evolution and classification of the millions of species on Earth. This chapter will give you the background you need to ace these questions. We'll be covering topics ranging from types of evidence for evolution to the taxonomic classification of various common species.

Origin of Life

The change in the genetic makeup of a population over time is termed *evolution*. Evolution can be explained as the constant propagation of new variations in the genes of a species, some of which impart an adaptive advantage. This process has led to the remarkable adaptations that many organisms have made to their environments.

Evidence for the theory of evolution comes from many sources. The fossils that archaeologists excavate support this hypothesis, as do present-day observations in the fields of embryology, anatomy, and biochemistry.

Put Things in Perspective

From an evolutionary perspective, humans have not been around for long at all. The first primitive humans appeared around one million years ago. But the oldest known animal fossil is an estimated 750,000,000 years old, and the earth is believed to have formed 4.5 billion years ago.

Evidence of Evolution

Scientists have uncovered various types of evidence that evolutionary change has occurred over time, some of which are outlined below.

Fossil Record

Fossils are the most direct evidence of evolutionary change. These ancient remains are generally found in sedimentary rocks. Unlike other varieties of rock, sedimentary rocks do not crush or melt foreign matter embedded in them. Although fossils are usually not the actual remains of organisms still alive today, they do represent the remains of extinct ancestors of these organisms.

Types of Fossils. Some varieties of fossils are presented below:

Actual remains. These include teeth, bones, etcetera, and are found in rock, tar pits, ice, and amber (the fossil resin of trees).

Petrifaction. Organisms that are covered by sediment and water for thousands of years eventually turn to stone; in other words, they become petrified. Petrifaction is the process by which mineral replaces the cells of an organism. Petrified fossils come to light when erosion wears away the sediments covering them and exposes them.

Imprints. Ancient organisms sometimes left impressions (footprints) in mud or sand, which subsequently dried out and turned to stone.

Molds. As organisms were buried in mud and decayed, the area that their bodies occupied was preserved in these organisms' original shapes. Molds may be defined as hollow impressions made in earth or rock by the outsides of fossil shells or other organic forms.

Casts. Decaying matter inside these molds was replaced by minerals, forming casts. Casts are fossil reproductions of natural objects; they can provide much information about the external appearance of ancient organisms.

Well-Known Examples of Fossil Remains. Certain types of fossil specimens have become particularly renowned. In our day, the *dinosaur* is perhaps the most familiar extinct organism that has been discovered in fossil form. This ancient organism, seemingly related to both reptiles and birds, was the dominant species during the Mezoic era. As was apparent in the film *Jurassic Park,* dinosaurs existed in many forms. Some were small, quick, and agile; others, mainly herbivores, were large, slow, and forced to live in water to support their immense weights. The Tyrannosaurus Rex was one of the largest predators in history, while the pterodactyl was a flying dinosaur whose shrieks probably filled the air during this era.

How About a Different Kind of Engagement Ring?

Specimens of amber that contain insect fossils are highly prized by jewellers.

Other types of fossil remains include the *trilobite,* a primitive crustacean (relative of the lobster) that was a dominant form during the early paleozoic era. The *eohippus* (dawn horse), meanwhile, was a primitive horse the size of a fox, with four toes and short teeth with pointed cusps for feeding on soft leaves. Fossil evidence indicates a gradual evolution of this creature into the modern horse. Today's horse has one toe (hoof) and two vestigial toes as side splints, flat teeth with ridges for grinding grain and tough prairie grass, and long legs for running.

The *woolly mammoth* (ancestor of the modern elephant) is another well-known prehistoric figure. This hairy animal was, however, found in Siberian ice rather than in the tropics. The *saber-tooth tiger,* ancestor of the big cats, has been preserved in asphalt tar pits, while various types of prehistoric insects have been preserved in amber.

Finally, the famous "missing link" between reptiles and birds has been identified as the *Archaeopteryx.* This organism possessed teeth and scales (like reptiles) and feathers (like birds).

Comparative Anatomy

When we compare the anatomies of two or more living organisms, we can not only form hypotheses about their common ancestors, but we can also glean clues that shed light upon the selective pressures that led to the development of certain adaptations, such as the ability to fly. Comparative anatomists study homologous and analagous structures in organisms.

Homologous Structures. Homologous structures have the same basic anatomical features and evolutionary origins. They demonstrate similar evolutionary patterns with late divergence of form due to differences in exposure to evolutionary forces. Examples of homologous structures include the wings of a bat, the flipper of a whale, the forelegs of a horse, and the arms of a human.

Analogous Structures. Analogous structures have similar functions but may have different evolutionary origins and entirely different patterns of development. The wings of a fly (membranous) and the wings of a bird (bony and covered in feathers) are analogous structures. Analogous organs demonstrate superficial resemblances that cannot be used as a basis for classification.

Don't Mix These Up on Test Day

Homologous structures share a common ancestry.

Analogous structures are not inherited from a common ancestor, but perform similar functions.

Comparative Embryology

Stages of embryonic development resemble the stages of a particular organism's evolutionary history. Hence close study of different organisms' embryos can tell us much about their evolutionary ancestry.

The life stages of the human embryo, for example, are similar to stages in the embryonic cycle of other organisms. This suggests a common ancestry and development history between humans and other organisms. It does *not* suggest that human early development is identical to that of these organisms. To be specific, the human embryo passes through the following stages:

- *Two-layer gastrula*, demonstrating common ancestry with the hydra.
- *Three-layer gastrula*, demonstrating common ancestry with the flatworm.
- *Gill slits*, demonstrating common ancestry with fish.

The earlier the stage at which embryotic development begins to diverge, the more dissimilar the mature organisms are. Thus, it is difficult to differentiate between the embryo of a human and that of an ape until relatively late in the development of each embryo, while human and flatworm embryos diverge much earlier.

Other embryonic evidence of evolution includes such characteristics as teeth in an avian embryo (recalling the reptile stage); the resemblance of the larvae of some mollusks (shellfish) to annelids (segmented worms), and the tail of the human embryo (indicating relationships to other mammals).

Comparative Biochemistry (Physiology)

Most organisms demonstrate the same basic needs and metabolic processes. They require the same basic nutrients and contain similar cellular organelles and energy storage forms (ATP). Respiratory processes, for example, are basically the same in most organisms. The similarity of the enzymes involved in these processes suggests that all organisms must contain some DNA sequences in common. The closer the organisms in the evolutionary scheme, the greater the similarity of their chemical constituents (enzymes, hormones, antibodies, and blood) and genetic information. We can therefore conclude that all organisms are descended from a common, primitive ancestral form.

The chemical similarity of the blood of different organisms very closely parallels the evolutionary pattern. A chimpanzee's blood is very similar to that of a human, but is quite different from that of a rabbit or fish. Thus, the more time that has elapsed since the divergence of two species, the more different their biochemical characteristics. This idea can be visualized as an "evolutionary clock."

When the Appendix Makes Its Presence Known

Although useless in present-day humans, the appendix can cause a lot of trouble if left unattended. If it becomes inflamed and bursts, it can cause illness and even death.

Vestigial Structures

Vestigial structures are structures that appear to be useless in the context of a particular modern-day organism's behavior and environment. It is apparent, however, that these structures used to have some function in an earlier stage of a particular organism's evolution. They serve as evidence of an organism's evolution over time, and can help scientists to trace its evolutionary path.

There are many examples of vestigial structures in humans, other animals, and plants. The appendix—small and useless in humans—assists digestion of cellulose in herbivores, indicating human's vegetarian ancestry, while the animal-like tail in humans is reduced to a few useless bones (coccyx) at the base of the spine. The splints on the legs of a horse are vestigial remains of the two side toes of the eohippus. Finally, the python has legs that are reduced to useless bones embedded in the sides of the adult python, as does the whale.

Barriers (Geographical Distribution)

Species multiplication is generally accompanied by migration, which lessens intraspecific competition. Separation of a widely distributed population by emerging geographic barriers increases the adaptations on either side of the barrier. Each population may evolve specific adaptations for the environment in which it lives, in addition to the accumulation of neutral (random, nonadaptive) changes. These adaptations will remain unique to the population in which they evolve, provided that interbreeding is prevented by the barrier. In time, genetic differences will reach the point where interbreeding becomes impossible and reproductive isolation would be maintained if the barrier were removed. In this manner, geographic barriers promote evolution.

Take marsupials, for example. A line of pouched mammals paralleling the development of placental mammals developed on the Australian side of a large water barrier. This geographic barrier protected the more primitive pouched mammals from competition with modern placental mammals. This resulted in the development of uniquely Australian plants and animals—like the kangaroo, the duck-billed platypus, pouched wolves, and the eucalyptus tree. (Opossums, however, did manage to cross the Bering Strait land bridge and complete successfully with placental mammals).

Darwin's finches of the Galapagos also illustrate this phenomenon. Over a comparatively short period of time, a single species of finch (birds) underwent adaptive radiation to form 13 different species of finches. Slight variations in the beak, for example, favored ground- or tree-feeding. Such adaptations minimized competition among the birds, enabling each emerging species to become firmly entrenched in its environmental niche.

Evolutionary History

Dissimilar species have often evolved from a common ancestor. Biologists seek to understand the evolutionary relationships between the species alive today; this evolutionary history is termed *phylogeny*. Evolutionary history may be visualized as a branching tree, where the common ancestor is found at the trunk and the modern species at the tips of the branches.

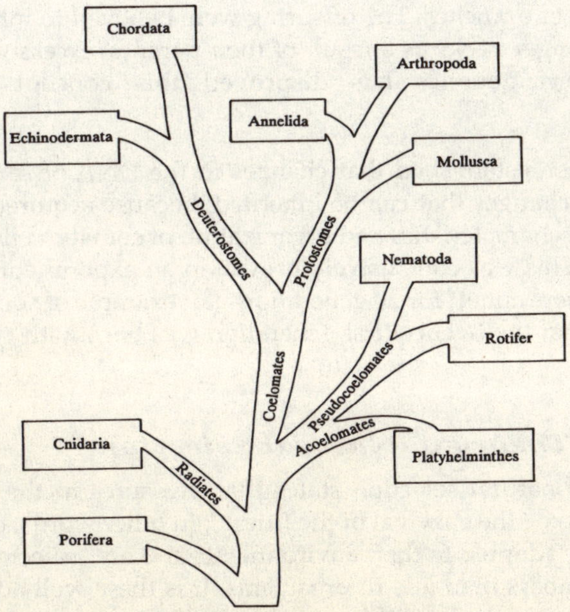

Phylogenetic Tree

Systematics

Creating phylogenetic trees is part of systematics, the field of biology that investigates the diversity of life.

It is interesting to note that groups within the branches of the phylogenetic tree develop in similar ways when exposed to similar environments. This is known as parallel evolution or convergent evolution. For example, whales, reptiles, and fish are only distantly related, yet each developed fishlike fins and a tail when exposed to water.

Theories of Evolution

Not Completely Wrong

Although Lamarck's theory of evolution has been disproven, he did correctly observe that evolution should not be envisioned as a ladder, but instead as a pathway with many branches.

For many years, scientists debated the forces of evolution. We knew we had evolved, but how? Early theories such as Lamarck's were founded on fallacies and eventually discredited. Then Charles Darwin proposed his theory of natural selection, which is still commonly accepted as the mechanism for evolution, although it has been modified considerably in recent decades.

Lamarckian Evolution

Until it was supplanted by Darwin's ideas, the scientist Lamarck's theory was one of the more widely accepted explanations of the mechanisms of evolution. The cornerstone of Lamarck's hypothesis was the principle of use and disuse. He asserted that organisms developed new organs, or changed their existing ones, in order to meet their changing needs. The amount of change that occurred was thought to be based on how much or little the organ in question was actually used.

Unfortunately for Lamarck, this theory of use and disuse was based upon a fallacious understanding of genetics. Any useful characteristic acquired in one generation was thought to be transmitted to the next. An oft-cited example was that of early giraffes, which stretched their necks to reach for leaves on higher branches. The offspring were believed to inherit the valuable trait of longer necks as a result of their parents' excessive use of their necks. Modern genetics has disproved this concept of acquired characteristics.

It has now been established that changes in the DNA of sex cells are the only types of changes that can be inherited; because acquired changes are changes in the characteristics and organization of somatic cells, they cannot be inherited. This was conclusively proven in an experiment in which the tails of mice were cut off for 20 generations (an example of somatic change), only to find that the twenty-first generation was born with tails of normal length.

Darwin's Theory of Natural Selection

The theory of natural selection states that pressures in the environment serve to promote "the survival of the fittest"; in other words, certain organisms are better adapted to their environments and are "selected" to survive in greater numbers than less fit organisms. It is these well-adapted organisms that reproduce and multiply, resulting in evolutionary change. Each of the following factors plays a role in the mechanism of natural selection:

Overpopulation. Too many offspring are produced, resulting in insufficient food, air, light, and space to support the entire population.

Variations. Parents and their offspring never have identical characteristics. Offspring naturally develop variations from their parents. Darwin did not know the reason for these differences. Dutch scientist Hugo DeVries—who verified Mendel's observations through his studies with the evening primrose—later suggested that *mutations* were the cause of variations. Some mutations are beneficial, although most are harmful.

Competition (The Struggle for Survival). The developing population must compete for the basic necessities of life. Many young die, and the number of adults in the population generally remains constant from generation to generation.

Natural Selection (Survival of the Fittest). Some organisms in a species have variations that give them an advantage over other members of the species. In the struggle for existence, these organisms might have adaptations that are advantageous for survival. For example, a giraffe with a variation of a longer neck would be able to get food from higher branches of a tree and thus be better fitted to survive than short-necked giraffes.

Inheritance of the Variations. The individuals that survive (those with the favorable variations) live to adulthood, reproduce their own kind, and thus transmit these favorable variations or adaptations to their offspring. These favored genes gradually dominate the gene pool.

Evolution of New Species. Over many generations of natural selection, the favorable changes (adaptations) are perpetuated in the species. The accumulation of these favorable changes eventually results in such significant changes of the gene pool that we can say a new species has evolved. These physical changes in the gene pool were perpetuated or selected for by environmental conditions.

The evolution of DDT-resistant insects is an example of this phenomenon. DDT-resistant mutants existed before DDT was introduced into the insect population. The introduction of this insecticide constitutes a favorable environmental change for these mutants, and a fatal environmental change for the rest of the insect population. Now, conditions select for the survival of DDT-resistant organisms, resulting in the evolution of a new DDT-resistant species.

Study Tip

Darwin's theory of evolution is a popular topic on the SAT II: Biology. Be prepared to answer questions on the following Darwinian themes:

- Overpopulation
- Variations
- Competition
- Natural selection
- Inheritance of variations
- Evolution of new species

Forces of Evolution

We know evolution occurs, but can we measure it? Once a new species has developed into a stable population, we are able to study it in more detail, using population genetics.

Test Strategy

All you need to know to solve any Hardy-Weinberg problem is the value of p (or p^2) and q (or q^2). Once you know these, you can calculate everything else using the formulas $p + q = 1$ and $p^2 + 2pq + q^2 = 1$.

Population Genetics

When studying populations, it is useful to be able to predict the frequencies of alleles and phenotypes that will appear in the offspring produced when two organisms reproduce, as well as in subsequent generations. To accomplish this, scientists use the equations of the Hardy-Weinberg principle.

Before we discuss the Hardy-Weinberg principle, we need to define some basic terms. A *population* includes all members of a particular species inhabiting a given locale. A *gene pool* of a population is the sum total of all the alleles for any given trait in the population. Finally, *gene frequency* is the decimal fraction representing the presence of an allele for all members of a population that have this particular gene locus. The letter p is used for the frequency of the dominant allele of a particular gene locus. The letter q represents the frequency of the recessive allele. For a given gene locus, $p + q = 1$.

The Hardy-Weinberg Principle. According to the Hardy-Weinberg principle, gene frequencies in a population remain constant from generation to generation. In this simple model, a population is maintained in equilibrium as long as the following conditions are met:

- Random mating (no isolation) must occur, so that no particular trait is favored. There can be no assortative mating (in other words, no organisms may select mating partners that resemble themselves).
- Migration—that is, immigration or emigration—cannot take place.
- There must be no total mutations.
- Large populations are required. As in all cases of probability, large samplings are needed to provide an accurate approximation of the expected occurrence. In addition, it must be highly unlikely that chance alone could significantly alter gene frequencies.

Under the above conditions, there is a free flow of genes between members of the same species, while the total gene content is continually being shuffled. A constant gene pool is nevertheless maintained for the entire species. The constancy of the gene pool is always threatened by changes in the environment (which would favor certain genes), mutations, migrations (new genes introduced), or reproductive isolation (lack of random mating favors certain genes).

Mathematical Demonstration of the Hardy-Weinberg Equilibrium Principle. We can cross two individuals to demonstrate mathematically that the gene pool frequencies remain constant generation after generation. Let us assume for the original gene pool that the gene frequency of the dominant allele for tallness, p, is $0.80T$, and the gene frequency of the recessive allele, q, is $0.20t$. Thus $p = 0.80$ and $q = 0.20$. The parents are crossed and their offspring frequencies are shown with a Punnett square.

The resulting gene frequency in the F_1 generation is: 64 percent *TT*, 16 percent + 16 percent or 32 percent *Tt*, and 4 percent *tt*.

		Possible Sperm 0.80*T*	0.20*t*
Possible Eggs	0.80*T*	0.64*TT* ($p^2 = 0.64$)	0.16*Tt* ($pq = 0.16$)
	0.20*t*	0.16*Tt* ($pq = 0.16$)	0.04*tt* ($q^2 = 0.04$)

If there are no mutations, no migrations, and no decrease in population size, the frequencies of the above F_1 generation are applicable in calculating the frequencies for the F_2 generation.

Some Mathematical Applications of the Hardy-Weinberg Principle. In working out population genetics problems, we must recall that p = gene frequency of the dominant allele, q = frequency of the recessive allele, and $p + q = 1$, since frequencies of the dominant and recessive alleles total 100 percent.

Now, using the Punnett square, we may calculate the offspring of the two individuals $p + q$. We have:

		Possible Sperm p	q
Possible Eggs	p	pp	pq
	q	pq	qq

Thus, when we cross $p + q$, we obtain $p^2 + 2pq + q^2 = 1$ (the "1" indicates that the total is 100 percent of the offspring). Note that the result is scientific confirmation of an obvious algebraic identity: $(p + q)(p + q) = p^2 + 2pq + q^2$. The official key for working out the following problem is as follows:

p = frequency of dominant allele (*C*)

q = frequency of recessive allele (*c*)

p^2 = frequency of homozygous dominant individuals (*CC*)

$2pq$ = frequency of heterozygous individuals (*Cc*)

q^2 = frequency of homozygous recessive individuals (*cc*)

Another mathematical application of the Hardy-Weinberg principle is explained below:

Problem: In a certain population, the frequency of homozygous curly hair (CC) is 64 percent. What percentage of the population has curly hair?

Solution: According to the key, p represents the frequency of the dominant allele (*C*), while q represents the frequency of the recessive allele (*c*). We are told that the *CC* frequency is 64 percent. This means that $p^2 = 0.64$ or $p = 0.8$. Since $p + q = 1$, $q = 1 - 0.8$, or 0.2. An individual with curly hair may be either *CC* or *Cc*. The frequency of each genotype = $p^2 + 2pq + q^2$.

$p^2 = 0.64$ or 64 percent homozygous curly hair

$2pq = 2(0.8)(0.2)$ or 32 percent heterozygous curly hair

$q^2 = 0.04$ or 4 percent homozygous straight hair

Therefore, the percentage of the population possessing curly hair is as follows:

$p^2 + 2pq = 64 \text{ percent} + 32 \text{ percent} = 96 \text{ percent}$

Instability in the Gene Pool

The following factors cause instability in the gene pool, resulting in the loss of Hardy-Weinberg equilibrium:

- Mutations
- Recombination
- Genetic drift
- Environment
- Isolation

Losing Hardy-Weinberg Equilibrium

The Hardy-Weinberg principle describes the stability of the gene pool. However, no population stays in Hardy-Weinberg equilibrium for very long, because the stable, ideal conditions needed to maintain it do not exist. As conditions change, the unstable gene pool that results produces changes in the population. Basic variations within a species also promote evolutionary trends. Various factors may be introduced into a population to cause change in its gene pool, and other factors influence the variations that are selected for in a given gene pool.

Factors Promoting Change in the Gene Pool. As we mentioned earlier, gene and chromosomal mutations may function to promote evolution. Certain mutations may provide a selective advantage for the organism.

Genes are sorted out and recombined during sexual reproduction. Recombination allows for new assortments of genes in a population, resulting in variations within that population.

Influx from other populations with different gene pools provides a constant source of new genes. In addition, small populations favor instability of the gene pool, since the laws of probability apply only to large samplings. Small populations undergo a random *genetic drift* (evolutionary change caused by chance), which causes an excess of one allele over another. The bottleneck effect (in which some form of disaster drastically reduces the size of a population) and the founder effect (in which a small section of a population becomes isolated from the rest of the population) are types of genetic drift. Genetic drift may cause a high degree of homozygosity, and may also result in a high incidence of genetic diseases.

Factors Influencing Variations Selected for in Gene Pool. The most important of these factors are the environment in which the population lives and this population's degree of isolation from other populations.

Environment. Traits that help the organism to adapt to its environment have high survival value, and the gene frequencies of these traits increase. Traits (alleles) with low adaptive values occur less frequently, and tend to disappear. For example, an allele for sharper teeth in an environment where large supplies of meat are available might give a selective advantage to organisms displaying this trait. In consequence, the gene frequency of this trait might rise. However, should the environment change to one that provides more grass and less meat, the gene frequency might change in favor of teeth adapted for grinding.

To summarize, a changing environment stimulates evolution, not by creating new traits, but by changing the frequency of genes already present in the population. As in natural selection, organisms with superior adaptations will have a higher tendency to reproduce and to pass their traits on to their offspring.

Isolation. Some members of a population become separated from the original population. When subdivisions of a population are isolated from each other, there is no random mating or intermingling of traits between the subdivisions. Any difference arising from mutations or new combinations of genes will be maintained. If the genetic differences become significant enough that future mating is not possible, we have developed a different species and established reproductive as well as geographic isolation.

When Different Species Interbreed

Members of different species can interbreed under certain circumstances, but the offspring they produce is frequently infertile. Hence the mule, which is the offspring of two different species, the horse and the ass, is usually sterile.

Speciation

A *species* is a group of populations that maintains reproductive communication by interbreeding. Members of different species generally do not interbreed. Several concepts that are basic to an understanding of the mechanisms of speciation, or the development of new species, are detailed below:

Demes. A deme is a small local population, such as all the beavers along a specific portion of a river. A specific species may encompass many demes. Members of a deme resemble one another more closely than they resemble the members of other demes for a couple of reasons. They are closely related genetically, since mating between members of the same deme occurs more frequently. In addition, they are influenced by similar environmental influences, making them subject to the same selection processes.

Changes Within a Species. Fluctuations in the environment alter the survival value of certain traits, and the gene frequencies for these traits change accordingly. The development of antibiotic-resistant strains of bacteria is an example.

Development of New Species. If the gene pools within a species become sufficiently different so that two individuals cannot mate and produce fertile offspring, two different species have developed. No gene flow is possible between two different species. Variation (mutation and recombination), changes in the environment, migration to new environments, adaptation to new environments, natural selection, and isolation from others who lack the variation are all factors in the development of new species.

Adaptive Radiation. Adaptive radiation is the production of a number of different species from a single ancestral species. *Radiation* refers to a branching out; *adaptive* refers to the hereditary change that allows a species to be more successful in its environment or to be successful in a new environment. Whenever two or more closely related populations occur together, natural selection favors evolution of different living habits. This results in the occupation of different niches by each population (this process is discussed in detail in our chapter on ecology). This divergent evolution through adaptive radiation has been an extremely frequent occurrence, as demonstrated by the famous example of Darwin's finches.

The concepts of adaptive radiation and phylogeny form the basis for the methods employed in developing a system for the classification of living things. This classification system was originally based upon evolutionary relationships between organisms.

The Heterotroph Hypothesis

We've discussed the mechanisms by which organisms evolve, but not where they came from in the first place, and the evolutionary stages that they went through early in their history. We will now touch upon one of the more widely accepted of the hypotheses that attempt to account for the origin and early evolution of life—the heterotroph hypothesis.

What Came First?

Heterotrophs were the first form of life to develop. As their need for nutrients surpassed the rate at which they were being spontaneously formed, autotrophs developed.

Development of the Heterotroph

The first forms of life lacked the ability to synthesize their own nutrients; they required access to preformed molecules outside their bodies. In order to survive, these "organisms" became heterotrophs that depended upon outside sources for food. Three things were necessary to facilitate the development of primitive heterotrophs: key raw material, energy, and synthesis of organic molecules.

Key Raw Material. The primitive seas contained simple inorganic and organic compounds such as salts, methane, ammonia, hydrogen, and water.

Energy Sources. These were present in heat, electricity, solar radiation (including X-rays and ultraviolet light), cosmic rays, and radioactivity.

Synthesis of Organic Molecules. The combined presence of both energy sources and simple inorganic and organic compounds created conditions which made possible the synthesis of simple organic compounds, such as sugars, amino acids, purines, and pyrimidines. These molecules dissolved in the "primordial soup," and, after many years, simple monomeric molecules combined to form a supply of macromolecules.

In 1953, Stanley L. Miller set out to demonstrate whether application of ultraviolet radiation, heat, or a combination of the two to a mixture of methane, hydrogen, ammonia, and water could result in the formation of complex organic compounds. Miller set up an apparatus in which the four gases were continuously circulated past electrical discharges from tungsten electrodes. After circulation of the gases for one week, Miller analyzed the liquid in the apparatus and found that an amazing variety of organic compounds, including urea, hydrogen cyanide, acetic acid, and lactic acid, had been synthesized.

Formation of Primitive Cells. It is known that colloidal protein molecules tend to clump together to form coacervate droplets (a cluster of colloidal molecules surrounded by a shell of water). These droplets tend to absorb and incorporate substances from the surrounding environment. In addition, the droplets tend to possess a definite internal structure.

Although these coacervate droplets were not living, scientists have hypothesized that coacervate systems arose in the seas of the early earth. Most of these systems were unstable; however, a few systems may have arisen that were stable enough to survive. A small percentage of the droplets that possessed favorable characteristics developed, eventually, into the first primitive cells. These first primitive cells probably possessed nucleic acid polymers, indicating that they were capable of reproduction.

Soon, the heterotrophs needed more nutrients than were being created spontaneously. Luckily for them (and us), *autotrophs* developed.

Development of Autotrophs

The primitive heterotrophs slowly evolved complex biochemical pathways enabling them to use a wider variety of nutrients. They also evolved anaerobic respiratory processes to convert nutrients into energy. However, these organisms required nutrients at a faster rate than they were being synthesized. Life would have ceased to exist if autotrophic nutrition had not developed. The pioneer autotrophs developed primitive photosynthetic pathways in which solar energy was captured and used in the synthesis of carbohydrates from carbon dioxide and water.

Anaerobe to Aerobe

The primitive autotrophs fixed carbon dioxide during the synthesis of carbohydrates and released molecular oxygen as a waste product. The addition of molecular oxygen to the atmosphere converted the *atmosphere* from a reducing to an oxidizing atmosphere. Some molecular oxygen was converted to ozone, which screens out high-energy radiation. Thus, living organisms destroyed the conditions that made possible the origin of life. Once molecular oxygen became a major component of the earth's atmos-

Study Tip

Know the following broad categories of organisms by heart when the time comes to take your SAT II: Biology exam:

- Heterotrophic aerobes (amoebae, earthworms, humans)
- Heterotrophic anaerobes (yeast)
- Autotrophic aerobes (green plants)
- Autotrophic anaerobes (chemosynthetic bacteria)

phere, both heterotrophs and autotrophs evolved the biochemical pathways of aerobic respiration.

Classification and Diversity

After the evolution of all the different organisms, scientists needed a way to tell them all apart. So they began to name them and to classify them in different groups, founding the discipline of *classification*.

Taxonomic Classification

The great diversity of living things is the product of billions of years of evolution. Scientists have attempted to sort out and discern relationships among the vast number of different types of organisms. This science of classification of living things, and the nomenclature it utilizes, is known as taxonomy.

Taxonomy

A modern classification system seeks to group organisms on the basis of evolutionary relationships. The bat, whale, horse, and human are placed in the same class of animals because they are believed to have descended from a common ancestor. Where possible, the taxonomist classifies all species known to have descended from the same common ancestor within the same taxonomic group.

Since much about early evolutionary history is not understood, there is some disagreement among biologists as to the best classification system to employ, particularly with regard to groups of unicellular organisms. Taxonomic organization proceeds from the largest, broadest group to the smaller, more specific subgroups. The largest group or kingdom is broken down into smaller and smaller subdivisions. Each smaller group has more specific characteristics in common. Furthermore, each subgroup is distinguishable from the next.

Classification and Subdivisions

The most modern scheme of taxonomy employs five kingdoms of living organisms. Each kingdom has several major phyla or divisions. The animal kingdom uses the term *phylum;* all other kingdoms employ the term *division*. A phylum or division has several subphyla or subdivisions, which are further divided into classes. Each class consists of many orders, and these orders are subdivided into families. Each family is made up of many genera. Finally, the species is the smallest subdivision. Organisms of the same species mate with one another to produce fertile offspring. The principal criterion in classifying organisms is the production of viable, fertile offspring.

Hence the order of classificatory divisions is as follows:

KINGDOM —> PHYLUM (DIVISION) —> SUBPHYLUM (SUBDIVISION) —> CLASS —> ORDER —> FAMILY —> GENUS —> SPECIES

The complete classification of humans is:

Kingdom:	Animalia
Phylum:	Chordata
Subphylum:	Vertebrata
Class:	Mammalia
Order:	Primates
Family:	Hominidae
Genus:	Homo
Species:	Sapiens

Assignment of Scientific Names

All organisms are assigned a scientific name consisting of the genus and species names of that organism. Thus, humans are *Homo sapiens*, and the common housecat is *Felis domestica*. This follows a pattern originated by the Swedish scientist Carl von Linné (Carolus Linnaeus).

Taxonomic nomenclature is based on diagnostic procedures applied to differentiate organisms, including anatomical and structural analysis. Modes of irritability, excretion, movement, digestion, and, where applicable, circulation and respiration, are also significant in determining nomenclature. Finally, the ratio of adenine-thymine bond pairs to guanine-cytosine bond pairs and the biochemical capabilities of the organism are also studied.

Classification of Living Things

Beginning with the simple virus, we will now take a closer look at how various species are classified.

The Virus

Viruses do not carry out physiological or biochemical processes outside of a host. For this reason they may be considered nonliving, although they are highly advanced parasites. Viruses infect cells and are capable of taking over the host's cellular machinery and directing the replication of the viral genome and protein coat. Viruses may be lytic (causing the cell to disintegrate) or lysogenic (remaining dormant in the cell). They contain either DNA or RNA and some essential enzymes surrounded by a protein coat.

Viruses that exclusively infect bacteria are termed bacteriophages. Bacteriophages resemble syringes in that they inject their nucleic acid (DNA or RNA) into the bacterium; the phage protein coat does not enter the bacterium.

Mnemonic

The following sentence will help you remember the order of classificatory divisions:

King **P**hillip **S**wiftly **C**ame **O**ver **F**or **G**ood **S**ushi.

Versatile Viruses

Viruses possess the characteristics of life only when they've infected a living host cell. Some examples of the wide range of diseases caused by viruses are:

- Chicken pox (caused by varicella zoster)
- AIDS (HIV)
- Colds (rhinoviruses)
- Cold sores (herpes simplex virus I)

Monera (Prokaryotes)

Monera incorporate two phyla: bacteria (schizomycetes) and blue-green algae (cyanophyta).

Blue-Green Algae. Blue-green algae are the only type of algae that are classified as prokaryotic. They possess a cell wall and photosynthetic pigments, but they have no flagella, true nuclei, chloroplasts, or mitochondria. Certain chemicals are unique to blue-green algae and to bacteria. They can withstand extreme temperatures, and are believed to be directly descended from the first organisms that developed photosynthetic capabilities.

Bacteria. The ubiquitous bacteria are single celled, lack true nuclei, and contain double-stranded circular chromosomal DNA that is not enclosed by a nuclear membrane. Almost all forms have cell walls. Many forms possess a wide variety of complex biochemical pathways. These creatures nourish themselves heterotrophically—either saprophytically or parasitically—or autotrophically, depending upon the species.

Bacteria are classified by their morphological appearance: *cocci* (round), *bacilli* (rods), and *spirilla* (spiral). Some forms are duplexes (*diplococci*), clusters (*staphylococci*), or chains (*streptococci*).

Protista

The kingdom Protista contains two phyla. The first includes mostly unicellular, animal-like organisms, while the second incorporates some of the algal species.

Protozoa may be unicellular, colonial, free living, or symbiotic. They are mostly heterotrophic, although some are photosynthetic (to be specific, euglenophytes and pyrrophytes). Protozoa proliferate in any sufficiently moist environment. Many possess contractile vacuoles, flagella, or cilia. Examples are the amoebae and paramecium. The different classes of protista are listed below:

- *Class Flagellata.* Flagella serve as the principal locomotor organelle in this class.
- *Class Sarcodina.* Examples are amoeboid protozoans.
- *Class Ciliata.* Cilia serve as the locomotor organelle in this class; an example is *paramecium.*
- *Class Sporozoa.* Organisms in this class are parasites—for example, *plasmodium* (malaria).

The Life of a Parasite

The *plasmodium,* which causes malaria, begins its life cycle when an infected mosquito bites a human. Sporozoites enter the human's bloodstream and infect the red blood cells. These red blood cells eventually lyse, causing severe anemia.

Fungi

This kingdom is divided into true fungi and slime molds. True fungi (Eumycophyte) are nongreen, plantlike, and reproduce via asexual sporulation or via intricate sexual processes. All true fungi are heterotrophic, parasites, or saprophytes; none are autotrophic. Some varieties of Eumycophyte utilize extracellular digestion. Notable types are molds, mushrooms, and yeast.

Meanwhile, slime molds (Myxomycophyta) are arranged in a coenocytic (that is, possessing many nuclei) mass of protoplasm. The slime mold undergoes a unique life cycle consisting of animallike and plantlike stages. These stages include fruiting bodies and unicellular flagellated spores. Slime molds reproduce asexually through sporulation.

Plantae (Plants) and Animalia (Animals)

Plants and animals (Metazoa) differ from protists (Protista) in several important ways. Plants possess specialized reproductive tissues and organs as well as other organs and organ systems. In addition, they develop via distinct embryonic stages.

Plants also differ from animals. Animals generally go through a larval stage, unlike plants. Plants are typically photosynthetic and sessile (i.e., they do not move), while animals are generally heterotrophic and motile. Plant structure is adapted for maximum exposure to light, air, and soil by extensive branching; animals, on the other hand, are adapted for minimum surface exposure. Finally, plant cells contain walls composed of cellulose.

Plant Phyla. Several important plant Phyla are listed below. These range from simple plants to complex ones.

The following are several of the more important varieties of algae:

- *Rhodophyta* (red algae)
- *Phaeophyta* (brown algae)
- *Chlorophyta* (green algae)

And here are a few of the higher plants:

- *Brophyta* (mosses and liverworts)
- *Tracheophyta* include several varieties, such as:
  —*Psilopsida* (Psilotum)
  —*Lycopsida* (club mosses)
  —*Sphenopsida* (horsetail)

Don't Mix These Up on Test Day

Most people can tell the difference between plants and animals. But you will need to keep in mind some less obvious differences when taking your biology test:

Animals have a larval stage, are heterotrophic and mobile, are adapted for minimum surface exposure, and have no cell walls.

Plants have no larval stage, are photosynthetic and sessile, have extended branching, and have cell walls made of cellulose.

The Sessile Sponge

Unlike most animals, sponges are sessile—they do not move, instead gathering food by filtering water through their bodies.

—*Pteropsida.* These encompass ferns (which produce monoploid spores that germinate to form gametophytes), gymnosperms (which produce naked seeds and include cycads, pines, spruce, and firs), and angiosperms (which produce covered seeds and include maples, carrots, buttercups, sugar cane, pineapple, and most other plants).

Animal Phyla. Animals can be broadly divided into invertebrates and vertebrates. *Invertebrates* do not have a backbone. The following are a few of the more common invertebrates:

Porifera (sponges). Porifera have pores and two layers of cells, are sessile, and have a low degree of cellular specialization.

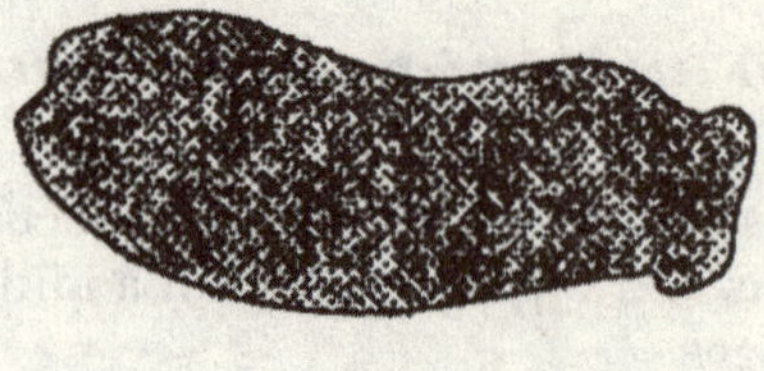

Sponge

Radiata (hydra, medusa, sea anemone). Coelenterates contain a digestive sac that is sealed at one end (gastrovascular cavity). Two layers of cells are present—the ectoderm and the endoderm. Coelenterates have many specialized features, such as tentacles, stinging cells (also known as *nematocysts*), and nerve nets.

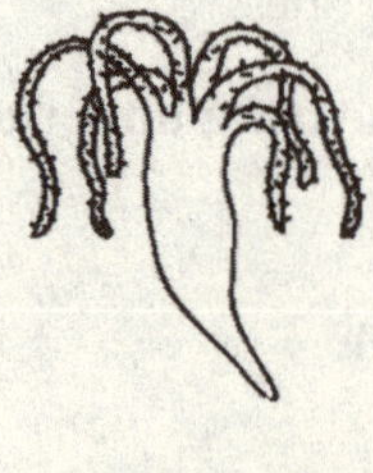

Coelenterate

Acoelomates (platylhelminthes). Flatworms are ribbonlike and bilaterally symmetrical. They possess three layers of cells, including a solid mesoderm. They do not have a circulatory system, and their nervous system consists of eyes, an anterior brain ganglion, and a pair of longitudinal nerve cords. Flatworms are capable of extensive regeneration.

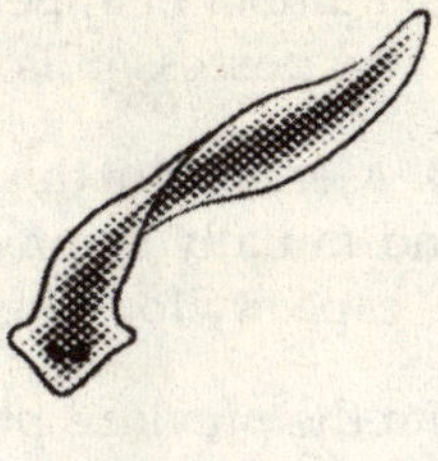

Flatworm

Pseudocoelomates (nematodes). Roundworms like hookworm and trichina possess long digestive tubes and an anus. A solid mesoderm is present; a circulatory system is not. They possess nerve cords and an anterior nerve ring.

Protostomes (mollusca, gastropoda). Protostomes, such as snails, clams, and squid, are soft bodied and possess mantles that often secrete calcareous (calcium carbonate) exoskeletons. They breathe through gills and have chambered hearts, blood sinuses, and a pair of ventral nerve cords.

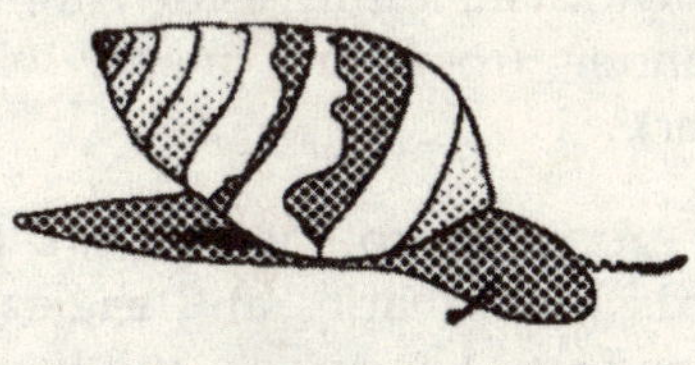

Mollusk

Annelida (segmented worms). Annelida (for example, earthworms) possess a coelem contained in the mesoderm. They have well-defined systems, including nervous, circulatory, and excretory systems.

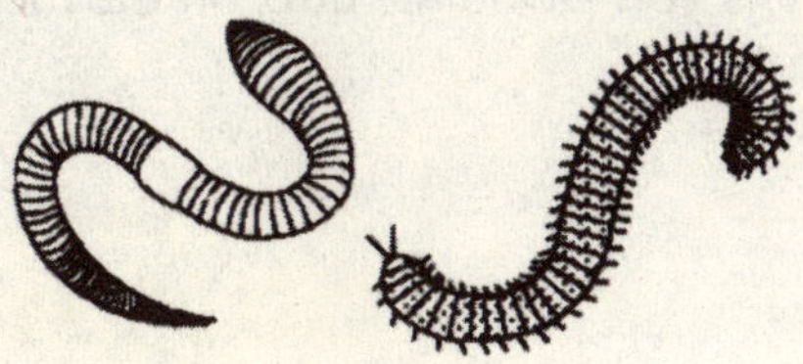

Annelids

Arthropods. Arthropods have jointed appendages, chitinous exoskeletons, and open circulatory systems. The three most important classes of arthropods are insects, arachnids, and crustaceans. Insects possess three pair of legs, spiracles, and tracheal tubes designed for breathing outside of an aquatic environment. Arachnids have four pair of legs and "book lungs"; examples include the scorpion and the spider. Lastly, crustaceans have seg-

Evolutionary Pathway to Humans

Porifera
Radiata
Acoelomates
Pseudocoelomates
Protostones
Annelida
Arthopoda
Deuterostomes
Chordata
Vertebrata
Mammalia
Homo sapiens

mented bodies with a variable number of appendages. Crustaceans like the lobster, crayfish, and shrimp also possess gills.

Deuterostomes (Echinodermata). Members of this group, which includes stars and sea urchins, are spiny and radially symmetrical, contain a water-vascular system, and possess the capacity for regeneration of parts.

Chordata. Chordata differ from the previous phyla in that a stiff, solid dorsal rod called the *notochord* is present at some stage of their embryologic development, as are paired gill slits. Chordata also have dorsal hollow nerve cords and tails extending beyond the anus at some point in their development.

The Chordata phylum encompasses all *Vertebrates*. It also includes amphioxus and tunicates. Chordata vertebrate fossils such as the diverse jawless creatures called agnathans are mostly extinct (the sea lamprey is an exception). Placodermi is another fossil chordata, a vertebrate fossil of armored fish called placoderms.

Chondrichthyes (cartilaginous fishes such as sharks and rays) and Osteichthyes (bony fishes such as trout and perch) are some marine creatures that belong to the chordata family, as do Amphibia (amphibious creatures, such as salamander, frogs, and toads, which divide their time between land and water).

On land, the Chordata group incorporates *Reptilia* (reptiles such as snakes and lizards); *Aves* (birds like canaries and eagles); *Mammalia* (examples include mammals like rodents, kangaroos, and horses); and human ancestors (primates such as humans and New and Old World monkeys).

Congratulations! You've made it through the chapter, and you should now be well versed in topics like theories of and evidence for evolution, speciation, the origin of early life, and classification and diversity. See just how much you've learned as you tackle the quiz on the following pages.

Evolution and Diversity Quiz

1. Nematocysts are characteristic of

   (A) Porifera
   (B) Protozoa
   (C) Coelenterata
   (D) Annelida
   (E) Echinodermata

2. According to the modern theory of evolution, which of the following evolved first?

   (A) the Krebs cycle
   (B) anaerobic respiration
   (C) autotrophic nutrition
   (D) photosynthesis
   (E) chemosynthesis

3. Which of the following is an INCORRECT association?

   (A) Porifera: sessile
   (B) Echinodermata: radial symmetry
   (C) Annelida: coelom
   (D) Platyhelminthes: anus
   (E) insects: tracheal tubes

4. Intestinal nematodes evolved from a free-living to a parasitic form through developing

   (A) special reproductive segments
   (B) a symbiotic relationship with intestinal bacteria
   (C) an external cuticle resistant to digestive enzymes
   (D) a long digestive tube
   (E) an intricate nervous system

5. Which statement about the phylum Echinodermata is false?

   (A) The phylum includes starfish and sea urchins.
   (B) Echinoderms reproduce sexually.
   (C) The phylum includes crayfish.
   (D) Echinoderms are heterotrophs.
   (E) Echinoderms are invertebrates.

6. Echinoderms are regarded as being closely related to chordates due to their

   (A) bilateral symmetry
   (B) form of circulatory system
   (C) endoskeleton
   (D) early embryonic development
   (E) possession of a notochord

7. Which of the following statements about evolution is correct?

   (A) Lamarck's theory of use and disuse adequately describes why giraffes have long necks.
   (B) Darwin's theory of natural selection relies solely on genetic mutation.
   (C) Darwin's theory explains the evolution of man from present-day apes.
   (D) Darwin's theory of natural selection relies solely on environmental conditions.
   (E) Natural selection is the process by which random mutations are selected for survival by the environment.

GO ON TO THE NEXT PAGE

8. Which of the following organisms is a chordate but NOT a vertebrate?

(A) shark
(B) lamprey eel
(C) turtle
(D) amphioxus
(E) none of the above

9. The notochord is

(A) present in all adult chordates
(B) present in all echinoderms
(C) present in chordates during embryological development
(D) always a vestigial organ in chordates
(E) part of the nervous system of all vertebrates

10. Which order of classificatory divisions is correct?

(A) kingdom, subphylum, genus, family
(B) phylum, class, order, genus
(C) species, order, family, phylum
(D) subphylum, kingdom, family, order
(E) genus, order, species, phylum

11. The hypothesis that chloroplasts and mitochondria were originally prokaryotic organisms living within eukaryotic hosts is supported by the fact that mitochondria and chloroplasts

(A) possess protein synthetic capability
(B) possess genetic material
(C) possess a plasma membrane
(D) possess characteristic ribosomes
(E) all of the above

12. Of the following forms of life, which is a member of the kingdom Protista?

(A) blue-green alga
(B) unicellular green alga
(C) anaerobic bacteria
(D) fungi
(E) mosses

13. Which of the following statement about viruses is NOT true?

(A) Their genetic material may be DNA or RNA.
(B) The virus may replicate in a bacterial host.
(C) The virus may replicate in a eukaryotic host.
(D) The virus may replicate autonomously in the absence of a host.
(E) The protein coat of the virus does not enter a host bacterial cell.

GO ON TO THE NEXT PAGE

14. Which of the following has a chitinous exoskeleton?

(A) oriole
(B) starfish
(C) clam
(D) honeybee
(E) earthworm

15. In the speculation concerning the origins of life, one theory states that purines, pyrimidines, sugars, and phosphates combined to form

(A) nucleotides
(B) nucleosides
(C) carbohydrates
(D) fats
(E) proteins

16. Compound eyes are found in

(A) Porifera
(B) Coelenterata
(C) Mollusca
(D) Arthropoda
(E) Annelida

17. Reptiles

(A) are homeothermic
(B) respire with gills
(C) must live in water at some stage of their life cycle
(D) possess notochords as adults
(E) lay leathery eggs

18. All viruses

(A) carry DNA
(B) carry RNA
(C) lack protein
(D) have chromosomes
(E) cannot reproduce outside of cells

19. Which of the following men might have explained the auk's loss of the ability to fly with the following hypothesis?

"Since the auk stopped using its wings, the wings became smaller, and this acquired trait was passed on to the offspring."

(A) Darwin
(B) Mendel
(C) de Vries
(D) Lamarck
(E) Morgan

20. All arthropods

(A) have a body consisting of three parts (the head, the thorax, and the abdomen)
(B) breathe through spiracles that lead to the tracheal tubes
(C) have a calcerous shell with a soft mantle
(D) have radial symmetry
(E) have jointed appendages and exoskeletons

Answers and Explanations for Evolution and Diversity Quiz

1. **(C)** Nematocysts are stinging cells found in coelenterates; they are used to immobilize prey. Coelenterates are organisms like hydra and jellyfish that are radially symmetrical and have one opening to their digestive tract that acts as both a mouth and an anus. Of the remaining choices, (A) Porifera are organisms such as sponges that have no defined tissue or organs, (B) Protozoa are one-celled organisms such as paramecium and amoebae, and (D) annelids are segmented invertebrates (such as earthworms) with a true body cavity (coelem) and two openings in their digestive tract. Lastly, echinoderms (E) are described in detail in the explanation to question 5 below.

2. **(B)** According to the evolutionary theory, large numbers of spontaneously formed organic molecules became simple organisms able to respire in the simplest way—anaerobically. These anaerobic heterotrophs developed first, followed by chemosynthetic bacteria. Next, photosynthetic bacteria developed and released O_2 as a byproduct, allowing the development of aerobic heterotrophs and the Krebs cycle.

3. **(D)** Platyhelminthes are ribbonlike, bilaterally symmetrical organisms with three layers of cells, including a solid mesoderm. They do not have a circulatory system and their nervous system consists of eyes, an anterior brain ganglion, and a pair of longitudinal nerve cords. They also have a primitive excretory system containing flame cells. They do not have an anus. Meanwhile, (A) porifera are sessile sponges that have two layers of cells, pores, and a low degree of cellular specialization. Echinoderms (B) are spiny, radially symmetrical, contain a water-vascular system, and possess the capacity for regeneration of parts, and (C) annelids are segmented worms which possess a true body cavity, the coelem, contained in the mesoderm. Annelids have well-defined nervous, circulatory, and excretory systems. As for (E), insects are arthropods and possess jointed appendages, chitinous exoskeletons, and open circulatory systems. They have three pairs of legs, spiracles, and tracheal tubes designed for breathing outside of an aquatic environment.

4. **(C)** This development was the most important evolutionary development for internal intestinal parasites, enabling them to avoid being digested by the host's acidic environment in the gut. (D) is false; as nematodes (roundworms) consume the already digested food of their host, they do not need developed digestive tracts. (B) and (E) have nothing to do with parasitism; nematodes actually have quite primitive nervous systems. And (A) is a characteristic of tapeworms, not roundworms. Roundworms actually reproduce by releasing eggs, so they can act as intestinal parasites without this ability.

5. **(C)** Echinoderms are the invertebrate predecessors of the chordates; because of this, they are a favorite question topic on the SAT II. Examples include the starfish, the sea urchin, and the sea cucumber. They are characterized by having a primitive vascular system known as the water vascular system. The adult echinoderm is radially symmetrical, while larvae are bilaterally symmetrical. They move with structures known as tube feet, have no backbone, and do not make their own food—in other words, they are heterotrophic. Crayfish are members of the phylum Arthropoda, not Echinodermata; their phylum includes insects, which are characterized by segmented bodies covered in a chitinous exoskeleton and jointed appendages.

6. **(D)** Echinoderms are considered our invertebrate predecessors because they are the only other organism besides chordates that are deuterostomes. In other words, the echinoderm's mouth is formed after the anus is formed from the blastopore, rather than before the anus is formed. In all other organisms, the blastopore forms the mouth first. As for (C), the endoskeleton of the echinoderm is made up of calcareous plates, unlike that of a chordate, and (E) echinoderms do not have a notochord in any

stage of their development. See explanation (5) above for more characteristics of the echinoderm.

7. **(E)** Natural selection is the process by which mutations are "tried out" in the environment; if they offer some degree of fitness, they will be passed on to the next generation. Lamarck's theory (A) asserted that an acquired trait was developed through use, such as a giraffe stretching its neck until it became longer, and that this acquired trait could be passed on to the next generation. We now know, however, that acquired characteristics do not affect genes and are therefore not passed on to the next generation (see the explanation to question 19 for a more detailed account of this theory). (B) Darwin's theory involves much more than mutation. It is contingent on overreproduction, which allows for genetic variation between the offspring. The most genetically fit of these offspring are then selected and pass on their genes to the next generation. (C) Darwin also believed that man and apes actually evolved from a common ancestor. And (D) is incorrect because natural selection includes selection pressures, but also needs a population with genetic variation in order to select the most fit organisms.

8. **(D)** Members of the phylum Chordata have a dorsal notochord, while members of the subphylum Vertebrata have a backbone. Nonvertebrate chordates include the amphioxus (lancet) or the unrelated tunicate worm. (A) and (B) are vertebrates with cartilaginous skeletons, while (C) has a bony skeleton.

9. **(C)** The notochord is a semirigid chord in the dorsal part of all chordates during a certain stage of embryonic development. In lower chordates, this chord remains as a semirigid chord, although in higher chordates it is seen only in the embryo and not as a vestigial organ. As for the other alternatives, (B) echinoderms do not possess a notochord, (D) the notochord remains in the lower chordates (such as the amphioxus and the tunicate worm), and (E) the notochord is not part of the nervous system.

10. **(B)** The correct order of classificatory divisions is: Kingdom, phylum, subphylum, class, order, family, genus, and species.

11. **(E)** The endosymbiotic hypothesis states that blue-green algae entered into a symbiotic arrangement with early eukaryotic plants to develop into chloroplasts, while bacteria entered into a similar arrangement with eukaryotic animal cells to become mitochondria. Chloroplasts and mitochondria have a plasma membrane (their inner membrane) and the ability to produce their own proteins without utilizing the cell's machinery (C). In addition, they have circular DNA, and their rRNA subunits are characteristic of prokaryotes.

12. **(B)** Protista are eukaryotes. They may be divided into two major categories, protozoa and algae. Protozoans are heterotrophic and resemble small animals. Algae, on the other hand, are photosynthetic, and resemble tiny plants; they are an important food source for marine animals. Blue-green algae, in contrast, are actually bacteria that are classified as Monera.

13. **(D)** A virus is a simple nonliving organism that takes on living characteristics when it enters cells that support its replication. The virus is made up of genetic material, either DNA or RNA, and a protein coat, and it is incapable of replicating autonomously; it must replicate within its host cell. A bacteriophage, for example, will inject its DNA into the bacterium, leaving its protein coat on the cell surface. In eukaryotes, however, the whole virus may enter the cell and not become unencapsulated until it enters the cytoplasm.

14. **(D)** Insects like the honeybee are characterized by having chitinous exoskeletons with three segmented bodies made up of a head, thorax, and abdomen. They have three pairs of jointed legs, two pairs of wings, and breathe through tracheal tubes. (A) Orioles, meanwhile, are vertebrates with bony endoskeletons, and (B) starfish are echinoderms whose endoskeletons are made up of calcareous

plates. (C) Clams are mollusks characterized by having a mantle secrete a calcium-mineral shell, and (E) earthworms are considered to have a hydrostatic skeleton of fluid surrounded by a layer of muscle.

15. (A) Nucleotides are made up of a nitrogenous base; guanine, uracil, adenine, cytosine, or thymidine; a sugar, either ribose or deoxyribose; and a phosphate group. Adenine and guanine are purines, while thymidine, cytosine, and uracil are pyrimidines. (B) Nucleosides do not contain the phosphate group, only the nitrogenous base and the sugar. Meanwhile, (C) carbohydrates are only sugars, while (D) fats are a combination of glycerols and three fatty acids. (E) Proteins are only linked amino acids and do not contain anything found in a nucleotide.

16. (D) Compound eyes are large groups of individual simple eyes, forming a type of composite eye characteristic of insects (phylum Arthropoda). Arthropods have jointed legs and an exoskeleton normally containing chitin. (A) The porifera, however, are sponges, extremely simple animal organisms that do not have eyes. (B) The coelentreates, including hydra and jellyfish, are radially symmetrical organisms which do not have eyes. Likewise, (C), the mollusks (snails, slugs, clams, and other shellfish) do not have compound eyes. Lastly, (E) annelids are segmented worms, such as the earthworm. They are invertebrate, have a fluid vascular skeleton which repeats segments, and do not have eyes.

17. (E) Reptiles lay soft, leathery eggs; they are water resistant, which implies that they tend to retain water on the inside. This is due to the fact that reptiles are terrestrial organisms. Reptiles are not homeothermic, or warm-blooded; they are poikilothermic or ectothermic, in other words cold-blooded. Their internal temperature is greatly dependent on the external environment. Reptiles do not respire with gills; fish and some larval amphibians, on the other hand, do. Reptiles also have functional lungs. They do not need to live in water at any stage of their life cycle; (C) refers to amphibians. Finally, reptiles do not possess notochords as adults; notochords are cartilaginous chords that form in the chordates during embryological development. In the chordates known as vertebrates, these are not present in the adult stage.

18. (E) Viruses must have a host cell in order to replicate, although this cell can be either a eukaryote or a prokaryote. (D) is incorrect because the genetic material of the virus does not take the form of a chromosome (the chromosome is actually a complex of nucleic acids and proteins known as a histone). Refer to the explanation for question 13 for more characteristics of the virus.

19. (D) Lamarck's theory of evolution states that new organs or changes in existing ones were believed to arise because of needs of the organism. The amount of change was thought to be based on the use or disuse of the organ. This theory was based on a fallacious understanding of genetics. Any useful characteristic acquired in one generation was thought to be transmitted to the next, but in actuality, only changes in the DNA of sex cells can be inherited. As for (A), Darwin's theory of natural selection states that pressures in the environment select the organism most fit to survive and reproduce. Darwin's theory incorporates a number of factors and stages, including overpopulation, variation, competition, natural selection, inheritance of variations, and evolution of a new species. (B) Mendel, meanwhile, defined modern genetics through experiments with pea plants, describing dominance, segregation, and independent assortment. (C) deVries confirmed Mendel's observations with results from experiments using a variety of different plant species. Finally, (E) Morgan induced mutation in Drosophila and studied the inheritance of these mutations. He also described sex-linked inheritance.

20. (E) Arthropods such as crustaceans, insects, spiders, and scorpions are characterized by a bilateral symmetry of jointed appendages, chitinous exoskeletons, and open circulatory systems. (A) and (B) refer specifically to insects, a class of arthropods.

They have a body consisting of a head, thorax, and an abdomen and a respiratory system consisting of spiracles and tracheal tubes. (C) Mollusks, such as snails, slugs, and clams, are characterized by a calcerous shell with a soft mantle; arthropods are not. Finally, coelenterates, not arthropods, are radially symmetrical organisms with a top and a bottom but no left and right (D).

STRESS MANAGEMENT

"She says the lava lamp relaxes her during the test."

CHAPTER EIGHT

THE KAPLAN ADVANTAGE™

The countdown has begun. Your date with THE TEST is looming on the horizon. Anxiety is on the rise. The butterflies in your stomach have gone ballistic. Perhaps you feel as if the last thing you ate has turned into a lead ball. Your thinking is getting cloudy. Maybe you think you won't be ready. Maybe you already know your stuff, but you're going into panic mode anyway. Worst of all, you're not sure of what to do about it.

Don't freak! It is possible to tame that anxiety and stress—before and during the test. We'll show you how. You won't believe how quickly and easily you can deal with that killer anxiety.

Avoid Must-y Thinking

Let go of "must-y" thoughts, those notions that you must do something in a certain way—for example, "I must get a great score, or else!" "I must meet Mom and Dad's expectations."

Make the Most of Your Prep Time

Lack of control is one of the prime causes of stress. A ton of research shows that if you don't have a sense of control over what's happening in your life, you can easily end up feeling helpless and hopeless. So, just having concrete things to do and to think about—taking control—will help reduce your stress. This section shows you how to take control during the days leading up to taking the test.

Identify the Sources of Stress

Jot down (in pencil) anything you identify as a source of your test-related stress. The idea is to pin down that free-floating anxiety so that you can take control of it. Here are some common examples to get you started:

- I always freeze up on tests.
- I'm nervous about trig (or functions, or geometry, etcetera).
- I need a good/great score to go to Acme College.
- My older brother/sister/best friend/girl- or boyfriend did really well. I must match their scores or do better.
- My parents, who are paying for school, will be really disappointed if I don't test well.

What Are "Signs of a Winner," Alex?

Here's some advice from a Kaplan instructor who won big on *Jeopardy!*™ In the green room before the show, he noticed that the contestants who were quiet and "within themselves" were the ones who did great on the show. The contestants who did not perform as well were the ones who were fact cramming, talking a lot, and generally being manic before the show. Lesson: Spend the final hours leading up to the test getting sleep, meditating, and generally relaxing.

- I'm afraid of losing my focus and concentration.
- I'm afraid I'm not spending enough time preparing.
- I study like crazy, but nothing seems to stick in my mind.
- I always run out of time and get panicky.
- I feel as though thinking is becoming like wading through thick mud.

Take a few minutes to think about the things you've just written down. Then rewrite them in some sort of order. List the statements you most associate with your stress and anxiety first, and put the least disturbing items last. Chances are, the top of the list is a fairly accurate description of exactly how you react to test anxiety, both physically and mentally. The later items usually describe your fears (disappointing Mom and Dad, looking bad, etcetera). As you write the list, you're forming a hierarchy of items so you can deal first with the anxiety provokers that bug you most. Very often, taking care of the major items from the top of the list goes a long way toward relieving overall testing anxiety. You probably won't have to bother with the stuff you placed last.

Take Stock of Your Strengths and Weaknesses

Take one minute to list the areas of the test that you are good at. They can be general ("algebra") or specific ("quadratic equations"). Put down as many as you can think of, and if possible, time yourself. Write for the entire time; don't stop writing until you've reached the one-minute stopping point.

Next, take one minute to list areas of the test you're not so good at, just plain bad at, have failed at, or keep failing at. Again, keep it to one minute, and continue writing until you reach the cutoff. Don't be afraid to identify and write down your weak spots! In all probability, as you do both lists, you'll find you are strong in some areas and not so strong in others. Taking stock of your assets and liabilities lets you know the areas you don't have to worry about, and the ones that will demand extra attention and effort.

Now, go back to the "good" list, and expand it for two minutes. Take the general items on that first list and make them more specific; take the specific items and expand them into more general conclusions. Naturally, if anything new comes to mind, jot it down. Focus all of your attention and effort on your strengths. Don't underestimate yourself or your abilities. Give yourself full credit. At the same time, don't list strengths you don't really have; you'll only be fooling yourself.

Every area of strength and confidence you can identify is much like having a reserve of solid gold at Fort Knox. You'll be able to draw on your reserves as you need them. You can use your reserves to solve difficult questions, maintain confidence, and keep test stress and anxiety at a distance. The encouraging thing is that every time you recognize another area of strength, succeed at coming up with a solution, or get a good score on a test, you increase your reserves. And, there is absolutely no limit to how much self-confidence you can have or how good you can feel about yourself.

What Do You Want to Accomplish in the Time Remaining?

The whole point to this next exercise is sort of like checking out a used car you might want to buy. You'd want to know up front what the car's weak points are, right? Knowing that influences your whole shopping-for-a-used-car campaign. So it is with your conquering-test-stress campaign: Knowing what your weak points are ahead of time helps you prepare.

So let's get back to the list of your weak points. Take two minutes to expand it just as you did with your "good" list. Be honest with yourself without going overboard. It's an accurate appraisal of the test areas that give you trouble. So, pick up your pencil, check the clock, and start writing.

Facing your weak spots gives you some distinct advantages. It helps a lot to find out where you need to spend extra effort. Increased exposure to tough material makes it more familiar and less intimidating. (After all, we mostly fear what we don't know and are probably afraid to face.) You'll feel better about yourself because you're dealing directly with areas of the test that bring on your anxiety. You can't help feeling more confident when you know you're actively strengthening your chances of earning a higher overall test score.

Imagine Yourself Succeeding

This next little group of exercises is both physical and mental. It's a natural follow-up to what you've just accomplished with your lists.

First, get yourself into a comfortable sitting position in a quiet setting. Wear loose clothes. If you wear glasses, take them off. Then, close your eyes and breathe in a deep, satisfying breath of air. Really fill your lungs until your rib cage is fully expanded and you can't take in any more. Then, exhale the air completely. Imagine you're blowing out a candle with your last little puff of air. Do this two or three more times, filling your lungs to their maximum and emptying them totally. Keep your eyes closed, comfortably but not tightly. Let your body sink deeper into the chair as you become even more comfortable.

With your eyes shut you can notice something very interesting. You're no longer dealing with the worrisome stuff going on in the world outside of you. Now you can concentrate on what happens inside you. The more you recognize your own physical reactions to stress and anxiety, the more you can do about them. You might not realize it, but you've begun to regain a sense of being in control.

Very Superstitious

Stress expert Stephen Sideroff, Ph.D., tells of a client who always stressed out before, during, and even after taking tests. Yet, she always got outstanding scores. It became obvious that she was thinking superstitiously—subconsciously believing that the great scores were a result of her worrying. She didn't trust herself, and believed that if she didn't worry she wouldn't study hard enough. Sideroff convinced her to take a risk and work on relaxing before her next test. She did, and her test results were still as good as ever—which broke her cycle of superstitious thinking.

Stress Tip

Don't work in a messy or cramped area. Before you sit down to study, clear yourself a nice, open space. And, make sure you have books, paper, pencils—whatever tools you will need—within easy reach before you sit down to study.

Let images begin to form on the "viewing screens" on the back of your eyelids. You're experiencing visualizations from the place in your mind that makes pictures. Allow the images to come easily and naturally; don't force them. Imagine yourself in a relaxing situation. It might be in a special place you've visited before or one you've read about. It can be a fictional location that you create in your imagination, but a real-life memory of a place or situation you know is usually better. Make it as detailed as possible, and notice as much as you can.

If you don't see this relaxing place sharply or in living color, it doesn't mean the exercise won't work for you. Some people can visualize in great detail, while others get only a sense of an image. What's important is not how sharp the details or colors, but how well you're able to manipulate the images. If you have only a faint sense of the images, that's okay—you'll still experience all the benefits of the exercise.

Think about the sights, the sounds, the smells, even the tastes and textures associated with your relaxing situation. See and feel yourself in this special place. Stay focused on the images as you sink farther back into your chair. Breathe easily and naturally. You might have the sensations of any stress or tension draining from your muscles and flowing downward, out your feet and away from you.

Take a moment to check how you're feeling. Notice how comfortable you've become. Imagine how much easier it would be if you could take the test feeling this relaxed and in this state of ease. You've coupled the images of your special place with sensations of comfort and relaxation. You've also found a way to become relaxed simply by visualizing your own safe, special place.

Now, close your eyes and start remembering a real-life situation in which you did well on a test. If you can't come up with one, remember a situation in which you did something (academic or otherwise) that you were really proud of—a genuine accomplishment. Make the memory as detailed as possible. Remember how confident you felt as you accomplished your goal. Now start thinking about the upcoming test. Keep your thoughts and feelings in line with that successful experience. Don't make comparisons between them. Just imagine taking the upcoming test with the same feelings of confidence and relaxed control.

This exercise is a great way to bring the test down to Earth. You should practice this exercise often, especially when the prospect of taking the exam starts to bum you out. The more you practice it, the more effective the exercise will be for you.

Link Your Thoughts

When you're committing new information to memory, link one fact to another, much as elephants are linked trunk to tail in a circus parade. Visualize an image (preferably a bizarre one) that connects the thoughts. You'll remember them in the same linked way, with one thought easily bringing the next to your mind.

Don't Force It

Never try to force relaxation. You'll only get frustrated and find yourself even more uptight. Be passive.

Exercise Your Frustrations Away

Whether it is jogging, walking, biking, mild aerobics, pushups, or a pickup basketball game, physical exercise is a very effective way to stimulate both your mind and body and to improve your ability to think and concentrate.

A surprising number of students get out of the habit of regular exercise, ironically because they're spending so much time prepping for exams. Also, sedentary people—this is a medical fact—get less oxygen to the blood and hence to the head than active people. You can live fine with a little less oxygen; you just can't think as well.

Any big test is a bit like a race. Thinking clearly at the end is just as important as having a quick mind early on. If you can't sustain your energy level in the last sections of the exam, there's too good a chance you could blow it. You need a fit body that can weather the demands any big exam puts on you. Along with a good diet and adequate sleep, exercise is an important part of keeping yourself in fighting shape and thinking clearly for the long haul.

There's another thing that happens when students don't make exercise an integral part of their test preparation. Like any organism in nature, you operate best if all your "energy systems" are in balance. Studying uses a lot of energy, but it's all mental. When you take a study break, do something active instead of raiding the fridge or vegging out in front of the TV. Take a 5- to 10-minute activity break for every 50 or 60 minutes that you study. The physical exertion gets your body into the act, which helps to keep your mind and body in sync. Then, when you finish studying for the night and hit the sack, you won't lie there, tense and unable to sleep because your head is overtired and your body wants to pump iron or run a marathon.

A warning about exercise, however: It's not a good idea to exercise vigorously right before you go to bed. This could easily cause sleep onset problems. For the same reason, it's also not a good idea to study right up to bedtime. Make time for a "buffer period" before you go to bed: For 30 to 60 minutes, just take a hot shower, meditate, simply veg out.

Ocean Dumping

Visualize a beautiful beach, with white sand, blue skies, sparkling water, a warm sun, and seagulls. See yourself walking on the beach, carrying a small plastic pail. Stop at a good spot and put your worries and whatever may be bugging you into the pail. Drop it at the water's edge and watch it drift out to sea. When the pail is out of sight, walk on.

Get High . . . Naturally

Exercise can give you a natural high, which is the only kind of high you should be aiming for. Using drugs (prescription or recreational) specifically to prepare for and take a big test is definitely self-defeating. (And if they're illegal drugs, you can end up with a bigger problem than the SAT on your hands.) Except for the drugs that occur naturally in your brain, every drug has major drawbacks—and a false sense of security is only one of them.

You may have heard that popping uppers helps you study by keeping you alert. If they're illegal, definitely forget about it. They wouldn't really work anyway, since amphetamines make it hard to retain information. Mild stimulants, such as coffee, cola, or over-the-counter caffeine pills can sometimes help as you study, since they keep you alert. On the down side, they can also lead to agitation, restlessness, and insomnia. Some people can drink a pot of high-octane coffee and sleep like a baby. Others have one cup and start to vibrate. It all depends on your tolerance for caffeine. Remember, a little anxiety is a good thing. The adrenaline that gets pumped into your bloodstream helps you stay alert and think more clearly. But, too much anxiety and you can't think straight at all.

Alcohol and other depressants are out, too. Again, if it's illegal, forget about it. Depressants wouldn't work anyway, since they lead to the inevitable hangover/crash, fuzzy thinking, and lousy sense of judgment. These would not help you ace the test.

Instead, go for endorphins—the "natural morphine." Endorphins have no side effects and they're free—you've already got them in your brain. It just takes some exercise to release them. Running around on the basketball court, bicycling, swimming, aerobics, power walking—these activities cause endorphins to occupy certain spots in your brain's neural synapses. In addition, exercise develops staying power and increases the oxygen transfer to your brain. Go into the test naturally.

Take a Deep Breath . . .

Here's another natural route to relaxation and invigoration. It's a classic isometric exercise that you can do whenever you get stressed out—just before the test begins, even during the test. It's very simple and takes just a few minutes.

Close your eyes. Starting with your eyes and—without holding your breath—gradually tighten every muscle in your body (but not to the point of pain) in the following sequence:

1. Close your eyes tightly.
2. Squeeze your nose and mouth together so that your whole face is scrunched up. (If it makes you self-conscious to do this in the test room, skip the face-scrunching part.)
3. Pull your chin into your chest, and pull your shoulders together.
4. Tighten your arms to your body, then clench your hands into tight fists.
5. Pull in your stomach.
6. Squeeze your thighs and buttocks together, and tighten your calves.
7. Stretch your feet, then curl your toes (watch out for cramping in this part).

At this point, every muscle should be tightened. Now, relax your body, one part at a time, in reverse order, starting with your toes. Let the tension drop out of each muscle. The entire process might take five minutes from start to finish (maybe a couple of minutes during the test). This clenching and unclenching exercise should help you to feel very relaxed.

The "New Age" of Relaxation

Here are some more tips for beating stress:

- Massage, especially shiatsu—see if it's offered through your school's phys ed department, or at the local "Y."
- Check out a book on acupressure, and find those points on your body where you can press a "relax button."
- If you're especially sensitive to smells, you might want to try some aromatherapy. Lavender oil, for example, is said to have relaxing properties. Health food stores, drug stores, and New Age bookstores may carry aromatherapy oils.
- Many health food stores carry herbs and supplements that have relaxing properties, and they often have a specialist on staff who can tell you about them.

And Keep Breathing

Conscious attention to breathing is an excellent way of managing test stress (or any stress, for that matter). The majority of people who get into trouble during tests take shallow breaths. They breathe using only their upper chests and shoulder muscles, and may even hold their breath for long periods of time. Conversely, the test taker who by accident or design keeps breathing normally and rhythmically is likely to be more relaxed and in better control during the entire test experience.

So, now is the time to get into the habit of relaxed breathing. Do the next exercise to learn to breathe in a natural, easy rhythm. By the way, this is another technique you can use during the test to collect your thoughts and ward off excess stress. The entire exercise should take no more than three to five minutes.

With your eyes still closed, breathe in slowly and deeply through your nose. Hold the breath for a bit, and then release it through your mouth. The key is to breathe slowly and deeply by using your diaphragm (the big band of muscle that spans your body just above your waist) to draw air in and out naturally and effortlessly. Breathing with your diaphragm encourages relaxation and helps minimize tension.

As you breathe, imagine that colored air is flowing into your lungs. Choose any color you like, from a single color to a rainbow. With each breath, the air fills your body from the top of your head to the tips of your toes. Continue inhaling the colored air until it occupies every part of you, bones and muscles included. Once you have completely filled yourself with the colored air, picture an opening somewhere on your body, either natural or imagined. Now, with each breath you exhale, some of the colored air will pass out the opening and leave your body. The level of the air (much like the water in a glass as it is emptied) will begin to drop. It will descend progressively lower, from your head down to your feet. As you continue to exhale the colored air, watch the level go lower and lower, farther and farther down your body. As the last of the colored air passes out of the opening, the level will drop down to your toes and disappear. Stay quiet for just a moment. Then notice how relaxed and comfortable you feel.

Thumbs Up for Meditation

Once relegated to the fringes of the medical world, meditation, biofeedback, and hypnosis are increasingly recommended by medical researchers to reduce pain from headaches, back problems—even cancer. Think of what these powerful techniques could do for your test-related stress and anxiety.

Effective meditation is based primarily on two relaxation methods you've already learned: body awareness and breathing. A couple of different meditation techniques follow. Experience them both, and choose the one that works best for you.

Stress Tip

Don't forget that your school probably has counseling available. If you can't conquer test stress on your own, make an appointment at the counseling center. That's what counselors are there for.

Take a Hike, Pal

When you're in the middle of studying and hit a wall, take a short, brisk walk. Breathe deeply and swing your arms as you walk. Clear your mind. (And, don't forget to look for flowers that grow in the cracks of the sidewalk.)

Stress Tip

If you want to play music, keep it low and in the background. Music with a regular, mathematical rhythm—reggae, for example—aids the learning process. A recording of ocean waves is also soothing.

Cyberstress

If you spend a lot of time in cyberspace anyway, do a search for the phrase *stress management.* There's a ton of stress advice on the Net, including material specifically for students.

Nutrition and Stress: The Dos and Don'ts

Do eat:

- Fruits and vegetables (raw is best, or just lightly steamed or nuked)
- Low-fat protein such as fish, skinless poultry, beans, and legumes (like lentils)
- Whole grains such as brown rice, whole wheat bread, and pastas (no bleached flour)

Don't eat:

- Refined sugar; sweet, high-fat snacks (simple carbohydrates like sugar make stress worse and fatty foods lower your immunity)
- Salty foods (they can deplete potassium, which you need for nerve functions)

Breath Meditation

Make yourself comfortable, either sitting or lying down. For this meditation you can keep your eyes opened or closed. You're going to concentrate on your breathing. The goal of the meditation is to notice everything you can about your breath as it enters and leaves your body. Take three to five breaths each time you practice the meditation; this set of breaths should take about a minute to complete.

Take a deep breath and hold it for 5 to 10 seconds. When you exhale, let the breath out very slowly. Feel the tension flowing out of you along with the breath that leaves your body. Pay close attention to the air as it flows in and out of your nostrils. Observe how cool it is as you inhale and how warm your breath is when you exhale. As you expel the air, say to yourself a cue word such as *calm* or *relax*. Once you've exhaled all the air from your lungs, start the next long, slow inhale. Notice how relaxed feelings increase as you slowly exhale and again hear your cue words.

Mantra Meditation

For this type of meditation experience you'll need a mental device (a mantra), a passive attitude (don't try to do anything), and a position in which you can be comfortable. You're going to focus your total attention on a mantra you create. It should be emotionally neutral, repetitive, and monotonous, and your aim is to fully occupy your mind with it. Furthermore, you want to do the meditation passively, with no goal in your head of how relaxed you're supposed to be. This is a great way to prepare for studying or taking the test. It clears your head of extraneous thoughts and gets you focused and at ease.

Sit comfortably and close your eyes. Begin to relax by letting your body go limp. Create a relaxed mental attitude and know there's no need for you to force anything. You're simply going to let something happen. Breathe through your nose. Take calm, easy breaths and as you exhale, say your mantra (*one, ohhm, aah, soup*—whatever is emotionally neutral for you) to yourself. Repeat the mantra each time you breathe out. Let feelings of relaxation grow as you focus on the mantra and your slow breathing. Don't worry if your mind wanders. Simply return to the mantra and continue letting go. Experience this meditation for 10 to 15 minutes.

Quick Tips for the Days Just Before the Exam

- The best test takers do less and less as the test approaches. Taper off your study schedule and take it easy on yourself. You want to be relaxed and ready on the day of the test. Give yourself time off, especially the evening before the exam. By then, if you've studied well, everything you need to know is firmly stored in your memory banks.

- Positive self-talk can be extremely liberating and invigorating, especially as the test looms closer. Tell yourself things such as, "I choose to take this test" rather than "I have to"; "I will do well" rather than "I hope things go well"; "I can" rather than "I cannot." Be aware of negative, self-defeating thoughts and images and immediately counter any you become aware of. Replace them with affirming statements that encourage your self-esteem and confidence. Create and practice visualizations that build on your positive statements.

- Get your act together sooner rather than later. Have everything (including choice of clothing) laid out days in advance. Most important, know where the test will be held and the easiest, quickest way to get there. You will gain great peace of mind if you know that all the little details—gas in the car, directions, etcetera—are firmly in your control before the day of the test.

- Experience the test site a few days in advance. This is very helpful if you are especially anxious. If at all possible, find out what room your part of the alphabet is assigned to, and try to sit there (by yourself) for a while. Better yet, bring some practice material and do at least a section or two, if not an entire practice test, in that room. In this situation, familiarity doesn't breed contempt, it generates comfort and confidence.

- Forego any practice on the day before the test. It's in your best interest to marshal your physical and psychological resources for 24 hours or so. Even race horses are kept in the paddock and treated like princes the day before a race. Keep the upcoming test out of your consciousness; go to a movie, take a pleasant hike, or just relax. Don't eat junk food or tons of sugar. And—of course—get plenty of rest the night before. Just don't go to bed too early. It's hard to fall asleep earlier than you're used to, and you don't want to lie there thinking about the test.

The Relaxation Paradox

Forcing relaxation is like asking yourself to flap your arms and fly. You can't do it, and every push and prod only gets you more frustrated. Relaxation is something you don't work at. You simply let it happen. Think about it. When was the last time you tried to force yourself to go to sleep, and it worked?

Dress for Success

On the day of the test, wear loose layers. That way, you'll be prepared no matter what the temperature of the room is. (An uncomfortable temperature will just distract you from the job at hand.)
And, if you have an item of clothing that you tend to feel "lucky" or confident in—a shirt, a pair of jeans, whatever—wear it. A little totem couldn't hurt.

Handling Stress During the Test

The biggest stress monster will be the test itself. Fear not; there are methods of quelling your stress during the test.

- Keep moving forward instead of getting bogged down in a difficult question. You don't have to get everything right to achieve a fine score. The best test takers skip difficult material temporarily in search of the easier stuff. They mark the ones that require extra time and thought. This strategy buys time and builds confidence so you can handle the tough stuff later.

- Don't be thrown if other test takers seem to be working more furiously than you are. Continue to spend your time patiently thinking through your answers; it's going to lead better results. Don't mistake

Stress Tip

Don't study on your bed, especially if you have problems with insomnia. Your mind might start to associate the bed with work, and make it even harder for you to fall asleep.

Stress Tip

A lamp with a 75-watt bulb is optimal for studying. But don't put it so close to your study material that you create a glare.

the other people's sheer activity as signs of progress and higher scores.

- Keep breathing! Weak test takers tend to forget to breathe properly as the test proceeds. They start holding their breath without realizing it, or they breathe erratically or arrhythmically. Improper breathing interferes with clear thinking.

- Some quick isometrics during the test—especially if concentration is wandering or energy is waning—can help. Try this: Put your palms together and press intensely for a few seconds. Concentrate on the tension you feel through your palms, wrists, forearms, and up into your biceps and shoulders. Then, quickly release the pressure. Feel the difference as you let go. Focus on the warm relaxation that floods through the muscles. Now you're ready to return to the task.

- Here's another isometric that will relieve tension in both your neck and eye muscles. Slowly rotate your head from side to side, turning your head and eyes to look as far back over each shoulder as you can. Feel the muscles stretch on one side of your neck as they contract on the other. Repeat five times in each direction.

With what you've just learned here, you're armed and ready to do battle with the test. This book and your studies will give you the information you'll need to answer the questions. It's all firmly planted in your mind. You also know how to deal with any excess tension that might come along, both when you're studying for and taking the exam. You've experienced everything you need to tame your test anxiety and stress. You're going to get a great score.

KAPLAN PRACTICE TESTS

- The two tests that follow offer realistic practice for the SAT II: Biology Subject Test. To get the most out of them, you should take them under testlike conditions.
- Take the tests in a quiet room with no distractions. Bring some No. 2 pencils.
- Time yourself. You should spend no more than one hour on the 95 questions on each test.
- Use the answer sheets provided on the pages before each test to mark your answers.
- Answers and explanations follow the test.
- Scoring instructions are in the "Compute Your Practice Test Score" sections immediately following the Answer Keys of each test.

ANSWER SHEET FOR PRACTICE TEST ONE

1 Ⓐ Ⓑ Ⓒ Ⓓ Ⓔ	26 Ⓐ Ⓑ Ⓒ Ⓓ Ⓔ	51 Ⓐ Ⓑ Ⓒ Ⓓ Ⓔ	76 Ⓐ Ⓑ Ⓒ Ⓓ Ⓔ
2 Ⓐ Ⓑ Ⓒ Ⓓ Ⓔ	27 Ⓐ Ⓑ Ⓒ Ⓓ Ⓔ	52 Ⓐ Ⓑ Ⓒ Ⓓ Ⓔ	77 Ⓐ Ⓑ Ⓒ Ⓓ Ⓔ
3 Ⓐ Ⓑ Ⓒ Ⓓ Ⓔ	28 Ⓐ Ⓑ Ⓒ Ⓓ Ⓔ	53 Ⓐ Ⓑ Ⓒ Ⓓ Ⓔ	78 Ⓐ Ⓑ Ⓒ Ⓓ Ⓔ
4 Ⓐ Ⓑ Ⓒ Ⓓ Ⓔ	29 Ⓐ Ⓑ Ⓒ Ⓓ Ⓔ	54 Ⓐ Ⓑ Ⓒ Ⓓ Ⓔ	79 Ⓐ Ⓑ Ⓒ Ⓓ Ⓔ
5 Ⓐ Ⓑ Ⓒ Ⓓ Ⓔ	30 Ⓐ Ⓑ Ⓒ Ⓓ Ⓔ	55 Ⓐ Ⓑ Ⓒ Ⓓ Ⓔ	80 Ⓐ Ⓑ Ⓒ Ⓓ Ⓔ
6 Ⓐ Ⓑ Ⓒ Ⓓ Ⓔ	31 Ⓐ Ⓑ Ⓒ Ⓓ Ⓔ	56 Ⓐ Ⓑ Ⓒ Ⓓ Ⓔ	81 Ⓐ Ⓑ Ⓒ Ⓓ Ⓔ
7 Ⓐ Ⓑ Ⓒ Ⓓ Ⓔ	32 Ⓐ Ⓑ Ⓒ Ⓓ Ⓔ	57 Ⓐ Ⓑ Ⓒ Ⓓ Ⓔ	82 Ⓐ Ⓑ Ⓒ Ⓓ Ⓔ
8 Ⓐ Ⓑ Ⓒ Ⓓ Ⓔ	33 Ⓐ Ⓑ Ⓒ Ⓓ Ⓔ	58 Ⓐ Ⓑ Ⓒ Ⓓ Ⓔ	83 Ⓐ Ⓑ Ⓒ Ⓓ Ⓔ
9 Ⓐ Ⓑ Ⓒ Ⓓ Ⓔ	34 Ⓐ Ⓑ Ⓒ Ⓓ Ⓔ	59 Ⓐ Ⓑ Ⓒ Ⓓ Ⓔ	84 Ⓐ Ⓑ Ⓒ Ⓓ Ⓔ
10 Ⓐ Ⓑ Ⓒ Ⓓ Ⓔ	35 Ⓐ Ⓑ Ⓒ Ⓓ Ⓔ	60 Ⓐ Ⓑ Ⓒ Ⓓ Ⓔ	85 Ⓐ Ⓑ Ⓒ Ⓓ Ⓔ
11 Ⓐ Ⓑ Ⓒ Ⓓ Ⓔ	36 Ⓐ Ⓑ Ⓒ Ⓓ Ⓔ	61 Ⓐ Ⓑ Ⓒ Ⓓ Ⓔ	86 Ⓐ Ⓑ Ⓒ Ⓓ Ⓔ
12 Ⓐ Ⓑ Ⓒ Ⓓ Ⓔ	37 Ⓐ Ⓑ Ⓒ Ⓓ Ⓔ	62 Ⓐ Ⓑ Ⓒ Ⓓ Ⓔ	87 Ⓐ Ⓑ Ⓒ Ⓓ Ⓔ
13 Ⓐ Ⓑ Ⓒ Ⓓ Ⓔ	38 Ⓐ Ⓑ Ⓒ Ⓓ Ⓔ	63 Ⓐ Ⓑ Ⓒ Ⓓ Ⓔ	88 Ⓐ Ⓑ Ⓒ Ⓓ Ⓔ
14 Ⓐ Ⓑ Ⓒ Ⓓ Ⓔ	39 Ⓐ Ⓑ Ⓒ Ⓓ Ⓔ	64 Ⓐ Ⓑ Ⓒ Ⓓ Ⓔ	89 Ⓐ Ⓑ Ⓒ Ⓓ Ⓔ
15 Ⓐ Ⓑ Ⓒ Ⓓ Ⓔ	40 Ⓐ Ⓑ Ⓒ Ⓓ Ⓔ	65 Ⓐ Ⓑ Ⓒ Ⓓ Ⓔ	90 Ⓐ Ⓑ Ⓒ Ⓓ Ⓔ
16 Ⓐ Ⓑ Ⓒ Ⓓ Ⓔ	41 Ⓐ Ⓑ Ⓒ Ⓓ Ⓔ	66 Ⓐ Ⓑ Ⓒ Ⓓ Ⓔ	91 Ⓐ Ⓑ Ⓒ Ⓓ Ⓔ
17 Ⓐ Ⓑ Ⓒ Ⓓ Ⓔ	42 Ⓐ Ⓑ Ⓒ Ⓓ Ⓔ	67 Ⓐ Ⓑ Ⓒ Ⓓ Ⓔ	92 Ⓐ Ⓑ Ⓒ Ⓓ Ⓔ
18 Ⓐ Ⓑ Ⓒ Ⓓ Ⓔ	43 Ⓐ Ⓑ Ⓒ Ⓓ Ⓔ	68 Ⓐ Ⓑ Ⓒ Ⓓ Ⓔ	93 Ⓐ Ⓑ Ⓒ Ⓓ Ⓔ
19 Ⓐ Ⓑ Ⓒ Ⓓ Ⓔ	44 Ⓐ Ⓑ Ⓒ Ⓓ Ⓔ	69 Ⓐ Ⓑ Ⓒ Ⓓ Ⓔ	94 Ⓐ Ⓑ Ⓒ Ⓓ Ⓔ
20 Ⓐ Ⓑ Ⓒ Ⓓ Ⓔ	45 Ⓐ Ⓑ Ⓒ Ⓓ Ⓔ	70 Ⓐ Ⓑ Ⓒ Ⓓ Ⓔ	95 Ⓐ Ⓑ Ⓒ Ⓓ Ⓔ
21 Ⓐ Ⓑ Ⓒ Ⓓ Ⓔ	46 Ⓐ Ⓑ Ⓒ Ⓓ Ⓔ	71 Ⓐ Ⓑ Ⓒ Ⓓ Ⓔ	
22 Ⓐ Ⓑ Ⓒ Ⓓ Ⓔ	47 Ⓐ Ⓑ Ⓒ Ⓓ Ⓔ	72 Ⓐ Ⓑ Ⓒ Ⓓ Ⓔ	
23 Ⓐ Ⓑ Ⓒ Ⓓ Ⓔ	48 Ⓐ Ⓑ Ⓒ Ⓓ Ⓔ	73 Ⓐ Ⓑ Ⓒ Ⓓ Ⓔ	
24 Ⓐ Ⓑ Ⓒ Ⓓ Ⓔ	49 Ⓐ Ⓑ Ⓒ Ⓓ Ⓔ	74 Ⓐ Ⓑ Ⓒ Ⓓ Ⓔ	
25 Ⓐ Ⓑ Ⓒ Ⓓ Ⓔ	50 Ⓐ Ⓑ Ⓒ Ⓓ Ⓔ	75 Ⓐ Ⓑ Ⓒ Ⓓ Ⓔ	

right

wrong

Use the answer key following the test to count up the number of questions you got right and the number you got wrong. (Remember to not count omitted questions as wrong.) The "Compute Your Score" section following the Answer Key will show you how to find your score.

Remove this answer sheet and use it to complete the Practice Test.

PRACTICE TEST ONE

Part A

<u>Directions:</u> Each question or incomplete statement below is followed by five possible answers or completions, lettered A–E. Choose the answer that is the best in each case. Fill in the corresponding oval on your answer sheet.

1. $CO_2 + H_2O \longrightarrow C_6H_{12}O_6 + O_2$

   This process is completed

   (A) in the cytoplasm
   (B) in the area of the cell membrane
   (C) in the chloroplast
   (D) in the mitochondria
   (E) in the area around the ribosomes

2. What are the most abundant molecules in living organisms?

   (A) proteins
   (B) NaCl
   (C) water
   (D) lipids
   (E) carbon dioxide

3. Which of the following is an adaptation by the fetus?

   I. ductus arteriosus
   II. foramen ovale
   III. fetal hemoglobin with a higher affinity for oxygen than adult hemoglobin

   (A) I only
   (B) II only
   (C) III only
   (D) I and III
   (E) I, II, and III

4. A bacteriophage which is present within a bacteria usually consists of

   (A) protein
   (B) DNA
   (C) a nucleus
   (D) specialized cells
   (E) other parasitic bacteria

5. Which of the following associations most closely corresponds to that of ectoderm: endoderm?

   (A) heart: stomach
   (B) retina: lungs
   (C) skeletal muscles: liver
   (D) skin: stomach muscle
   (E) taste buds: uterus

6. In the pyramid of energy, the least amount of stored chemical energy is found in

(A) primary producers
(B) decomposers
(C) primary consumers
(D) secondary consumers
(E) tertiary consumers

7. A movement of muscle that tends to bend one part upon the other is termed

(A) flexion
(B) insertion
(C) tonus
(D) diastole
(E) extension

8. Which of the following is true regarding DNA replication?

I. DNA replication is semi-conservative.
II. DNA replication occurs during Prophase.
III. Okasaki fragments are formed.
IV. Purines bind to pyrimidines.

(A) I and II
(B) I, II, and III
(C) I, III, and IV
(D) I, II, III, and IV
(E) I and IV

9. Down's syndrome is a human genetic disorder that can be caused by the nondisjunction of chromosome 21 during meiosis. During which phase of meiosis does nondisjunction occur?

(A) Metaphase II
(B) Prophase I
(C) Metaphase I
(D) Anaphase I
(E) Telophase I

10. Which hormone triggers your body to retain NaCl, especially during periods of excessive heat?

(A) aldosterone
(B) progesterone
(C) ACTH
(D) epinephrine
(E) testosterone

11. If one parent is homozygous dominant and the other is homozygous recessive, which of the following might appear in an F_2 generation, but not in an F_1 generation?

I. heterozygous genotype
II. dominant phenotype
III. recessive phenotype

(A) I only
(B) II only
(C) III only
(D) I and II
(E) II and III

GO ON TO THE NEXT PAGE

12. Which of the following statements about oxidative phosphorylation is NOT correct?

(A) It occurs in the inner membrane of the mitochondrion.
(B) It involves O_2 as the final electron acceptor.
(C) It produces 2 ATPs for each $FADH_2$.
(D) It can occur under anaerobic conditions.
(E) It involves a cytochrome electron transport chain.

13. Which of the following pairs of differentiated tissue might be malformed if only one of the primary germ layers failed to properly differentiate during development?

(A) brain and skeleton
(B) muscle and kidneys
(C) kidneys and brain
(D) intestinal epithelium and kidneys
(E) bladder and skin

14. The components of blood include

(A) plasma
(B) red blood cells
(C) white blood cells
(D) cell fragments known as platelets
(E) all of the above

15. The climate with the shortest growing season would be located in the

(A) taiga
(B) tropical rain forest
(C) deciduous forest
(D) savanna
(E) steppe

16. A factor that tends to keep the gene pool constant is

I. nonrandom mating
II. freedom to migrate
III. no net mutations
IV. large populations

(A) I and II
(B) III and IV
(C) I, III, and IV
(D) II, III, and IV
(E) I, II, III, and IV

17. A person takes a large overdose of antacid. As a result, the activity of which of the following enzymes would be most affected?

(A) maltase
(B) lactase
(C) lipase
(D) pepsin
(E) sucrase

GO ON TO THE NEXT PAGE

18. During the process of oxidative phosphorylation, oxygen serves as

(A) the initial acceptor of H electrons
(B) the final acceptor of H electrons
(C) a high-energy intermediate
(D) a phosphorylating agent
(E) a reducing agent

19. Which of the following is not a characteristic of the kingdom Protista?

(A) Members can be photosynthetic.
(B) Members can be free living.
(C) Some members move via flagella.
(D) Some members are shaped like rods and termed bacilli.
(E) Some members spend part of their life cycle inside insects.

20. Which of the following would most likely cause malabsorption in the small intestine?

(A) the presence of extremely dilute chyme in the small intestine
(B) the destruction of the intestinal villi
(C) decreased secretion of HCl in the stomach
(D) liver failure
(E) high glucose levels in the nephron

21. Which enzyme digests disaccharides to monosaccharides?

(A) lactase
(B) kinase
(C) zymogen
(D) lipase
(E) phosphorylase

22. The vessel with the LEAST oxygenated blood is the

(A) pulmonary vein
(B) aorta
(C) renal artery
(D) pulmonary artery
(E) superior vena cava

23. Bacteria have all of the following organelles EXCEPT

(A) ribosomes
(B) a cell membrane
(C) a cell wall
(D) a nucleus
(E) a flagellum

GO ON TO THE NEXT PAGE

24. From which germ layer do the kidneys form?

    I. ectoderm
    II. endoderm
    III. mesoderm

    (A) I only
    (B) II only
    (C) III only
    (D) I and II
    (E) I and III

25. Which of the following has a chambered heart and breathes through gills?

    (A) cardinal
    (B) sea urchin
    (C) snail
    (D) praying mantis
    (E) earthworm

26. A single nondisjunction may cause all of the following except

    (A) spontaneous miscarriage of the fetus
    (B) 47 chromosomes
    (C) 45 chromosomes
    (D) congenital disorders such as Down's syndrome
    (E) breakage near the centromere

27. When a tick bird consumes insect pests from a rhino's back, which of the following processes is taking place?

    (A) mutualism
    (B) parasitism
    (C) saprophytism
    (D) commensalism
    (E) autotrophism

28. Which organelle is chiefly responsible for digestive breakdown of the cell during autolysis?

    (A) pinocytic vesicle
    (B) Golgi body
    (C) ribosome
    (D) mitochondria
    (E) lysosome

29. A person's anterior pituitary gland is removed by surgery. Plasma concentrations of which of the following hormones would be LEAST affected?

    (A) GH
    (B) LH
    (C) FSH
    (D) insulin
    (E) ACTH

GO ON TO THE NEXT PAGE

30. The membrane that functions in respiration in the embryo is the

(A) amnion
(B) allantois
(C) chorion
(D) umbilical cord
(E) yolk sac

31. Which organism is correctly matched to its trophic level?

(A) shark: primary consumer
(B) cattle: primary consumer
(C) fungi: producer
(D) cyanobacteria: decomposer
(E) butterflies: scavenger

32. How many different types of gametes would be produced by an organism of genotype *AabbCcDD* if all of the genes were to assort independently?

(A) 4
(B) 6
(C) 8
(D) 10
(E) 12

33. Ribosomes function in aggregates called

(A) histones
(B) nucleoli
(C) endoplasmic reticulum
(D) the Golgi complex
(E) polysomes

34. Red is dominant over white in a certain flower. A red flower was crossed with a white flower. Both red and white offspring were produced. This demonstrates

(A) the law of segregation
(B) incomplete dominance
(C) linkage
(D) mutation
(E) the law of independent assortment

35. Phagocytosis is found in

(A) paramecium
(B) erythrocytes
(C) hydra
(D) liver cells
(E) flame cells

36. Salmon return to their specific home stream to spawn. This is an example of

(A) pheromones
(B) signal stimuli
(C) imprinting
(D) classical conditioning
(E) circadian rhythms

GO ON TO THE NEXT PAGE

37. If two animals mate and produce viable, fertile offspring under natural conditions, we can conclude that

(A) they both have diploid somatic cells
(B) they are both from the same species
(C) for any given allele, they both have the same gene
(D) their blood types are compatible
(E) none of the above

38. In an emergency, an individual with type AB antigen in his red blood cells may receive a transfusion of

I. type O blood
II. type A blood
III. type B blood

(A) I only
(B) II only
(C) II and III
(D) III only
(E) I, II, and III

39. The gene for color blindness is X-linked. If normal parents have a color-blind son, what is the probability that he inherited the gene for color blindness from his mother?

(A) 0%
(B) 25%
(C) 50%
(D) 75%
(E) 100%

40. Which of the following is a correct association?

(A) mitochondria: transport of materials form the nucleus to the cytoplasm
(B) Golgi apparatus: modification and glycosylation of proteins
(C) endoplasmic reticulum: selective barrier for the cell
(D) ribosomes: digestive enzymes most active at acidic pH
(E) lysosomes: membrane-bound organelles that convert fat into sugars

41. Spermatogenesis and oogenesis differ in that

I. spermatogenesis is continuous; oogenesis is discontinuous
II. spermatogenesis only occurs at puberty
III. spermatogenesis produces four haploid sperm cells, while oogenesis produces one egg cell and two or more polar bodies

(A) I only
(B) II only
(C) III only
(D) I and III
(E) I, II, and III

GO ON TO THE NEXT PAGE

42. What would be the sequence of mRNA transcribed from the DNA segment ACGTCA?

(A) UGCAGU
(B) TGCAGT
(C) ACGTCA
(D) ACTGCA
(E) ACUGCA

43. Which of the following foods contains the greatest amount of energy per gram?

(A) sugar
(B) starch
(C) fat
(D) proteins
(E) vitamins

44. Which statement about human gamete production is false?

(A) In the testes, sperm develop in the seminiferous tubules.
(B) In the ovaries, eggs develop in the ovarian follicles.
(C) FSH stimulates gamete production in both sexes.
(D) Gametes arise via meiosis.
(E) The result of meiosis in females is the production of four egg cells.

45. The hormone progesterone

(A) stimulates follicle growth
(B) shuts off FSH production
(C) is solely responsible for the maintenance of secondary sex characteristics
(D) is produced by the anterior pituitary
(E) readies the uterus for implantation

46. In a particular population, for a trait with two alleles, the frequency of the recessive allele is 0.6. What is the frequency of individuals expressing the dominant phenotype?

(A) 0.16
(B) 0.36
(C) 0.48
(D) 0.64
(E) 0.6

47. The first major part of the photosynthetic process involves

(A) the conversion of water and carbon dioxide into starch and oxygen
(B) the conversion of sugar into water and CO_2
(C) the splitting of water into oxygen, hydrogen, and electrons
(D) the chaining of CO_2 into a bicarbonate ion
(E) the splitting of CO_2 into carbon, oxygen, and an electron

GO ON TO THE NEXT PAGE

Questions 48–52 refer to the following diagram.

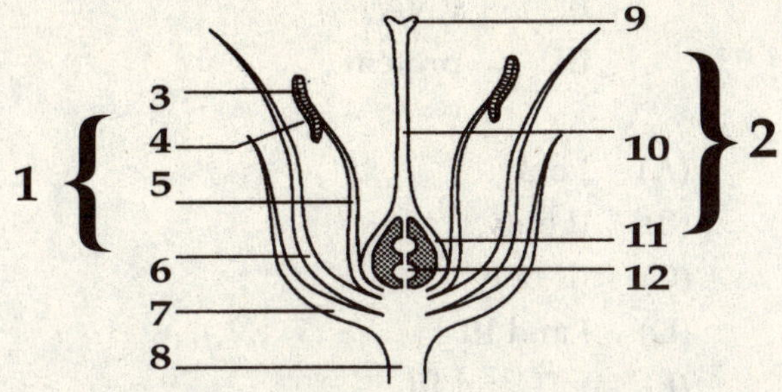

48. Which structure is the female gametophyte?

(A) 1
(B) 6
(C) 7
(D) 9
(E) 12

49. Which is the part of the flower that catches the pollen?

(A) 3
(B) 4
(C) 9
(D) 5
(E) 6

50. Where do pollen grains develop?

(A) 4
(B) 5
(C) 6
(D) 7
(E) 8

51. A pollen grain fuses with the two polar bodies to form

(A) the flower
(B) the endosperm
(C) the pistil
(D) the sepal
(E) the stamen

52. Which structure contains the pollen tube and allows the pollen grain to reach the ovules?

(A) 1
(B) 3
(C) 4
(D) 5
(E) 10

GO ON TO THE NEXT PAGE

Questions 53–56 refer to the figure below.

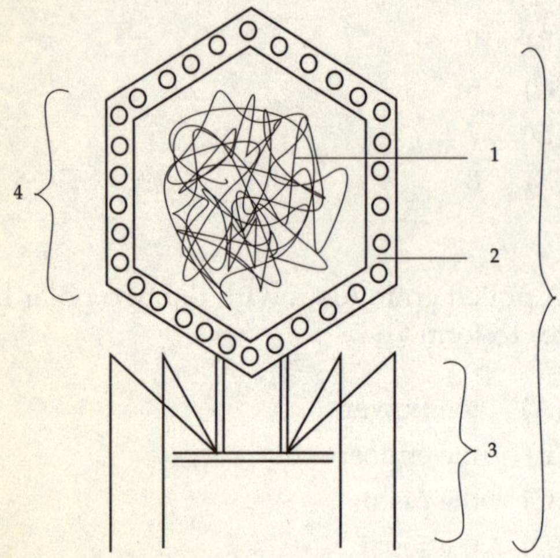

53. This is a figure of a

(A) bacteriophage
(B) eukaryotic virus
(C) bacteria
(D) spirochete
(E) none of the above

54. Which structure attaches this organism to the cell surface?

(A) 1
(B) 2
(C) 3
(D) 4
(E) 5

55. Structure 1 can be made up of

I. DNA
II. RNA
III. protein

(A) I only
(B) II only
(C) III only
(D) I and III
(E) I, II, and III

56. Structure 2 can be made up of

I. DNA
II. RNA
III. protein

(A) I only
(B) II only
(C) III only
(D) I and II
(E) I, II, and III

GO ON TO THE NEXT PAGE

Part B

Each set of choices A–E below should be compared to the numbered statements that follow it. Choose the lettered choice that best matches each numbered statement. Fill in the correct oval on your answer sheet. Remember that a choice may be used once, more than once, or not at all in each set.

Questions 57–60:

(A) 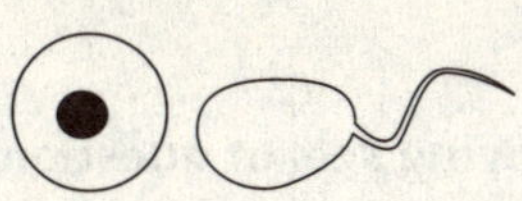

(B) 

(C) 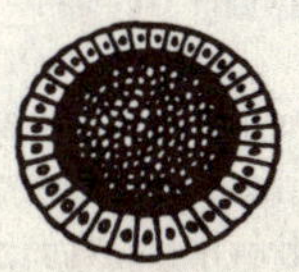

(D) 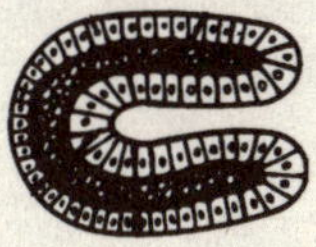

(E) 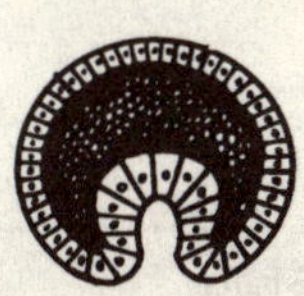

57. contains endoderm, mesoderm, and ectoderm
58. is a gamete
59. has cells that will ultimately differentiate into muscle and circulatory systems
60. is a blastula

Questions 61–64:

(A) ductus arteriosus
(B) pulmonary artery
(C) foramen ovale
(D) superior/inferior vena cava
(E) aorta

61. diverts blood from the right atrium to the left atrium
62. shunts blood away from the fetal lungs by funneling it into the aorta
63. in adults, carries the most oxygenated blood of all those listed above
64. collects blood from all over the body

Questions 65–68:

(A) simplex reflex
(B) complex reflex
(C) fixed action patterns
(D) behavioral cycles
(E) environmental rhythms

GO ON TO THE NEXT PAGE

65. startle response

66. retrieval and maintenance of eggs of their species by female birds

67. simple response to simple stimuli

68. circadian rhythms

Questions 69–73:

(A) simple diffusion
(B) nephridia
(C) tracheae and spiracles
(D) flame cells
(E) nephrons

69. grasshopper

70. man

71. planaria

72. earthworm

73. protozoa

Questions 74–78:

(A) fission
(B) budding
(C) spore formation
(D) vegetative propagation
(E) parthenogenesis

74. reproduction of bacteria

75. possible production of seedless fruit

76. reproduction of yeast cells

77. development of an egg without a sperm

78. an example is an underground stem with buds

Part C

Each of the following sets of questions is based on a laboratory or experimental situation. Begin by studying the description of each situation. Next, choose the best answer to each of the questions that follow it. Fill in the corresponding oval on your answer form.

A researcher grew *E. coli* bacteria that either had a plasmid (an extranuclear piece of DNA) or did not have a plasmid. These two types of *E. coli* were grown in two different growth mediums—a normal nutrient growth medium, and a normal nutrient growth medium with tetracycline, an antibiotic. The results of these experiments are summarized in the following graphs of growth of bacteria versus time.

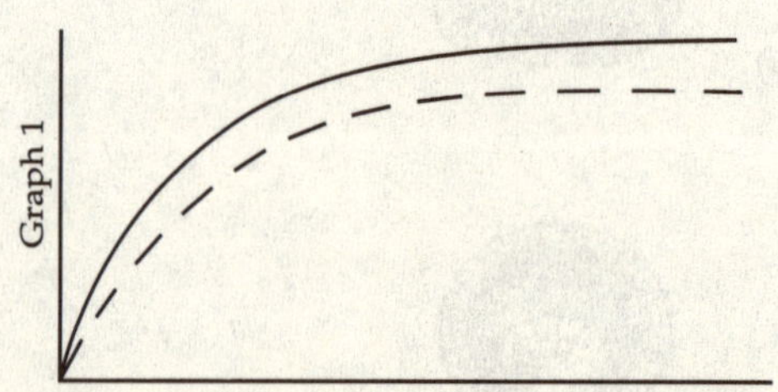

GO ON TO THE NEXT PAGE

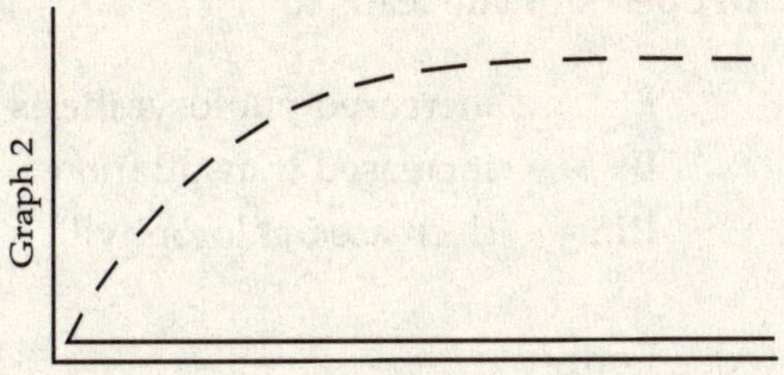

Use this data to answer questions 79–83.

79. When the two different *E. coli* strains, with and without the plasmid, were grown in nutrient media without antibiotics, which of the following occurred?

(A) The strain without the plasmid did not reach maximum growth potential.
(B) Both strains reached maximum growth potential.
(C) The strain with the plasmid did not reach maximum growth potential.
(D) Neither strain reached maximum growth potential.
(E) The strain with the plasmid had a higher population density than the strain without the plasmid.

80. When the two strains were cultured in nutrient media with the antibiotic tetracycline

(A) the strain without the plasmid was able to reach maximum population potential
(B) both strains reached maximum population potential
(C) the strain with the plasmid was able to reach maximum population potential
(D) neither strain was able to reach maximum population potential
(E) tetracycline did not affect either bacterial colony

81. The plasmid found in this *E. coli* strain codes for an enzyme that

(A) allows *E. coli* to grow in the presence of tetracycline
(B) digests the nutrient media
(C) causes the cells to autolyse
(D) makes this *E. coli* strain virulent
(E) none of the above

82. Which of the following would not be found in the plasmid?

(A) thymine
(B) adenine
(C) uracil
(D) cytosine
(E) guanine

GO ON TO THE NEXT PAGE

83. According to Darwin's theory of natural selection, the *E. coli* strain with the plasmid

(A) is more fit in regular media than the strain without the plasmid
(B) is a result of genetic mutation
(C) is less fit than the strain without the plasmid in the antibiotic media
(D) has a competitive advantage when grown in regular media
(E) has a competitive advantage when grown in the media that contains tetracycline

The following experiment describes the cyclical opening and closing of the stomata of a plant during a 24-hour period. Use this information to answer questions 84–87.

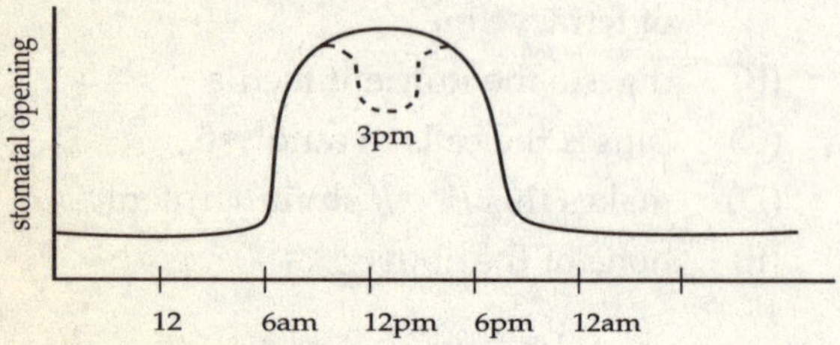

84. The dotted line represents stomatal closing during the hottest part of the day. This is because

(A) the dark reactions begin
(B) chlorophyll begins to act
(C) increased CO_2 concentration occurs
(D) transpiration exceeds the uptake of water
(E) none of the above

85. An open stomate leads to

I. increased photosynthesis
II. increased transpiration
III. decreased chlorophyll

(A) I only
(B) II only
(C) III only
(D) I and II
(E) I, II, and III

86. Guard cells open at dawn when light hits the guard cells. What would happen if you kept the plant in a dark room until 2 hours after dawn?

(A) The stomates would open anyway.
(B) The stomates would stay closed until they received the light stimulus.
(C) Stomates would stay closed all day.
(D) Photosynthesis would increase.
(E) Chlorophyll would degrade.

87. Stomate opening and closing is an approximately 24-hour cycle. This is an example of

(A) a dark reaction
(B) circadian rhythms
(C) environmental rhythms
(D) habituation
(E) response to pheromones

GO ON TO THE NEXT PAGE

A scientist was studying differential gene expression in a variety of tissues from the same organism. He investigated 8 different genes. The following chart describes which genes are turned on and producing their respective proteins in each of the different tissue types.

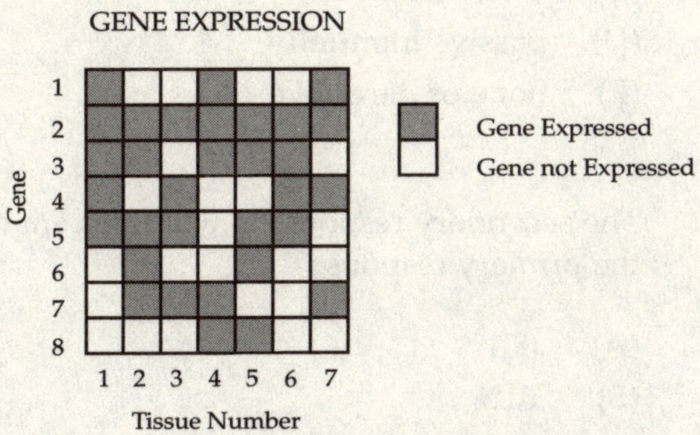

The chart will help you answer questions 88–91.

88. Which two cell types are the closest in the differentiation pathway?

(A) 1 and 2
(B) 2 and 5
(C) 3 and 4
(D) 4 and 5
(E) 5 and 6

89. Which two are the farthest apart in the differentiation pathway?

(A) 2 and 5
(B) 2 and 3
(C) 1 and 7
(D) 5 and 7
(E) 4 and 5

90. Gene 8 is a protein that digests fats. The cell type that expresses this gene is probably

(A) gall bladder
(B) stomach
(C) liver
(D) pancreas
(E) salivary gland

91. Which is a housekeeping gene (a gene involved in protein synthesis in the ribosome that all cells must express to survive)?

(A) 1
(B) 2
(C) 3
(D) 4
(E) 5

GO ON TO THE NEXT PAGE

The graph below illustrates the data from an experiment where antibody levels for two different antigens, A and B, were measured following exposure to these antigens. Antibodies play a crucial role in the human body's defense against disease.

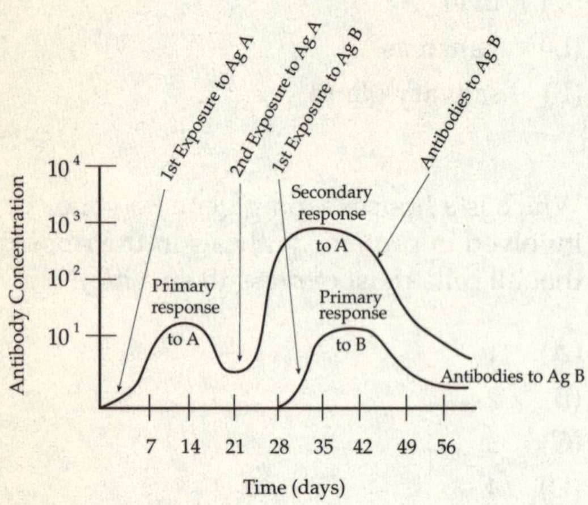

Questions 92–95 refer to this graph.

92. Antibodies are produced by

(A) B cells
(B) T cells
(C) macrophages
(D) neutrophils
(E) natural killer cells

93. A vaccination is defined by which part of the chart?

(A) Antigen A days 28–56
(B) Antigen A days 0–56
(C) Antigen A days 0–21
(D) Antigen B days 0–56
(E) Antigen B days 0–21

94. The large peak upon secondary stimulation is due to

(A) primed memory cells secreting antibody
(B) more antigen in the organism
(C) T cell activation
(D) passive immunity
(E) none of the above

95. The secondary response is what percent of the primary response?

(A) .001%
(B) .01%
(C) 10%
(D) 100%
(E) 1000%

Answer Key

1.	**C**	**20.**	**B**	**39.**	**E**	**58.**	**A**	**77.**	**E**
2.	**C**	**21.**	**A**	**40.**	**B**	**59.**	**E**	**78.**	**D**
3.	**E**	**22.**	**D**	**41.**	**D**	**60.**	**C**	**79.**	**B**
4.	**B**	**23.**	**D**	**42.**	**A**	**61.**	**C**	**80.**	**C**
5.	**B**	**24.**	**C**	**43.**	**C**	**62.**	**A**	**81.**	**A**
6.	**E**	**25.**	**C**	**44.**	**E**	**63.**	**E**	**82.**	**C**
7.	**A**	**26.**	**E**	**45.**	**E**	**64.**	**D**	**83.**	**E**
8.	**C**	**27.**	**A**	**46.**	**D**	**65.**	**B**	**84.**	**D**
9.	**C**	**28.**	**E**	**47.**	**C**	**66.**	**C**	**85.**	**D**
10.	**A**	**29.**	**D**	**48.**	**E**	**67.**	**A**	**86.**	**B**
11.	**C**	**30.**	**B**	**49.**	**C**	**68.**	**D**	**87.**	**B**
12.	**D**	**31.**	**B**	**50.**	**A**	**69.**	**C**	**88.**	**B**
13.	**B**	**32.**	**A**	**51.**	**B**	**70.**	**E**	**89.**	**D**
14.	**E**	**33.**	**E**	**52.**	**E**	**71.**	**D**	**90.**	**D**
15.	**A**	**34.**	**A**	**53.**	**A**	**72.**	**B**	**91.**	**B**
16.	**C**	**35.**	**A**	**54.**	**C**	**73.**	**A**	**92.**	**A**
17.	**D**	**36.**	**C**	**55.**	**A**	**74.**	**A**	**93.**	**C**
18.	**B**	**37.**	**B**	**56.**	**C**	**75.**	**D**	**94.**	**A**
19.	**D**	**38.**	**E**	**57.**	**E**	**76.**	**B**	**95.**	**E**

Compute Your Practice Test Score

Step 1: Figure out your raw score. Refer to your answer sheet for the number right and the number wrong on the practice test you're scoring. (If you haven't checked your answers, do that now, using the answer key that follows the test.) You can use the chart below to figure out your raw score. Multiply the number wrong by 0.25 and subtract the result from the number right. Round the result to the nearest whole number. This is your raw score.

SAT II: Biology Practice Test

Number right		Number wrong		Raw score
☐	– (	0.25 x ☐)	=	☐ (rounded)

Step 2: Find your practice test score. Find your raw score in the left column of the table below. The score in the right column is your Practice Test score.

Find Your Practice Test Score

Raw	Scaled	Raw	Scaled	Raw	Scaled	Raw	Scaled	Raw	Scaled	Raw	Scaled
95	800	78	690	61	590	43	490	26	390	9	290
94	800	77	680	59	580	42	480	25	380	8	290
93	790	76	670	58	570	41	470	24	380	7	280
92	780	75	670	57	570	40	470	23	370	6	280
91	770	74	660	56	560	39	460	22	370	5	270
90	760	73	660	55	550	38	460	21	360	4	260
89	750	72	650	54	550	37	450	20	360	3	260
88	740	71	640	53	540	36	450	19	350	2	250
87	740	70	640	52	540	35	440	18	350	1	240
86	730	69	630	51	530	34	430	17	340	0	240
85	730	68	630	50	530	33	430	16	330	–1	230
84	720	67	620	49	520	32	420	15	330	–2	230
83	720	66	620	48	510	31	420	14	320	–3	220
82	710	65	610	47	510	30	410	13	320	–4	220
81	710	64	610	46	500	29	410	12	310	–5	210
80	700	63	600	45	500	28	400	11	300	–6	210
79	690	62	590	44	490	27	400	10	300	–7 – –9	200

Practice Test Score

☐

A note on your practice test scores: Don't take these scores too literally. Practice test conditions cannot precisely mirror real test conditions. Your actual SAT II: Biology Subject Test score will almost certainly vary from your practice test scores. Your scores on the practice tests will give you a rough idea of your range on the actual exam.

Explanations to Practice Test One

1. **(C)** This equation signifies the formation of carbohydrates from carbon dioxide and water by plants. This process occurs in the chloroplasts.

2. **(C)** All organisms are made up of 50–70 percent water, which means that water is the most abundant molecule in living things.

3. **(E)** The fetus has the ductus arteriosus, which shunts blood from the pulmonary artery to the aorta. The foramen ovale, meanwhile, shunts blood from the right atrium to the left atrium. Both of these keep blood from traveling to the underdeveloped lungs. Fetal hemoglobin has a higher affinity for oxygen than adult hemoglobin, enabling it to become oxygenated in the placenta from the mother's blood.

4. **(B)** Bacteriophages are made up of DNA and a protein coat. The protein coat works like a syringe that injects DNA into the cell.

5. **(B)** The ectoderm develops into the skin, eyes, and the nervous system. The endoderm develops into the lungs, the gastrointestinal tract, and the bladder lining. Finally, the mesoderm develops into everything else, including the musculoskeletal system, the circulatory system, and the reproductive organs.

6. **(E)** Each level in the pyramid of energy loses some energy from the level before it, due to the loss of heat and energy costs of maintenance of the organism at each level. Hence the least amount of energy would be found at the highest point of the chain.

7. **(A)** Flexors bend a joint to an acute angle, while extensors extend or straighten the bones at a joint.

8. **(C)** DNA replication is semiconservative in that the two daughter strands receive one strand from the parent strand. DNA replication occurs at Interphase and Okasaki fragments are formed. Purines (such as adenine and guanine) always bind with pyrimidines (like thymine, uracil, and cytosine).

9. **(C)** Disjunction, the separation of sister chromatids during meiosis, occurs during Metaphase I. Failure of this to occur (nondisjunction) will lead to either one extra or one missing chromosome in the daughter cell.

10. **(A)** Aldosterone causes the active absorption of sodium ions in the nephron. Chloride ions and water passively follow.

11. **(C)** In the *AA* x *aa* cross, the F_1 generation will be 100 percent *Aa*. In the F_2 generation, there will be a 1:2:1 ratio of *AA:Aa:aa*. Therefore, the F_1 generation is entirely dominant heterozygous. Answer choices I and II can be eliminated, leaving only III.

12. **(D)** The electron transport chain is a complex carrier mechanism located on the inside of the inner mitochondrial membrane. During oxidative phosphorylation, ATP is produced when high energy potential electrons are transferred from NADH and $FADH_2$ to oxygen by a series of carriers located in the inner mitochondrial membrane. As the electrons are transferred from carrier to carrier, free energy is released, which is then used to form ATP. The last carrier of the electron transport chain, cytochrome a3, passes its electron to the final electron acceptor, O_2. In addition to the electrons, O_2 picks up a pair of hydrogen ions from the surrounding medium, forming water. This process cannot occur without oxygen.

13. (B) See the explanation to question 5 for descriptions of the ectoderm, endoderm, and mesoderm.

14. (E) Blood is made up of three major components. Plasma is the liquid part of blood, containing dissolved nutrients, wastes, proteins, hormones, and fibrinogen. Red blood cells are anucleated, biconcave discs filled with hemoglobin which unite with oxygen to form oxyhemoglobin. Finally, the cells of the body's defense system are called white blood cells. These include phagocytes, which engulf bacteria; lymphocytes, which are involved in the specific immune response; and platelets, which are cell fragments that aid in blood clotting.

15. (A) The taiga have the most northern latitude of the choices available. Since growing season decreases as latitude increases, it must also have the shortest growing season.

16. (C) The Hardy-Weinberg Law states that gene ratios and allelic frequencies remain constant through the generations in a nonevolving population. Four criteria must be met in order for this to occur: random mating, a large population, no migration into or out of the population, and a lack of mutation. If all four of these criteria are met, gene frequencies will remain constant. Any time these four are not met, gene frequencies will change and evolution may occur.

17. (D) When a person takes too much antacid, HCl in the stomach is neutralized and pH rises. When pH is raised, pepsin becomes nonfunctional. Therefore, pepsin activity would be most affected by an antacid overdose.

18. (B) See the explanation to question 12 for a description of the electron transport chain and the process of oxidative phosphorylation.

19. (D) Protistans may be unicellular or colonial, free living or symbiotic. They also can be heterotrophic or photosynthetic. Flagellates move via flagella, while sporozoans are responsible for malaria and spend part of their life cycle in mosquitoes. Protistans cannot, however, be rod-shaped. Bacteria are defined by their shape, and rod-shaped bacteria are termed *bacilli*.

20. (B) The millions of small villi that cover the entire surface of the small intestine are crucial to its absorptive capacity. The villi greatly enhance the surface area of the small intestine's absorptive membrane. Extremely dilute chyme presents little problem for the small intestine, because it is readily transported through the intestinal mucosa into the blood of the villi when osmotic pressure is sufficient. HCl is secreted in the stomach, not the small intestine. If decreased secretion of HCl in the stomach causes the chyme entering the small intestine to be less acidic, this would not adversely affect intestinal absorption. In fact, the digestive enzymes of the small intestine operate best at a slightly basic pH. The presence of alkaline fluid in the small intestine is the norm, due to secretion of bicarbonate ion into the small intestine.

21. (A) Lactase breaks lactose into glucose and galactose. In (B), kinase is an enzyme that phosphorylates its substrate. As for (C), zymogen is an enzyme that is secreted in an inactive form. The zymogen is cleaved under certain physiological conditions to the active form of the enzyme. Important examples of zymogens include pepsinogen, trypsinogen, and chymotrypsinogen, which are cleaved in the digestive tract to yield the active enzymes pepsin, trypsin, and chymotrypsin. Finally, in (D), lipase breaks down lipids into free fatty acids, while in (E), phosphorylase removes a phosphate from its substrate.

22. (D) The least oxygenated blood would be found in the pulmonary artery. This artery transports blood to the lungs to become oxygenated.

23. (D) Bacteria are prokaryotes, and do not have membrane organelles like nuclei, mitochondria, and lysosomes.

24. (C) See the explanation for question 5 for definitions of the ectoderm, the endoderm, and the mesoderm.

25. (C) Snails are mollusks. Their characteristic mantle secretes a calcium-mineral shell. Snails also breathe through gills and have chambered hearts, blood sinuses, and a pair of ventral nerve cords. As for (D), insects are characterized by having chitinous exoskeletons with three segmented bodies made up of a head, thorax, and abdomen. They have three pairs of jointed legs, two pairs of wings, and breathe through tracheal tubes. Therefore, the praying mantis would have a chitinous exoskeleton. Cardinals (A), on the other hand, are vertebrates, with a bony endoskeleton, while starfish (B) are echinoderms with calcareous plates making up their endoskeletons. Finally, earthworms (D) are believed to have a hydrostatic skeleton of fluid surrounded by a layer of muscle.

26. (E) Nondisjunction is a failure of some homologous pairs of chromosomes to separate following meiotic synapsis. The result is an extra chromosome or a missing chromosome for a given pair. For example, Down's syndrome can be attributed to an extra chromosome number 21. The number of chromosomes in a case of single nondisjunction is $2n + 1$ or $2n - 1$. In Down's syndrome, it is 47. Most of these embryos are aborted early in their development and only a few, like Down's syndrome Trisomy 21, Trisomy 13, and Trisomy 18, make it to term, albeit with developmental disorders. Hence breakage near the centromere might be induced by environmental factors such as mutagens, but would not be caused by nondisjunction.

27. (A) The tick bird helps the rhino by keeping it clean of parasites, eating the insects that infect the rhino, while the rhino in its turn provides food for the tick bird. This +/+ relationship is known as mutualism.

28. (E) The lysosome is a small, single-membrane-enclosed vesicle, containing intracellular digestive enzymes. These enzymes are sometimes used to digest worn-out organelles within the cell. In some cells, such as phagocytic cells, they may be used to help digest ingested bacteria and foreign material, and in large numbers, the lysosomes may break down the complete interior of a cell, causing cell death. This will occur if many of the lysosomes release their enzymatic contents simultaneously. This type of cell death is known as autolysis.

A pinocytotic vesicle (A) is a small vesicle formed by the cell membrane in order to take into the cell a small particle or drop of liquid. As for (B), a Golgi body, normally located at the end of a section of endoplasmic reticulum, functions to package and modify proteins. In (C), ribosomes produce proteins under the direction of messenger RNAs, using amino acids that are present in the cytoplasm. Finally, mitochondria (D) are the site of respiration and all energy-producing processes within the cell.

29. (D) GH, FSH, LH, and ACTH are all secreted by the anterior pituitary, and would therefore by affected by its surgical removal. Insulin is produced by the pancreas.

30. (B) The allantois is found in the eggs of birds and reptiles and is used as a receptacle for nitrogenous wastes. The vessels of this structure also lie close enough to the surface of the shell to enable the exchange of gases. In choice (A), the amnion is the innermost fluid-filled embryonic membrane; it forms a protective sac surrounding the embryos of birds, reptiles, and mammals. The chorion (C) is the outermost extra-embryonic membrane that separates the embryo from the albumin of reptiles and birds. The umbilical cord (D), meanwhile, connects the embryo to the placenta. Gas exchange occurs in the placenta. And the yolk sac (E) contains food for the developing embryo.

31. (B) Cattle, being herbivorous, are primary consumers; sharks are secondary consumers or tertiary consumers. Fungi are decomposers and cyanobacteria are producers. Butterflies, meanwhile, are primary consumers.

32. (A) You can disregard *bb* and *DD*; all gametes will be *bD*. Your remaining options are now *ACbD*, *AcbD*, *aCbD*, and *acbD*, giving you a total of four different gametes.

33. (E) Polysomes are defined as a group of ribosomes that attach to a strand of mRNA and simultaneously translate it. Histones are attached to the DNA, allowing it to coil tightly in the nucleus, and the nucleoli is the region of the nucleus where rRNA is produced.

34. (A) The law of segregation states that when gametes are formed, the two alleles for a particular trait will separate or segregate into the gametes, so that each of the gametes contains only one of the alleles for a given trait. So if red is dominant over white, and both red and white offspring were produced, then the red flower must be a hybrid or heterozygous flower. This means that the genotype of the red flower contains both one red allele and one white allele. The white flower contains two white alleles. When the gametes are formed for this mating, the two alleles in the red flower, the red and white, will segregate into the gametes, forming red-containing gametes and white-containing gametes. When these meet and fertilize the white-containing gametes from the other flower, half the offspring produced will be red because they are the result of the red gamete's fertilization of the white gamete, and the other half will result from the white gamete's fertilization of the other white gamete.

35. (A) Phagocytosis is the means by which paramecium engulf food by extending their cell membrane to encapsulate food particles. Phagosomes are formed and fuse with lysosomes to degrade the contents.

36. (C) Imprinting is a process in which environmental patterns or objects presented to a developing organism during a brief "critical period" in early life become accepted as permanent elements in the organism's life. In this case, the location at which the salmon was born becomes a permanent behavioral element, demonstrated when the salmon returns to this location to spawn.

37. (B) The definition of a species states that members of a particular species are able to mate and produce fertile, viable offspring. When animals of different species mate, their offspring is usually infertile, as when the horse and the donkey produce the sterile ass.

38. (E) AB is known as the universal acceptor. It does not have antibodies to either the A or B antigens. Therefore, AB patients can receive blood from A, B, AB, or O people.

39. (E) The important thing to remember about questions dealing with *X*-linked traits is that a given female has two *X* chromosomes, one inherited from her mother and one inherited from her father, while a given male has one *X* chromosome inherited from his mother and one *Y* chromosome inherited from his father. So if a male expresses an *X*-linked trait, he must have inherited it from his mother. Hence if normal parents have a color-blind son, he *must* have inherited the color blind gene, which is *X*-linked, from his mother. His mother *must* be a carrier of the color blind allele. The probability that a color blind son inherited the gene for color blindness from his mother is therefore 100 percent.

40. (B) The Golgi apparatus consists of a stack of membrane-enclosed sacs. The Golgi receives vesicles and their contents from the smooth ER, modifies them (as in glycosylation), repackages them into vesicles, and distributes them. In (A), mitochondria are involved in cellular respiration, and in (C), the ER transports polypeptides around the cell and to the Golgi apparatus for packaging. The ribosome (D) is the site of protein synthesis, while lysosomes

(E) are membrane-bound organelles which contain digestive enzymes and typically have a low pH.

41. (D) Spermatogenesis and oogenesis are both examples of gametogenesis in that both produce haploid gametes through reductional division (meiosis) of diploid cells. These processes occur in the gonads. They differ in that in spermatogenesis, the cytoplasm is equally divided during meiosis and four viable sperm are produced from one diploid cell. In oogenesis, on the other hand, the cytoplasm is divided unequally, and only one ovum, with the bulk of the cytoplasm, is produced in addition to two or three inert polar bodies. Spermatogenesis is also continuous, meaning that it occurs throughout life and not only during puberty. Meanwhile, oogenesis freezes at the end of Meiosis I and does not continue throughout Meiosis II until fertilization.

42. (A) If the DNA sequence is ACGTCA, the complementary mRNA sequence must be UGCAGU. Remember that thymine is replaced by uracil in RNA and becomes the base pair for adenine.

43. (C) Fat contains approximately 9 calories/gram, while carbohydrates and proteins contain only 4 calories/gram. Sugar and starch are two forms of carbohydrates, and vitamins are coenzymes that are typically not metabolized.

44. (E) Oogenesis produces only one viable egg and two or three polar bodies. This is a result of unequal distribution of the cytoplasm during meiosis. Interstitial cells (A) are stimulated by LH to produce testosterone. FSH and testosterone then initiate the development of sperm in the seminiferous tubules. As for (B), eggs develop in follicles in the ovaries under the control of FSH. It is obvious in (C) that FSH plays a role in gamete production in both sexes. Finally, in (D), gametes become haploid through reductional division (meiosis) in which a diploid cell gives rise to either four haploid sperm or one haploid egg and two or three polar bodies.

45. (E) Progesterone readies the uterus for implantation by thickening and vascularizing the uterine lining. In (A), FSH secreted by the anterior pituitary, not progesterone, stimulates follicle growth. As for (B), it is the production of estrogen by the ovaries that shuts off FSH secretion, rather than progesterone. Testosterone and estrogen are responsible for the secondary sex characteristics of males and females respectively (D). As for the final choice, (E), FSH and LH are produced by the anterior pituitary.

46. (D) To answer this question, we must use the Hardy Weinberg equation, $p^2 + 2pq + q^2 = 1$, where p equals the gene frequency for the dominant allele, and q equals the gene frequency of the recessive allele. Hence, p^2 is the frequency of homozygous dominants in the population, $2pq$ is the frequency of heterozygotes, and q^2 is the frequency of homozygous recessives. For a trait with only two alleles, $p + q$ must equal 1, since the combined frequencies of the alleles must total 100 percent. In this problem, we are told that the frequency of the recessive allele for a particular trait is 0.6; hence q=0.6. Since $p + q = 1$, $p = 0.4$. You're asked to determine the frequency of individuals expressing the dominant phenotype, not the dominant genotype. So, you're looking for $p^2 + 2pq$; p^2 = 0.4 squared = 0.16. $2pq$ = 2(0.6)(0.4) = 0.48. Finally, 0.16 + 0.48 = 0.64, which means that 64 percent of the individuals in the question express the dominant phenotype.

47. (C) In the light reaction, chlorophyll absorbs energy to split water into excited electrons, H^+ and O_2. The excited electrons make ATP through photophosphorylation. In the dark reaction, carbohydrates are produced in the absence of sunlight by CO_2; H^+ ions in the form of $NADPH_2$ and ATP are other products of this reaction. (B), meanwhile, describes cellular respiration, not photosynthesis. (D) does not occur at all, while in (E) H_2O is split, not CO_2.

48. (E) The ovules are contained in the enlarged base of the pistil known as the ovary. Each ovule contains a monoploid egg nucleus.

49. (C) The stigma is the sticky top part of the flower that catches pollen.

50. (A) Pollen grains develop in the sac (called the anther) at the top of the thin, stalklike filament.

51. (B) One pollen grain (sperm) fuses with the egg to form the zygotes, while another fuses with two polar bodies to form the endosperm that nourishes the growing zygote.

52. (E) The style is a tubelike structure connecting the stigma to the ovary at the base of the pistil. It allows sperm to reach the ovules.

53. (A) A bacteriophage is a virus that infects bacteria.

54. (C) Tail fibers attach the bacteriophage to the cell surface, where it injects DNA into the cell.

55. (A) Bacteriophages contain only DNA, although some other viruses, such as HIV, use RNA as their genetic material.

56. (C) The viral capsid is always made up of protein.

57. (E) The three-layer gastrula is the first stage in the developmental pathway that contains all three germs layers—the endoderm, the ectoderm, and the mesoderm.

58. (A) Eggs and sperm are known as gametes. Sperm move via flagellar tails that propel them toward the eggs.

59. (E) Since it is the first stage in which mesoderm is present, the three-layer gastrula is also the first stage that has cells that will ultimately differentiate into the musculoskeletal, circulatory, and reproductive systems.

60. (C) The blastula is a hollow ball of cells with a fluid-filled center.

61. (C) The foramen ovale is a hole in the fetal heart between the right and left atria that shunts blood from the right atrium to the left atrium. This prevents the blood from entering the developing lungs.

62. (A) The ductus arteriosus has the same function as the foramen ovale, although it shunts blood away from the lungs by diverting it from the pulmonary artery into the aorta.

63. (E) In an adult, the most oxygenated blood in the body flows through the pulmonary veins. This blood has just left the lungs. The next most oxygenated blood in the body would flow through the left atrium, followed by the blood that flows through the left ventricle and the aorta.

64. (D) The vena cavas collect blood from all over the body and transport it to the right atrium, allowing the blood to be oxygenated and renewing the cycle of systemic circulation.

65. (B) Complex reflexes involve neural integration at higher levels such as the brain stem or the cerebrum. An example of this kind of reflex is the startle response that is provoked when you hear your name called.

66. (C) Fixed action patterns are complex, coordinated, innate behavioral responses to specific patterns or stimuli from the environment. Female birds will take care of eggs of their own species more vigilantly than they would an egg that doesn't resemble one of their own.

67. (A) Simple reflexes are simple, automatic responses to simple stimuli; one example is the knee-jerk reflex.

68. (D) Daily behavioral cycles are called circadian rhythms. These cycles are initiated intrinsically but modified by external factors such as day length.

69. (C) Grasshoppers, like all insects, remove wastes through tracheae and spiracles.

70. (E) Humans remove waste from their blood stream through filtration in the nephron, the functional unit of the kidney.

71. (D) Planaria have a primitive excretory system that consists of flame cells.

72. (B) Earthworms excrete carbon dioxide directly through their moist skin. Two pairs of nephridia in each body segment excrete water, mineral salts, and urea wastes.

73. (A) Because protozoans are in contact with their external aqueous environment, ammonia and carbon dioxide exit via simple diffusion through the cell membrane.

74. (A) Fission may be defined as a method of cell division in which an equal division of cytoplasm and nucleus occurs. It is found in most one-celled organisms.

75. (D) Vegetative propagation is when undifferentiated meristem cells provide for the development of a new plant from the parent. Plants that practice this form of propagation are able to produce seedless fruit, and daughter organisms are genetically identical to their parents. Examples of vegetative propagation include underground stems with buds (found in potatoes) and runners, which are stems running above or below the ground found in strawberries and some grasses.

76. (B) Yeast cells and some hydra reproduce by budding. This involves the unequal division of cytoplasm, although equal division of the genetic material is maintained.

77. (E) Parthenogenesis occurs in some insects when the eggs develop without sperm. These organisms are haploid and contain only half of the normal genetic component.

78. (D) See the explanation for question 75 for a description of vegetative propagation.

79. (B) Without antibiotics, both strains reached their maximum growth potential.

80. (C) With the antibiotics in Graph 2, only the *E. coli* strain that has the plasmid reached its maximum growth potential. The strain that did not have the plasmid could not grow.

81. (A) The plasmid must allow the bacteria to grow in the presence of an antibacterial compound such as tetracycline. Therefore, it must either inactivate the antibiotics or circumvent the pathway that they affect.

82. (C) Plasmids are extra chromosomal DNA, and uracil is a component of RNA not found in DNA.

83. (E) Natural selection states that fitter organisms have a competitive advantage over unfit organisms.

Therefore, a bacteria that can grow in the presence of an antibiotic such as tetracycline is more fit and has a competitive advantage over a strain that does not contain this plasmid when both are grown in the presence of tetracycline.

84. (D) During the hottest part of the day, the loss of water (transpiration) through stomata exceeds the uptake of water through the roots. Therefore, to conserve water, the plant closes its stomata.

85. (D) An open stomate allows carbon dioxide to reach the chloroplasts so that photosynthesis can occur. This would therefore increase the rate of photosynthesis. The open stomata also leads to the loss of water through the opening, but does not affect chlorophyll at all.

86. (B) Stomata open when they receive light stimulus; therefore, they would stay closed in the dark room until they received light stimulus when put back into the sun.

87. (B) Twenty-four-hour daily cycles, such as sleep patterns and the opening and closing of stomata, are examples of circadian rhythms.

88. (B) Tissue types 2 and 5 express the same genes, except that 2 expresses one more; therefore, 2 and 5 are probably very similar cell types.

89. (D) Tissue types 5 and 7 express completely opposite genes, except for gene 2, which all cells express. Therefore, based on their protein secretion, 5 and 7 are probably very different cells.

90. (D) The pancreas produces lipase, the enzyme that digests fats.

91. (B) Gene 2 is expressed in all cell types and is probably associated with something that all cells do, such as protein synthesis.

92. (A) B cells produce antibodies; T cells, meanwhile, may be either cytotoxic (that is, they kill other cells) or serve as helper T cells that secrete proteins to stimulate other cell types.

93. (C) When you are vaccinated, you receive a pathogen for the first time in an attenuated or weakened form, so that when you are exposed to this pathogen for the second time (in the form of the disease) you are able to make a potent protective response.

94. (A) The large peak the second time you are exposed to an antigen or pathogen is due to the memory response of cells that have been primed for that antigen.

95. (E) The secondary response contained 10,000 antibodies per unit of blood, compared with 10 antibodies per unit of blood in the primary response. So, 10,000/10 = 1,000 percent of the primary response.

ANSWER SHEET
FOR PRACTICE TEST TWO

Remove this answer sheet and use it to complete the Practice Test.

1 Ⓐ Ⓑ Ⓒ Ⓓ Ⓔ	26 Ⓐ Ⓑ Ⓒ Ⓓ Ⓔ	51 Ⓐ Ⓑ Ⓒ Ⓓ Ⓔ	76 Ⓐ Ⓑ Ⓒ Ⓓ Ⓔ
2 Ⓐ Ⓑ Ⓒ Ⓓ Ⓔ	27 Ⓐ Ⓑ Ⓒ Ⓓ Ⓔ	52 Ⓐ Ⓑ Ⓒ Ⓓ Ⓔ	77 Ⓐ Ⓑ Ⓒ Ⓓ Ⓔ
3 Ⓐ Ⓑ Ⓒ Ⓓ Ⓔ	28 Ⓐ Ⓑ Ⓒ Ⓓ Ⓔ	53 Ⓐ Ⓑ Ⓒ Ⓓ Ⓔ	78 Ⓐ Ⓑ Ⓒ Ⓓ Ⓔ
4 Ⓐ Ⓑ Ⓒ Ⓓ Ⓔ	29 Ⓐ Ⓑ Ⓒ Ⓓ Ⓔ	54 Ⓐ Ⓑ Ⓒ Ⓓ Ⓔ	79 Ⓐ Ⓑ Ⓒ Ⓓ Ⓔ
5 Ⓐ Ⓑ Ⓒ Ⓓ Ⓔ	30 Ⓐ Ⓑ Ⓒ Ⓓ Ⓔ	55 Ⓐ Ⓑ Ⓒ Ⓓ Ⓔ	80 Ⓐ Ⓑ Ⓒ Ⓓ Ⓔ
6 Ⓐ Ⓑ Ⓒ Ⓓ Ⓔ	31 Ⓐ Ⓑ Ⓒ Ⓓ Ⓔ	56 Ⓐ Ⓑ Ⓒ Ⓓ Ⓔ	81 Ⓐ Ⓑ Ⓒ Ⓓ Ⓔ
7 Ⓐ Ⓑ Ⓒ Ⓓ Ⓔ	32 Ⓐ Ⓑ Ⓒ Ⓓ Ⓔ	57 Ⓐ Ⓑ Ⓒ Ⓓ Ⓔ	82 Ⓐ Ⓑ Ⓒ Ⓓ Ⓔ
8 Ⓐ Ⓑ Ⓒ Ⓓ Ⓔ	33 Ⓐ Ⓑ Ⓒ Ⓓ Ⓔ	58 Ⓐ Ⓑ Ⓒ Ⓓ Ⓔ	83 Ⓐ Ⓑ Ⓒ Ⓓ Ⓔ
9 Ⓐ Ⓑ Ⓒ Ⓓ Ⓔ	34 Ⓐ Ⓑ Ⓒ Ⓓ Ⓔ	59 Ⓐ Ⓑ Ⓒ Ⓓ Ⓔ	84 Ⓐ Ⓑ Ⓒ Ⓓ Ⓔ
10 Ⓐ Ⓑ Ⓒ Ⓓ Ⓔ	35 Ⓐ Ⓑ Ⓒ Ⓓ Ⓔ	60 Ⓐ Ⓑ Ⓒ Ⓓ Ⓔ	85 Ⓐ Ⓑ Ⓒ Ⓓ Ⓔ
11 Ⓐ Ⓑ Ⓒ Ⓓ Ⓔ	36 Ⓐ Ⓑ Ⓒ Ⓓ Ⓔ	61 Ⓐ Ⓑ Ⓒ Ⓓ Ⓔ	86 Ⓐ Ⓑ Ⓒ Ⓓ Ⓔ
12 Ⓐ Ⓑ Ⓒ Ⓓ Ⓔ	37 Ⓐ Ⓑ Ⓒ Ⓓ Ⓔ	62 Ⓐ Ⓑ Ⓒ Ⓓ Ⓔ	87 Ⓐ Ⓑ Ⓒ Ⓓ Ⓔ
13 Ⓐ Ⓑ Ⓒ Ⓓ Ⓔ	38 Ⓐ Ⓑ Ⓒ Ⓓ Ⓔ	63 Ⓐ Ⓑ Ⓒ Ⓓ Ⓔ	88 Ⓐ Ⓑ Ⓒ Ⓓ Ⓔ
14 Ⓐ Ⓑ Ⓒ Ⓓ Ⓔ	39 Ⓐ Ⓑ Ⓒ Ⓓ Ⓔ	64 Ⓐ Ⓑ Ⓒ Ⓓ Ⓔ	89 Ⓐ Ⓑ Ⓒ Ⓓ Ⓔ
15 Ⓐ Ⓑ Ⓒ Ⓓ Ⓔ	40 Ⓐ Ⓑ Ⓒ Ⓓ Ⓔ	65 Ⓐ Ⓑ Ⓒ Ⓓ Ⓔ	90 Ⓐ Ⓑ Ⓒ Ⓓ Ⓔ
16 Ⓐ Ⓑ Ⓒ Ⓓ Ⓔ	41 Ⓐ Ⓑ Ⓒ Ⓓ Ⓔ	66 Ⓐ Ⓑ Ⓒ Ⓓ Ⓔ	91 Ⓐ Ⓑ Ⓒ Ⓓ Ⓔ
17 Ⓐ Ⓑ Ⓒ Ⓓ Ⓔ	42 Ⓐ Ⓑ Ⓒ Ⓓ Ⓔ	67 Ⓐ Ⓑ Ⓒ Ⓓ Ⓔ	92 Ⓐ Ⓑ Ⓒ Ⓓ Ⓔ
18 Ⓐ Ⓑ Ⓒ Ⓓ Ⓔ	43 Ⓐ Ⓑ Ⓒ Ⓓ Ⓔ	68 Ⓐ Ⓑ Ⓒ Ⓓ Ⓔ	93 Ⓐ Ⓑ Ⓒ Ⓓ Ⓔ
19 Ⓐ Ⓑ Ⓒ Ⓓ Ⓔ	44 Ⓐ Ⓑ Ⓒ Ⓓ Ⓔ	69 Ⓐ Ⓑ Ⓒ Ⓓ Ⓔ	94 Ⓐ Ⓑ Ⓒ Ⓓ Ⓔ
20 Ⓐ Ⓑ Ⓒ Ⓓ Ⓔ	45 Ⓐ Ⓑ Ⓒ Ⓓ Ⓔ	70 Ⓐ Ⓑ Ⓒ Ⓓ Ⓔ	95 Ⓐ Ⓑ Ⓒ Ⓓ Ⓔ
21 Ⓐ Ⓑ Ⓒ Ⓓ Ⓔ	46 Ⓐ Ⓑ Ⓒ Ⓓ Ⓔ	71 Ⓐ Ⓑ Ⓒ Ⓓ Ⓔ	
22 Ⓐ Ⓑ Ⓒ Ⓓ Ⓔ	47 Ⓐ Ⓑ Ⓒ Ⓓ Ⓔ	72 Ⓐ Ⓑ Ⓒ Ⓓ Ⓔ	
23 Ⓐ Ⓑ Ⓒ Ⓓ Ⓔ	48 Ⓐ Ⓑ Ⓒ Ⓓ Ⓔ	73 Ⓐ Ⓑ Ⓒ Ⓓ Ⓔ	
24 Ⓐ Ⓑ Ⓒ Ⓓ Ⓔ	49 Ⓐ Ⓑ Ⓒ Ⓓ Ⓔ	74 Ⓐ Ⓑ Ⓒ Ⓓ Ⓔ	
25 Ⓐ Ⓑ Ⓒ Ⓓ Ⓔ	50 Ⓐ Ⓑ Ⓒ Ⓓ Ⓔ	75 Ⓐ Ⓑ Ⓒ Ⓓ Ⓔ	

right

wrong

Use the answer key following the test to count up the number of questions you got right and the number you got wrong. (Remember to not count omitted questions as wrong.) The "Compute Your Score" section that follows the Answer Key will show you how to find your score.

Practice Test Two

Part A

Directions: Each question or incomplete statement below is followed by five possible answers or completions, lettered A–E. Choose the answer that is the best in each case. Fill in the corresponding oval on your answer sheet.

1. An organism is heterozygous with respect to three pairs of genes, named *Aa*, *Bb*, and *Cc*. How many different types of gametes can be formed?

   (A) 2
   (B) 4
   (C) 6
   (D) 8
   (E) 9

2. Which of the following is found in all forms of life?

   I. genetic material
   II. protein
   III. water

   (A) I only
   (B) II only
   (C) III only
   (D) I and III
   (E) I, II, and III

3. A defect in the ectoderm would lead to

   (A) a malformed bladder
   (B) malformed reproductive organs
   (C) a malformed gastrointestinal tract
   (D) a malformed nervous system
   (E) a malformed skeletal system

4. Which of the following is NOT essential for blood clotting?

   (A) sodium ions
   (B) calcium ions
   (C) prothrombin
   (D) vitamin K
   (E) platelets

5. Which substance does NOT act as a digestive enzyme for proteins?

   (A) pepsin
   (B) trypsin
   (C) carboxypeptidase
   (D) chymotrypsin
   (E) gastrin

6. Which of the following is a function of bone?

   I. formation of blood cells
   II. protection of vital organs
   III. framework for movement

   (A) I only
   (B) II only
   (C) III only
   (D) I and II
   (E) I, II, and III

7. The hypothesis that chloroplasts and mitochondria were originally prokaryotic organisms living within eukaryotic hosts is supported by the fact that mitochondria and chloroplasts

   I. possess protein synthetic capability
   II. possess genetic material
   III. possess a plasma membrane
   IV. possess characteristic ribosomes

   (A) II only
   (B) IV only
   (C) II and IV
   (D) III and IV
   (E) I, II, III, and IV

8. Which of the following situations is NOT an example of environmental factors affecting the way a gene is expressed?

   (A) Drosophila with a given set of genes have crooked wings at low temperatures and straight wings at high temperatures.
   (B) The hair color of the Himalayan hare changes after the naturally warm portions of the animal are cooled.
   (C) Heat shock proteins are produced in a cell after an increase in temperature.
   (D) A decrease in temperature causes shivering.
   (E) all of the above

9. A culture of white blood cells is grown on a nutrient media containing dinitrophenol, which is a poison that blocks the electron transport chain. Under these conditions

   (A) ATP production will remain the same
   (B) ATP production will decrease
   (C) oxygen consumption will increase
   (D) ethanol production will increase
   (E) none of the above

GO ON TO THE NEXT PAGE

10. Many animals utilize panting as a means of cooling themselves down. The mechanism behind panting is to

(A) rapidly increase carbon dioxide expiration
(B) moisten the mucosa of the respiratory passages
(C) minimize the movement of respiratory muscles
(D) decrease body heat via evaporation
(E) none of the above

11. Compounds from the glomerular filtrate are partially resorbed in the

(A) Bowman's capsule
(B) glomerulus
(C) proximal convoluted tubule
(D) villi
(E) ureter

12. A climax community

(A) is independent of the environment
(B) will have a large number of lichens
(C) consists of fungi that decay plant matter
(D) is relatively stable within a given climate
(E) is populated mainly by pioneer organisms

13. How many genetically different gametes can be formed by a mouse that is genotypically *AABbCc*? (Assume that there is no linkage between these genes and that none are lethal.)

(A) 4
(B) 6
(C) 7
(D) 8
(E) 9

14. Fetal lungs are supplied with only enough blood to nourish the lung tissue itself, since they are nonfunctional prior to birth. Which of the following shunts blood away from the fetal lungs?

(A) aorta
(B) ductus arteriosus
(C) ductus venosus
(D) pulmonary artery
(E) pulmonary vein

15. Which are correctly related?

(A) white blood cell: no nucleus
(B) smooth muscle cell: multinuclear
(C) smooth muscle: voluntary action
(D) cardiac muscle: involuntary action
(E) smooth muscles: striations

GO ON TO THE NEXT PAGE

16. A process that cannot take place in a haploid cell is

(A) mitosis
(B) meiosis
(C) cell division
(D) growth
(E) digestion

17. To test whether a tall plant in which the tallness trait is dominant is homozygous or heterozygous, you could

I. cross it with a tall plant that had a short parent
II. cross it with a tall plant that had two tall parents
III. cross it with a short plant

(A) I only
(B) II only
(C) III only
(D) I and II
(E) I and III

18. Strontium is preferentially incorporated into growing long bone. Therefore, if a child were exposed to strontium, where would the highest concentration of strontium most likely be found?

(A) in the cartilage lining the joints
(B) in the center of long bones
(C) in the skull bones
(D) near the epiphyseal plates of long bones
(E) evenly distributed throughout the body

19. Venous blood en route from the kidneys to the heart must pass through the

(A) iliac vein
(B) inferior vena cava
(C) liver
(D) hepatic vein
(E) pulmonary vein

GO ON TO THE NEXT PAGE

20. Humans cannot produce Vitamin K. However, even when their diet is lacking in Vitamin K, they have plenty of it in their bloodstream. How can this be possible?

    (A) Vitamin K is synthesized in the liver using hormones absorbed from ingested plant cells.
    (B) Vitamin K is a byproduct of protein degradation during digestion.
    (C) Vitamin K is synthesized by bacteria that inhabit the colon.
    (D) Vitamin K can be absorbed from sunlight.
    (E) none of the above

21. Which statement about glycolysis is NOT true?

    (A) Glycolysis converts a single molecule into two molecules of pyruvate.
    (B) Glycolysis can produce a net total of 2 ATPs.
    (C) The end-product of glycolysis can form ethanol, lactate, or acetyl CoA.
    (D) During glycolysis, $FADH_2$ is produced.
    (E) During glycolysis, NADH is produced.

22. The bicarbonate ion in the digestive tract

    (A) neutralizes stomach acid
    (B) promotes phagocytosis by white blood cells
    (C) carries oxygen to the lungs
    (D) functions in the blood clotting mechanism
    (E) none of the above

23. Which of the following is NOT characteristic of fermentation?

    (A) It is anaerobic.
    (B) It requires glucose.
    (C) It produces energy.
    (D) It requires oxygen.
    (E) It produces ethanol.

24. If a tracer substance is injected into a patient's superior vena cava, which of the following structures would the tracer reach last?

    (A) the right ventricle
    (B) the left ventricle
    (C) the pulmonary veins
    (D) the left atrium
    (E) the right atrium

GO ON TO THE NEXT PAGE

25. Which of the following is the pacemaker of the heart?

(A) the foramen ovale
(B) the sinoatrial node
(C) the ductus arteriosus
(D) the bundle of His
(E) the vagus nerve

26. $CO_2 + H_2O \longrightarrow C_6H_{12}O_6 + O_2$

The above reaction is catalyzed by

(A) light
(B) ADP
(C) CO_2
(D) chlorophyll
(E) none of the above

27. If a diabetic accidentally overdosed on insulin, which of the following would be likely to occur?

(A) increased levels of glucose in the blood
(B) increased glucose concentration in urine
(C) dehydration due to increased urine excretion
(D) increased conversion of glycogen to glucose
(E) increased conversion of glucose to glycogen

28. The best description of identical twins is that they are

(A) twins of the same sex
(B) twins from a single egg
(C) twins from two eggs that have been fertilized by the same sperm
(D) twins from two eggs fertilized by two separate sperm
(E) twins from a single egg fertilized by two separate sperm

29. Which of the following lacks a backbone?

(A) the duckbill platypus
(B) the turtle
(C) the amphioxus
(D) the trout
(E) the rabbit

30. A stable ecosystem

I. is self-sustaining
II. cycles materials between biotic and abiotic components
III. requires a constant energy source
IV. will have a high mutation rate

(A) I only
(B) I and II
(C) I, II, and III
(D) I, II, and IV
(E) I, II, III, and IV

GO ON TO THE NEXT PAGE

31. Ingestion of the insecticide Parathion, which blocks acetylcholinesterase function, would cause

(A) a decrease in postsynaptic signals
(B) a halt to all synaptic nervous transmissions
(C) an increase in acetylcholine concentration in synapses
(D) a decrease in acetylcholine concentration in synapses
(E) levels of acetylcholine to remain the same

32. A cell is placed in a medium containing radioactively labeled thymine. If the cell undergoes two rounds of replication while in this medium, the radioactivity will appear

(A) in all the strands of DNA
(B) in half of the strands of DNA
(C) in the proteins produced
(D) in the mRNA
(E) in the rRNA

33. Straight tail (*T*) is dominant over bent tail (*t*) in mice, and long-tailed mice (*L*) are dominant over short-tailed mice (*l*). Which cross must produce all straight, long-tailed mice?

(A) *TtLl* x *TtLl*
(B) *Ttll* x *TTLl*
(C) *TtLL* x *ttLL*
(D) *TtLl* x *TTLL*
(E) none of the above

34. Which of the following are ectothermic animals?

(A) birds
(B) cats
(C) dogs
(D) humans
(E) none of the above

35. In adult humans, red blood cells

(A) carry oxygen to the tissues
(B) are replaced in the liver
(C) are outnumbered by white blood cells in the circulatory system
(D) are made in the spleen
(E) are the sites of rapid protein synthesis

36. Which of the following biomes are correctly paired?

(A) tundra: treeless frozen plain
(B) taiga: higher temperatures, torrential rains
(C) savanna: less than ten inches of rain per year
(D) desert: cold winters, warm summers, moderate rainfall
(E) none of the above

GO ON TO THE NEXT PAGE

37. Which statement regarding protein synthesis is false?

(A) tRNA molecules shuttle amino acids that are incorporated into the protein.
(B) Proteins are formed on the ribosomes.
(C) mRNA is not necessary for proper protein synthesis.
(D) Ribosomal RNA is needed for proper binding of the mRNA message.
(E) Ribosomes are found either in the cytoplasm or attached to the endoplasmic reticulum.

38. The notochord is

(A) present in all adult chordates
(B) present in all echinoderms
(C) present in chordates during embryonic development
(D) always a vestigial organ in chordates
(E) part of the nervous system of all vertebrates

39. Which of the following is an acceptable nitrogen base composition for double-stranded DNA?

(A) 31% A; 19% T; 31% C; 19% G
(B) 36% A; 36% U; 24% C, 24% G
(C) 48% A; 48% T; 52% C; 52% G
(D) 31% A; 31% T; 19% C; 19% G
(E) 24% A; 24% U; 36% C; 36% G

40. A typical human gamete

I. contains a haploid number of genes
II. will always contain an *X* or a *Y* chromosome
III. is a result of mitosis
IV has undergone genetic recombination

(A) I and II
(B) I and III
(C) II and III
(D) II, III, and IV
(E) I, II, and IV

41. Which of these ecosystems has the largest population of primary producers?

(A) tundra
(B) stagnant pond
(C) rocky meadow
(D) deciduous forest
(E) desert

42. The absorption of oxygen in the capillaries of the lungs takes place in the

(A) pulmonary artery
(B) pulmonary vein
(C) alveoli
(D) trachea
(E) bronchi

GO ON TO THE NEXT PAGE

43. Which of the following is a substance secreted by a member of a species that affects other members of the same species?

(A) gender-specific proteins
(B) lacrimal fluid
(C) enzymes
(D) hormones
(E) pheromones

44. The rate of breathing is controlled by involuntary centers in the

(A) cerebrum
(B) cerebellum
(C) medulla oblongata
(D) spinal cord
(E) hypothalamus

45. Enzymes

I. are proteins
II. typically work best at pH 7.2
III. are changed during a reaction
IV. are found in the nucleus only

(A) I only
(B) II only
(C) I and II
(D) I, II, and III
(E) I, II, III, and IV

46. Which of the following occurs in the cell nucleus?

I. RNA synthesis
II. protein synthesis
III. DNA synthesis

(A) I only
(B) II only
(C) III only
(D) I and III
(E) I, II, and III

47. The genetic code is considered degenerate because

(A) more than one codon can code for a single amino acid
(B) one codon can code for multiple amino acids
(C) more than one anticodon can bind to a given codon
(D) only one anticodon can bind to a given codon
(E) none of the above

GO ON TO THE NEXT PAGE

48. Which statement about the plasma membrane is false?

(A) It serves as a selectively permeable barrier to the external environment.
(B) It serves as a mediator between the internal and external environments.
(C) In eukaryotes, it contains the cytochrome chain of oxidative phosphorylation.
(D) It contains phospholipids as a structural component.
(E) It contains proteins which in some cases span the membrane.

49. Which of the following is a type of genetic mutation?

(A) point
(B) silent
(C) insertion
(D) frameshift
(E) all of the above

50. Pancreatic exocrine secretions contain all of the following EXCEPT

(A) proteases
(B) lipases
(C) amylases
(D) glucagon
(E) bicarbonate ions

Questions 51–55 refer to the figures below, which depict the most significant stages of embryonic development.

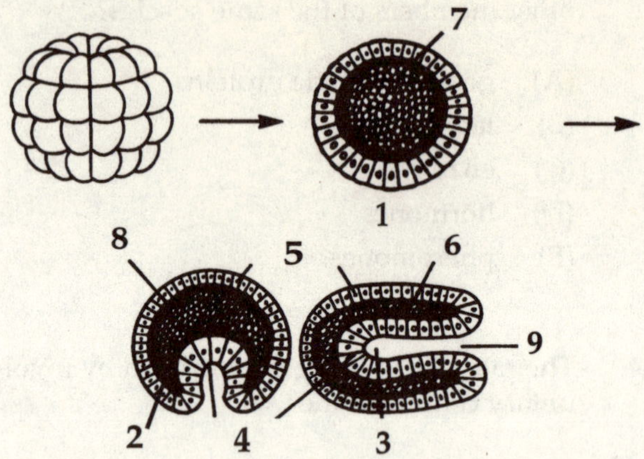

51. Which structure develops into the skin and nervous systems?

(A) 2
(B) 3
(C) 4
(D) 5
(E) 6

52. Which structure develops into the circulatory and muscular systems?

(A) 2
(B) 3
(C) 4
(D) 5
(E) 6

GO ON TO THE NEXT PAGE

53. Which structure is referred to as the blastocoel?

(A) 1
(B) 2
(C) 5
(D) 6
(E) 7

54. A problem with which structure would affect lung development?

(A) 2
(B) 3
(C) 4
(D) 5
(E) 6

55. Which structure has only two germ layers present?

(A) 1
(B) 2
(C) 3
(D) 5
(E) 6

Questions 56–59 refer to the figure of the vascular system of a plant below.

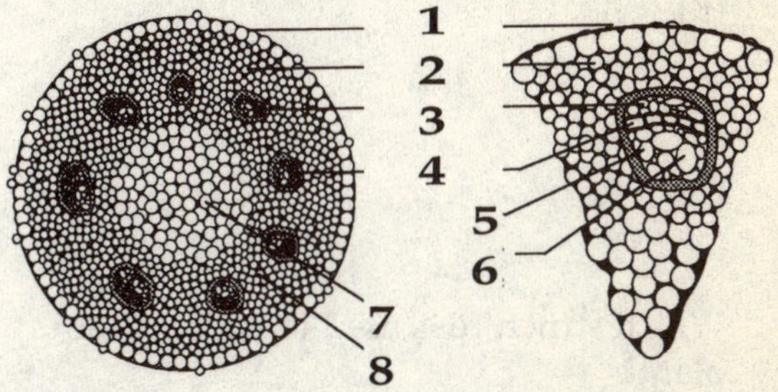

56. Which structure is responsible for the transport of nutrients?

(A) 2
(B) 3
(C) 4
(D) 5
(E) 6

57. Which structure is responsible for the transport of water and minerals?

(A) 2
(B) 3
(C) 4
(D) 5
(E) 6

GO ON TO THE NEXT PAGE

58. Which structure is made up of rapidly dividing, undifferentiated cells?

(A) 2
(B) 3
(C) 4
(D) 5
(E) 6

59. Which structure is used to support the plant?

(A) 4
(B) 5
(C) 6
(D) 7
(E) 8

Part B

Each set of choices A–E below should be compared to the numbered statements that follow it. Choose the lettered choice that best matches each numbered statement. Fill in the correct oval on your answer sheet. Remember that a choice may be used once, more than once, or not at all in each set.

Questions 60–63:

(A) virus
(B) bacteria
(C) amoeba
(D) planaria
(E) sponge

60. may contain plasmid DNA

61. has a cell membrane but lacks a true nucleus

62. genetic material can be either DNA or RNA

63. bilaterally symmetrical

Questions 64–68:

(A) glycolysis
(B) fermentation
(C) Krebs cycle
(D) electron transport chain
(E) photosynthesis

64. utilizes a proton pump

65. occurs in the mitochondrial matrix

66. is an anaerobic process that forms NAD^+

67. occurs in the inner membrane of the mitochondria

68. creates byproducts of ethanol or lactic acid

Questions 69–72:

(A) mitochondria
(B) lysosome
(C) Golgi apparatus
(D) endoplasmic reticulum
(E) nucleolus

GO ON TO THE NEXT PAGE

69. the site of rRNA production

70. thought to be a result of the symbiotic relationship between eukaryotic cells and bacterial cells

71. digestive vacuole of the cell

72. packages and modifies secretory proteins

Questions 73–76:

(A) musculoskeletal system
(B) endocrine system
(C) circulatory system
(D) nervous system
(E) reproductive system

73. transports respiratory gases, nutrients, and wastes

74. enables organisms to receive and respond to stimuli

75. is a basic internal framework

76. performs internal communication and coordinates activities of the organ systems

Questions 77–80:

(A) adenine
(B) guanine
(C) thymine
(D) cytosine
(E) uracil

77. purine that binds with three hydrogen bonds

78. found only in RNA

79. pyrimidine that binds with three hydrogen bonds

80. purine found in both DNA and RNA that binds with two hydrogen bonds

Part C

Each of the following sets of questions is based on a laboratory or experimental situation. Begin by studying the description of each situation. Next, choose the best answer to each of the questions that follow it. Fill in the corresponding oval on your answer form.

In 1952, Alfred Hershey and Martha Chase conducted a set of experiments to determine whether DNA or proteins were the genetic material of living organisms. The original experiments took advantage of the fact that sulfur is not a component of DNA but is found in most proteins, while phosphorous is a component of DNA and not of proteins. Variations of the Hershey-Chase experiments have been performed by other researchers, as described below. These experiments used lambda bacteriophage, which is a DNA virus. Lambda bacteriophage has both a lytic cycle and a lysogenic cycle.

Experiment 1:

The protein coat of lambda bacteriophage was labeled with radioactive sulfur, ^{35}S. This phage culture was then allowed to infect a culture of *E. coli*. The phage carcasses, called ghosts, were separated from the bacterial cells before any virions could be produced. The bacterial cells were separated by centrifugation and the radioactivity was measured.

GO ON TO THE NEXT PAGE

Experiment 2:

The phage DNA was labeled with radioactive phosphorus, ^{32}P. This phage culture was then allowed to infect *E. coli.* The phage carcasses, called ghosts, were separated from the bacterial cells before any virions could be produced. The bacterial cells were separated by centrifugation and the radioactivity was measured.

Use this information to answer questions 81–85 below.

81. In Experiment 1, the radio-labeled sulfur subsequently appeared in

(A) *E. coli* proteins
(B) *E. coli* chromosomes
(C) lambda ghost proteins
(D) lambda ghost DNA
(E) none of the above

82. In Experiment 2, most of the radioactivity was found in

(A) lambda ghosts
(B) *E. coli* cells
(C) *E. coli* progeny
(D) lambda progeny
(E) none of the above

83. Suppose a new organism was discovered that contained no sugar molecules, either as free molecules or as chemical building blocks, but was similar in all other regards to a typical eukaryotic cell. Which of the following conclusions could be drawn about this novel organism?

(A) The organism's genetic information is not encoded in DNA.
(B) The organism does not have any membranes.
(C) The organism cannot use facilitated diffusion as a mechanism of membrane transport.
(D) The organism uses ATP to store chemical energy.
(E) The organisms uses RNA to code for proteins.

GO ON TO THE NEXT PAGE

84. ^{32}P-labeled phage infected a culture of *E. coli* and entered a lysogenic cycle. The radioactivity of a 1 mL sample of the culture with a concentration of 1 x 105 cells/mL was measured. The *E. coli* were then allowed to undergo 3 rounds of replication. What fraction of the initial radioactivity would be present in a 1 mL sample, which contains cells diluted to a final concentration of 1 x 105 cells/mL, from the culture containing the final generation of *E. coli*?

(A) $\frac{1}{8}$
(B) $\frac{1}{4}$
(C) $\frac{1}{10}$
(D) $\frac{1}{3}$
(E) $\frac{1}{2}$

85. After centrifugation of the *E. coli* using standard laboratory protocols, all of the following structures would be found EXCEPT

(A) ribosomes
(B) DNA
(C) mitochondria
(D) cell walls
(E) RNA

A stable population exists in Hardy-Weinberg equilibrium with two alleles, *T* and *t*. *TT* and *Tt* individuals have the ability to curl their tongues, while *tt* individuals cannot curl their tongues. The allele frequency of *T* is 0.8.

The data above will help you answer questions 86–89.

86. What is the allele frequency of *t*?

(A) 0.04
(B) 0.2
(C) 0.8
(D) 0.16
(E) 0.64

87. What is the percentage of heterozygotes?

(A) 4%
(B) 32%
(C) 64%
(D) 50%
(E) 75%

88. What is the percentage of individuals that can curl their tongues?

(A) 32%
(B) 4%
(C) 64%
(D) 96%
(E) 50%

GO ON TO THE NEXT PAGE

89. Which of the following will keep this population in Hardy-Weinberg equilibrium?

(A) a large population
(B) geographic isolation
(C) random mating
(D) no net mutations
(E) all of the above

A scientist grew her bacterial colonies in a variety of different concentrations of nutrients to determine at which concentration they grew best. The results of her experiments are summarized in the following bar graph.

GROWTH BAR CHART

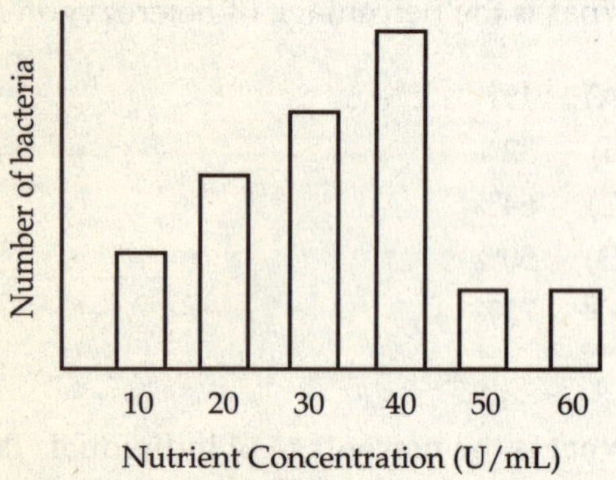

Now answer questions 90–91.

90. What is the optimum concentration of nutrient supplement?

(A) 10 U/mL
(B) 20 U/mL
(C) 30 U/mL
(D) 40 U/mL
(E) 50 U/mL

91. At 50 U/mL and greater, what happens to the bacteria?

I. They begin to mutate.
II. They lose water and shrivel up, due to the high osmotic potential of the concentrated nutrient solution.
III. They enter dormancy.

(A) I only
(B) II only
(C) III only
(D) I and II
(E) I, II, and III

A mouse was put in a maze and timed to determine how long it took him to find the food pellet at the end. A total of eight trials were done until the researchers were confident that the mouse had learned the maze. The data follows in both tabular form in the table and graphical form.

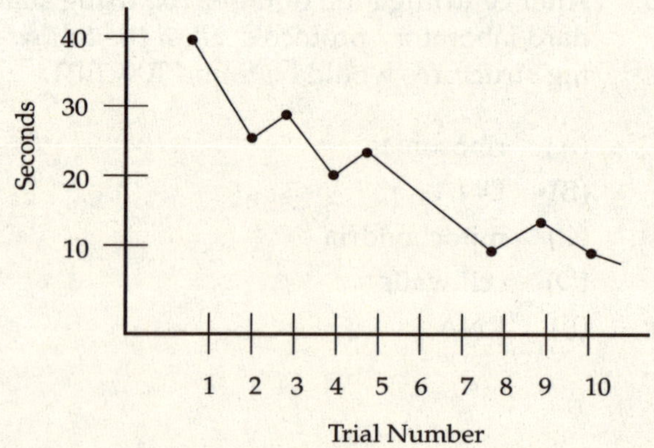

GO ON TO THE NEXT PAGE

Trial Number	Number of Seconds
1	40
2	20
3	25
4	16
5	17
6	10
7	12
8	10

Use this data to answer questions 92–93.

92. The greatest improvement is shown in the

(A) first trial
(B) second trial
(C) third trial
(D) fourth trial
(E) fifth trial

93. The mouse learned the maze after trial number

(A) 4
(B) 5
(C) 6
(D) 7
(E) 8

The solid line in the figure below shows the course of the reaction A + B —> C + D in the absence of a catalyst.

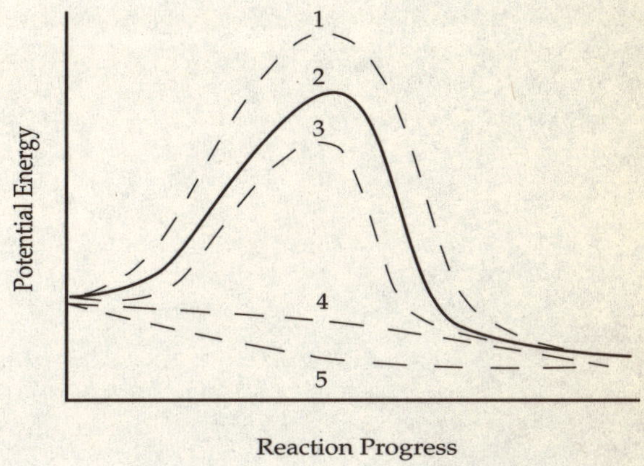

Questions 94–95 refer to the data above.

94. Which of the dotted lines would represent the course of the same reaction in the presence of a catalyst?

(A) 1
(B) 2
(C) 3
(D) 4
(E) 5

95. Which line would represent changing the pH from 7.2 to 5.0?

(A) 1
(B) 2
(C) 3
(D) 4
(E) 5

Answer Key

1. D
2. E
3. D
4. A
5. E
6. E
7. E
8. D
9. B
10. D
11. C
12. D
13. A
14. B
15. D
16. B
17. E
18. D
19. B
20. C
21. D
22. A
23. D
24. B
25. B
26. D
27. E
28. B
29. C
30. C
31. C
32. A
33. D
34. E
35. A
36. A
37. C
38. C
39. D
40. E
41. D
42. C
43. E
44. C
45. C
46. D
47. A
48. C
49. E
50. D
51. D
52. E
53. E
54. C
55. B
56. A
57. D
58. C
59. D
60. B
61. B
62. A
63. D
64. D
65. C
66. B
67. D
68. B
69. E
70. A
71. B
72. C
73. C
74. D
75. A
76. B
77. B
78. E
79. D
80. A
81. C
82. A
83. A
84. A
85. C
86. B
87. B
88. D
89. E
90. D
91. B
92. B
93. C
94. C
95. A

Compute Your Practice Test Score

Step 1: Figure out your raw score. Refer to your answer sheet for the number right and the number wrong on the practice test you're scoring. (If you haven't checked your answers, do that now, using the answer key that follows the test.) You can use the chart below to figure out your raw score. Multiply the number wrong by 0.25 and subtract the result from the number right. Round the result to the nearest whole number. This is your raw score.

SAT II: Biology Practice Test

Number right		Number wrong		Raw score
☐	–	(0.25 x ☐)	=	☐ (rounded)

Step 2: Find your practice test score. Find your raw score in the left column of the table below. The score in the right column is your Practice Test score.

Find Your Practice Test Score

Raw	Scaled	Raw	Scaled	Raw	Scaled	Raw	Scaled	Raw	Scaled	Raw	Scaled
95	800	78	690	61	590	43	490	26	390	9	290
94	800	77	680	59	580	42	480	25	380	8	290
93	790	76	670	58	570	41	470	24	380	7	280
92	780	75	670	57	570	40	470	23	370	6	280
91	770	74	660	56	560	39	460	22	370	5	270
90	760	73	660	55	550	38	460	21	360	4	260
89	750	72	650	54	550	37	450	20	360	3	260
88	740	71	640	53	540	36	450	19	350	2	250
87	740	70	640	52	540	35	440	18	350	1	240
86	730	69	630	51	530	34	430	17	340	0	240
85	730	68	630	50	530	33	430	16	330	–1	230
84	720	67	620	49	520	32	420	15	330	–2	230
83	720	66	620	48	510	31	420	14	320	–3	220
82	710	65	610	47	510	30	410	13	320	–4	220
81	710	64	610	46	500	29	410	12	310	–5	210
80	700	63	600	45	500	28	400	11	300	–6	210
79	690	62	590	44	490	27	400	10	300	–7– –9	200

Practice Test Score

☐

A note on your practice test scores: Don't take these scores too literally. Practice test conditions cannot precisely mirror real test conditions. Your actual SAT II: Biology Subject Test score will almost certainly vary from your practice test scores. Your scores on the practice tests will give you a rough idea of your range on the actual exam.

Explanations to Practice Test Two

1. (D) The potential gametes are *ABC, ABc, AbC, Abc, aBC, aBc, abC*, and *abc*—a total of eight gametes.

2. (E) All forms of life, from bacteria to man, have certain things in common. They all have some form of genetic material as well as proteins, and they are composed largely of water.

3. (D) A defect in the ectoderm would affect the development of the skin, eyes, and nervous system.

4. (A) Clotting occurs when platelets in an open wound release thromboplastin, which initiates a series of reactions ultimately leading to the formation of a fibrin clot. Thromboplastin, with the aid of calcium and vitamin K as cofactors, converts inactive plasma prothrombin to the active form thrombin in a series of steps. Thrombin in its turn converts fibrinogen (dissolved in plasma) into the fibrinous protein fibrin. Finally, threads of fibrin trap red blood cells to form clots.

5. (E) Pepsin, trypsin, carboxypeptidase, and chymotrypsin all digest proteins. Gastrin is a hormone released by the pyloric mucosa when food enters the stomach. It stimulates the secretion of gastric juices.

6. (E) The bony skeleton serves as a framework within all vertebrate organisms. Muscles are attached to the bones, permitting movement. The skeleton also provides protection for vital organs. For example, the rib cage protects the heart and the lungs, while the skull and vertebral column protect the brain and the spinal cord. The hollow cavity formed within each bone is subsequently filled with bone marrow, which is the site of formation of blood cells. Although all hematopoietic cells are formed in the bone marrow, *T* lymphocytes are matured in the thymus.

7. (E) The endosymbiotic hypothesis states that blue-green algae entered into a symbiotic arrangement with early eukaryotic plants to develop into chloroplasts, while bacteria entered into a similar arrangement with eukaryotic animal cells to become mitochondria. Chloroplasts and mitochondria have a plasma membrane (their inner membrane) and the ability to produce their own proteins without utilizing the cell's machinery. In addition, they have circular DNA, and their rRNA subunits are characteristic of prokaryotes.

8. (D) Genes supply the hereditary background, but the environment determines the way in which a certain gene is expressed. Thus, genetic and environmental factors interact to produce the phenotype of the individual. (A), (B), and (C) are all examples of the environment affecting what genes are transcribed. (D), on the other hand, is an example of a change in environment causing homeostasis to be maintained. In shivering, muscles contract rapidly to generate heat.

9. (B) If the electron transport chain is blocked in a cell that utilizes cellular respiration to produce ATP, then the amount of ATP that can be produced is decreased. This is because the cell is only able to produce ATP through glycolysis and through the Krebs cycle. The NADH and $FADH_2$ produced by the Krebs cycle will not lead to the production of any ATP.

10. (D) The large amounts of air that come into the upper respiratory passages permit water evaporation from the mucosal surfaces, thereby allowing for heat loss. Panting is turned on by the thermoregulatory centers of the brain that monitor blood temperatures.

11. (C) As blood passes through the capillary tuft of the glomerulus, it is filtered as it passes through the capillary walls. Only blood cells and protein remain in the blood, while water, salts, glucose, and amino acids are filtered into the Bowman's capsule. Resorption of amino acids, glucose, and salts occurs predominantly in the proximal convoluted tubules. The cells lining the tubules actively transport these material out of the tubular lumen and into the peritubular capillary network. Movement of these materials produces an osmotic gradient that drives some water out of the tubules through simultaneous passive diffusion. In (D), villi are small projections in the walls of the small intestine that increase the surface area to facilitate absorption of nutrients. As for (E), the ureter is a duct that carries urine from the kidneys to the bladder.

12. (D) A climax community is the final succession in an ecosystem. It will be stable within a given environment, as long as too much interference from human or natural forces, such as pollution or forest fires, does not take place.

13. (A) In this question, you can disregard the *A* allele because all gametes will be *A*. This means that there are a total of four potential gametes—*A* in combination with *BC*, *Bc*, *bC*, and *bc*.

14. (B) The ductus arteriosus shunts blood from the pulmonary artery to the aorta in order to divert this blood away from the developing fetal lungs.

15. (D) Cardiac cells have intercalated disk connections, termed protoplasmic connections, between cells. Although they have some striations, they are not voluntary. Smooth muscle, on the other hand, has no striations, is mononuclear, and is involuntary. White blood cells have nuclei. In adult humans, red blood cells lack nuclei, in order to make room for as much hemoglobin as possible.

16. (B) A cell that is n (haploid) cannot undergo meiosis to become $\frac{1}{2}n$. (A), (C), and (D) are all incorrect, meanwhile, because there are a number of organisms that are haploid, and these organisms undergo mitosis to divide and grow. An example of such an organism is a braconid wasp. These animals in the haploid form are males (n); females ($2n$) are only formed when a female mates. As for (E), an organism, whether diploid or haploid, must be able to digest to maintain life.

17. (E) This question illustrates two types of test cross. A test cross is performed to determine if a particular dominant individual's phenotype is a homozygous or heterozygous genotype. In this case, there are two possible tall genotypes—*TT*, the pure homozygous tall, and *Tt*, the hybrid heterozygous tall. These two individuals would have the same phenotype. In order to determine the genotype, the unknown tall plant would be mated with a recessive, short plant. If the tall plant and the short plant produce only tall offspring, then it can be assumed that the original tall plant is homozygous. If the mating of the unknown tall organism and the short organism produces any short offspring at all, then we know that the original unknown tall was heterozygous. This is because short offspring can be produced only when one short gamete is produced from each parent. In I, crossing an unknown with a *Tt* will also give you the information you seek, because if any short progeny are produced at all, your unknown must be heterozygous. If all of the offspring are tall, your unknown must be homozygous dominant.

18. (D) The question is simply asking you where growth occurs in long bone. The answer is in the epiphyseal plates. The epiphyseal plates are regions of cartilaginous cells separating the shaft of the long bone, called the diaphysis, from its two dilated ends, called the epiphyses. The epiphyseal plates are located at either end of long bone, which are referred to as the proximal and distal ends. Since growth occurs only at the epiphyseal plates, the strontium would be incorporated near the plates.

19. (B) Blood leaving the kidneys travels through the inferior vena cava before entering the right atrium. All of the blood from the lower half of the body is collected in the inferior vena cava. This vessel merges with the superior vena cava (which collects blood from the upper half of the body) immediately before it enters the right atrium. In (A), the iliac veins circulate blood in each leg and return it to the more central portions of the body before joining to become the common iliac vein, which then enters the inferior vena cava. As for (C) and (D), blood enters the liver through the hepatic portal artery after absorbing monosaccharides and amino acids from the small intestine. It leaves the liver through the hepatic portal vein after filtration occurs. Finally, in (E), the pulmonary vein brings blood back from the lungs to the left atrium. This blood has already reached the heart via the right atrium.

20. (C) The vitamin K we need for blood coagulation is synthesized by the intestinal bacteria residing in our colons. Vitamin K is required for the liver to synthesize some important blood factors. Deficiency of Vitamin K leads to defective blood clotting.

21. (D) The first set of respiratory reactions are anaerobic and involve the breakdown of glucose into pyruvate (A). Two ATP molecules are required and four are produced, leaving a net total of 2 ATP (B). Pyruvic acid will either become acetyl CoA and enter the Krebs cycle during aerobic respiration, or will become lactic acid or ethanol in anaerobic respiration (C). Two NADH molecules are produced by glycolysis; these will donate their electrons to compound Q in the electron transport chain (E). $FADH_2$, however, is not produced until the Krebs cycle.

22. (A) Bicarbonate ion acts as a buffer to maintain the pH of the blood. For example, metabolic alkalosis will cause hyperventilation, which will effectively increase the concentration of CO_2 in the blood. This in its turn will lead to the production of more carbonic acid, decreasing the pH of the blood. Bicarbonate ion also serves as a mechanism for transporting CO_2 from the tissue to the lungs for exhalation, but it does not carry O_2 (C). As for the remaining choices, (B) and (D) are not affected by levels of bicarbonate in the blood, while (E) is untrue because (A) is true.

23. (D) Fermentation is a process that occurs during anaerobic respiration in organisms such as yeast. After the yeast begins the respiration process by changing glucose to pyruvic acid, gaining some net ATPs in the process, the pyruvic acid is then changed into ethyl alcohol, a waste product of the fermentation process. Fermentation does produce energy, as described in the first set of reactions in which glucose is changed into pyruvic acid. It does not, however, require oxygen. The oxygen-requiring reactions of respiration occur in aerobic respiration, not in fermentation, and they occur as NADH and $FADH_2$ molecules produced in the Krebs cycle are sent to the electron-transport chain for the production of ATPs. The final electron acceptor in these reactions is oxygen. However, in fermentation, there is no aerobic stage, and ATP is not produced through a Krebs cycle and electron transport chain mechanism.

24. (B) This question is a matter of knowing the pathway that blood travels in the heart. Deoxygenated blood drains into the right atrium from both the inferior vena cava and the superior vena cava. From the right atrium, the blood flows into the right ventricle, which then pumps it to the lungs via the pulmonary arteries. Carbon dioxide is exchanged for oxygen in the alveoli of the lungs. Oxygenated blood is returned to the left atrium via the pulmonary veins. From the left atrium, the blood drains into the left ventricle, which pumps it into the aorta for circulation throughout the body. Hence if a tracer substance is injected into the superior vena cava, it would take the longest amount of time to reach the left ventricle.

25. (B) The heartbeat begins in the sinoatrial node located in the wall of the right atrium at the approximate point at which the vena cava enters and travels through the atria. It is then picked up by the AV nodes and carried to the AV bundle (also known as the bundle of His) and transported through the ventricles through the Purkinje fibers. (A) and (C) are structures in the fetal heart that ensure that blood is shunted away from the fetus' developing lungs. The vagus nerve (E) is a cranial nerve that regulates the heartbeat in accordance with signals from the parasympathetic nervous system, although it does not determine the heartbeat. The SA node is able to maintain the heartbeat without any stimulation from the nervous system.

26. (D) Chlorophyll is a green pigment that performs essential functions in photosynthesis, donating electrons and capturing light. When photons of light strike chlorophyll molecules, those molecules transfer the energy of the light to their electrons. This energy is then transferred to ADP to form ATP, to NADP to form NADPH, and to H_2O, which splits to form $2H^+$ and $\frac{1}{2}$ O_2.

27. (E) Insulin is the hormone secreted by the beta cells of the pancreas in response to high blood glucose levels. Insulin decreases blood glucose by stimulating cells to uptake glucose, and by stimulating the conversion of glucose into its storage form, glycogen, in the liver and muscle cells. An overdose of insulin can, and often does, lead to a sharp decrease in blood glucose concentration.

28. (B) Identical twins are produced when a zygote formed by one egg and one sperm splits during the four- or eight-cell stage to develop into two genetically identical organisms. Identical twins (A) will always be of the same sex, but fraternal twins can also be of the same sex. The twins in (D) are termed fraternal twins and are no more genetically alike than siblings. (C) and (E), meanwhile, are impossible events.

29. (C) An amphioxus is a chordate but not a vertebrate. Chordates have a stiff dorsal rod called the notochord during a certain period of their embryological development, and amphioxus and tunicate worms do not lose their notochords. Chordates also have paired gill slits, a dorsal hollow nerve cord, and a tail extending beyond the anus at some point during development. Vertebrates, meanwhile, have bones called vertebrae that form the backbone. Bony vertebrae replace the notochord of the embryo and protect the nerve cord. Mammals, amphibians, reptiles, birds, and fish all possess these vertebrae. Let's touch on the other choices here: the duckbilled platypus (A) is an example of an egg-laying mammal; the turtle (B) is an example of a reptile; the trout (D) is a fish, and rabbits (E) are mammals.

30. (C) A stable ecosystem is by nature self-sustaining (I). It needs no outside input of materials. It cycles its components, such as nitrogen and carbon, between its biotic and abiotic components as both elemental and "fixed" components (II). It also requires a constant energy source, such as the sun (III). Mutation rates, however, are not affected by an ecosystem's stability or instability (IV).

31. (C) Acetylcholine is a neurotransmitter, which causes depolarization of the postsynaptic membrane of one neuron when released by the presynaptic terminal of another neuron. Acetylcholine is removed from the synapse by way of the enzyme acetyl cholinesterase. Anticholinesterases like parathion disrupt this activity, meaning that the acetylcholine can't be degraded. And this in its turn implies that the concentration of acetylcholine in the synapse will increase.

32. (A) DNA replication is semiconservative. In consequence, after the DNA has been replicated in the medium containing a radioactive tracer, every helix of DNA will have at least one strand that will have incorporated some of the radioactivity.

33. (D) If a homozygous dominant, *TTLL*, is crossed with any genotype, the offspring will be heterozygous and have the dominant phenotype. All the other options have potential for some homozygous recessives.

34. (E) Poikilothermic animals, also known as ectothermic or cold-blooded animals, do not possess the ability to maintain a consistent internal body temperature. Poikilothermic animals include fish, reptiles, amphibians, and most of the lower animals. They are sensitive to the outside environmental temperature, and are much more active in warmer temperatures and less active in colder temperatures. Any animals that are not ectothermic are called warm-blooded (alternatively, endothermic or homeothermic). These include all mammals and birds.

35. (A) Red blood cells are produced in the bone marrow. They lose their nuclei to make room for more hemoglobin, which means that they cannot reproduce, repair themselves, or make proteins. Red blood cells actually greatly outnumber leukocytes (white blood cells) (C); as for (D), the spleen stores a reservoir of red blood cells and acts as a biological and physical filter for the blood, but it does not make red blood cells.

36. (A) The tundra is a treeless frozen plain between the taiga and the northern ice sheets. The taiga (B) is a spruce-filled forest inhabited by moose and bear, while the savanna (C) is a grassland with low rainfall and populated with hoofed herbivores and carnivorous predators. Finally, the desert (D) receives less than ten inches per year of rainfall and is populated by small plants and animals.

37. (C) Protein synthesis does require energy. tRNA (A) brings the amino acid to the ribosome, where it interacts with the mRNA that has the appropriate sequence. As for (B), tRNA molecules do in fact have an amino acid bound to their 3′ end. The mRNA is read from 5′ to 3′ as the ribosome moves along the message. (D) and (E) are also both true.

38. (C) The notochord appears as a semirigid chord in the dorsal part of all chordates sometime during embryonic development. In lower chordates, this chord remains as a semirigid chord, although in higher chordates it is seen only in the embryo and not as a vestigial organ. Echinoderms (B) do not possess a notochord. As for (D), the notochord remains in the lower chordates, such as the amphioxus and the tunicate worm, while the notochord (E) is not part of the nervous system.

39. (D) Because DNA is double stranded and because of the rules of complementary base pairing, the quantity of adenine must equal the quantity of thymine, and the quantity of cytosine must equal the quantity of guanine. The reason for this is that in DNA, adenine will bind only with thymine, while cytosine binds only with guanine. After eliminating the answer choices that do not follow this rule, you can eliminate any answer choices that mention uracil, since it is found only in RNA. Finally, to reach the correct answer, you must add up your percentages and make sure that they equal 100 percent of the bases in the DNA strand.

40. (E) During meiosis, the gamete reduces its genetic component from $2n$ to n, resulting in a haploid cell with half the normal chromosome number. When a haploid egg and sperm unite, they form a diploid organism known as a zygote. All ova will contain an X chromosome and all sperm will contain either an X or a Y chromosome. These gametes are formed during the two reductional divisions called meiosis. During Metaphase I of Meiosis I, tetrads form, and sister chromatids undergo the homologous recombination known as crossing over.

41. (D) The deciduous forest, full of trees, shrubs, and mosses, would have the greatest number of photosynthetic plants. A stagnant pond (B) lacks enough oxygen to reach its growth potential, while the other options are terrains not conducive to plant growth.

42. (C) Alveoli are thin air sacs that act as the sites of air exchange between the environment and the blood via passive diffusion. The bronchi (E) are the two main branches of the air intake pathway. One bronchus goes to each lung, and each bronchus is divided into smaller sections termed bronchioles.

The trachea (D) is the region of the air intake pathway located between the glottis and the bronchi. It is also known as the windpipe.

43. (E) Pheromones like primer or releaser pheromones are substances secreted by organisms that result in long-term change in the behavior of other members of that organism's species.

44. (C) The breathing center in the medulla oblongata monitors the increase in CO_2 through its sensory cells. It will also detect a decrease in pH in the blood, which is indicative of an increase of CO_2 levels in the blood. A decrease in O_2 is monitored peripherally by chemoreceptors, located in the carotid bodies in the carotid arteries and in the aortic bodies in the aorta. In (A), the cerebrum is involved in sensory interpretation, memory, and thought, while the cerebellum (B) is involved in fine motor coordination, balance, and equilibrium. Finally, the spinal cord (D) relays sensory and motor information to and from the brain, and the hypothalamus (E) regulates hunger, thirst, body temperature, sex drive, and emotion.

45. (C) Enzymes are necessary to catalyze the reactions that allow the breakdown of glucose into ATP. Enzymes are typically proteins (I), although they may rarely be RNA, and they typically act most effectively at a physiological pH of 7–7.4 (except for the enzymes which break down protein in the stomach, which act most effectively in an acidic environment) (II). Enzymes do not always require an coenzyme to function. They usually require a cofactor, however, and that cofactor may be either an organic molecule coenzyme or a metal ion such as copper or iron. Enzymes are never changed during a reaction (III). They increase the rate of a reaction, but are not themselves affected by that reaction.

46. (D) In the nucleus, DNA is produced during cell division, while RNA is transcribed. mRNA travels from the nucleus into the cytoplasm, where it is translated into polypeptides on the ribosomes.

47. (A) Genetic codes are referred to as "degenerate" because they are not precise or specific. More than one codon can code for each of the different amino acids. For example, both UAU and UAC code for tyrosine. Since there are three bases in a codon, there are 64 possible codons and only 20 amino acids. Obviously, there will be some repetition of amino acids among the codons. (B) is the opposite of answer choice (A) and therefore untrue. It is essential that the anticodon be the exact base pair complement of the codon to maintain protein integrity. (D) is true, but it does not account for the degeneracy of the genetic code.

48. (C) The plasma membrane separates the cellular contents from the environment. It is responsible for the permeability of the membrane—in other words, for what is allowed in and out. The fluid mosaic model of the plasma membrane states that this membrane is a bilayer of phospholipid interspersed with proteins acting as receptors, pores, and channels. The pores and channels cross the entire membrane. The cytochrome chain referred to in the correct answer is actually located in the cristae (the inner membrane) of the mitochondria.

49. (E) A point mutation changes the sequence of one nucleotide; a silent mutation (B), on the other hand, affects the DNA sequence, but does not affect the protein produced from that DNA sequence. An insertion (C) is an addition of nucleotides which often leads to a frameshift mutation (D) in which the reading frame of the DNA is altered and the protein produced is rendered nonfunctional.

50. (D) As an exocrine gland, the pancreas secretes proteases, lipases, and amylases that aid in the digestion of food, as well as bicarbonate ion that buffers the pH of the chyme coming from the stomach. Glucagon, meanwhile, is an endocrine gland secreted by the pancreas in response to a low blood glucose level. It causes an increase in the levels of glucose by degrading glycogen and decreasing the uptake of glucose by muscles.

51. (D) The skin, nervous system, and eyes develop from the ectoderm, the outside cell layer in both the two- and three-layer gastrula.

52. (E) The circulatory, musculoskeletal, and reproductive systems develop from the mesoderm, the middle layer of cells in the three-layer gastrula.

53. (E) The blastocoel is the hollow, fluid-filled center of the blastula.

54. (C) The lungs develop from the endoderm, the cell layer in the invagination of the two- and three-layer gastrula.

55. (B) The two-layer gastrula is the stage between the fluid-filled blastula and the three-layer gastrula. It has two germ layers.

56. (A) The phloem, the thin-walled cells on the outside of the vascular bundle, transport nutrients down the stem of the plant.

57. (D) The xylem, the thick-walled, usually hollow cells on the inside of the vascular bundles, transport water and minerals up the plant stem.

58. (C) The cambium is two cells thick and located in between the xylem and phloem. These actively dividing, undifferentiated cells give rise to the xylem and the phloem.

59. (D) The pith is the innermost tissue layer that is used for the storage of nutrients and plant support.

60. (B) Bacteria contain extrachromosal circular DNA called plasmids. They often code for genes that in their turn code for antibiotic resistance.

61. (B) Bacteria may also be characterized as prokaryotes. These organisms have a plasma membrane. They lack membrane-bound organelles and a true nucleus, although they may have a nucleiod region.

62. (A) Viruses are composed of genetic material, either DNA or RNA, enclosed in a protein coat called a capsid. They are obligate parasites in that they cannot replicate unless they have infected a host cell.

63. (D) Planaria are bilaterally symmetrical, which implies that they have two sides that are mirror images of each other.

64. (D) The electron transport chain uses a proton pump to drive the ATP synthase that produces ATP. It is found in the inner mitochondrial membrane of the mitochondria. The energy of the proton gradient is harvested by the cell through the production of large quantities of ATP.

65. (C) The Krebs cycle occurs in the mitochondrial matrix.

66. (B) Fermentation is anaerobic and regenerates NAD^+ for glycolysis.

67. (D) See the explanation to question 64.

68. (B) The byproducts of fermentation are ethanol (in prokaryotes) and some yeast and lactic acid (in most of the other eukaryotes, including humans). This accounts for the soreness of your muscles after a tough workout.

69. (E) rRNA is produced in the region of the nucleus known as the nucleolus.

70. (A) Mitochondria and chloroplasts are thought to have arisen after bacteria were phagocytosed by eukaryotic cells, initiating a symbiotic relationship between the two.

71. (B) The lysosome is a membrane-bound organelle with a low pH that acts as a digestive system for the cell by degrading proteins and other organic molecules.

72. (C) The Golgi apparatus receives translated proteins from the endoplasmic reticulum, modifies them (e.g., by adding sugars to produce glycoproteins), and packages them for secretion.

73. (C) The function of a circulatory system is to transport gases, nutrients, and wastes throughout the body.

74. (D) The nervous system receives input from the five senses. This input is analyzed by the central nervous system and acted upon by the motor neurons. This allows organisms to receive and respond to stimuli.

75. (A) The musculoskeletal system forms the basic framework for the human body. Everything else is either built upon this framework or protected within it.

76. (B) The hormones secreted by the various endocrine organs allow tissues and organs to communicate. This is crucial for the maintenance of homeostasis, reproduction, and other life functions.

77. (B) Guanine (along with adenine) is a purine that binds cytosine (a pyrimidine) with three hydrogen bonds.

78. (E) Uracil is a pyrimidine found only in RNA.

79. (D) See the explanation for question 77.

80. (A) Adenine (along with guanine) is a purine that binds thymine with two hydrogen bonds in DNA and uracil with two hydrogen bonds in RNA.

81. (C) When the ^{32}P labeled phage were used in Experiment 2, most of the radioactivity ended up inside the bacterial cells, indicating that the phage DNA entered the cells. If the phage had been allowed to enter a lytic cycle, the ^{32}P would also have been recovered in the phage progeny. In Experiment 1, when the ^{35}S-labeled phage were used, most of the radioactive material ended up in the phage ghosts, indicating that the phage protein never entered the bacterial cell. The conclusion to be drawn from these experiments is that DNA is the hereditary material, and that phage proteins are mere structural packaging that is discarded after delivering the viral DNA to the bacterial cell. The entire "life purpose" of a virus is devoted to finding a host cell and getting its nucleic acid inside it. This is crucial for a virus, since it must use the genetic machinery of a host cell to replicate. Since the protein was labeled with ^{35}S, you would expect to find this isotope of sulfur only in the viral ghost, which you're told is the phage carcass. And since the DNA was labeled with ^{32}P, you would expect to find this radioactive isotope only in the bacterial cells.

82. (A) See the explanation to question 81 above.

83. (A) DNA contains the sugar deoxyribose; remember that the acronym DNA stands for deoxyribonucleic acid. So, if this novel organism does not have any sugar molecules, then it follows that it can't have any DNA. If there isn't any DNA in this hypothetical organism, yet it somehow still exists, then it must be concluded that this organism's genetic information is not encoded in DNA.

84. (A) We know that the radio labeled phage DNA has been inserted into the bacterial chromosome and will undergo replication just as if it were native bacterial DNA. During DNA replication, the double helix unwinds, and each strand acts as a template for complementary base-pairing in the synthesis of two new daughter helices. Each new daughter helix contains an intact strand from the parent helix, which in this problem contains radioactivity, and a newly synthesized strand, which contains no radioactivity, since all of its nucleotides were synthesized by the bacterial cell. Thus DNA replication is semiconservative—half of the original DNA is conserved from one generation to the next. Since every bacterial cell division doubles the number of cells and the total amount of DNA, three rounds of replication will increase the total number of cells and the total amount of DNA by a factor of eight, while the amount of radioactivity remains constant. In essence, the ^{32}P is diluted with each round of replication. This means that the fraction of radioactivity in a 1 mL sample taken from the culture of the final generation will be one-eighth the amount of radioactivity taken from a culture of the parental generation.

85. (C) This questions tests your ability to differentiate between prokaryotic and eukaryotic cells. A prokaryote is a unicellular organism that lacks membrane-bound organelles. Mitochondria, which are membrane bound, would not be found in a prokaryotic cell.

86. (B) In order for Hardy-Weinberg equilibrium to be maintained, a large population, no net mutations, no migration, and random mating must all occur. If these criteria are met, then the allele frequencies for two alleles must equal 1, $p + q = 1$. The phenotypic frequencies must also equal one in the following equation: $p^2 + 2pq + q^2 = 1$. In this case, $T = 0.8$, so $p = 0.8$. $0.8 + q = 1$, so $q = 0.2$. Therefore, the allele frequency of *t* is 20 percent, or 0.2.

87. (B) The percentage of heterozygous individuals is defined by the $2pq$ term. So, $2(0.8)(0.2)$ = the percentage of heterozygous individuals, which comes to 0.32 or 32 percent.

88. (D) Individuals that can curl their tongues have at least one dominant allele, but cannot have two. This means that you are looking for the number of p^2 individuals, who have two dominant alleles, and the number of $2pq$ individuals, who have only one dominant allele. These two numbers added together will give you the number of individuals that can curl their tongues. Therefore, $(0.8)2 + 2(0.8)(0.2) = 0.96$ or 96 percent.

89. (E) The four criteria for Hardy Weinberg equilibrium are (i) a large population, (ii) no net mutations, (iii) geographic isolation and no migration, and (iv) random mating.

90. (D) The bacteria that have the largest population are growing most optimally. In this experiment, the optimum concentration is at 40 U/mL.

91. (B) At high nutrient concentrations, there are so many particles on the outside of the cell compared to the inside that osmotic potential drives water out of the cell into the media, causing the cells to shrivel and die.

92. (B) The second time the mouse tried the maze, he cut his time in half, from 40 to 20 seconds. The next largest drop in time is after the fourth trial, when the mouse shaved 9 seconds off his previous time.

93. (C) By the sixth time the mouse did the maze, the time required was at a minimum, implying that the mouse had learned the maze.

94. (C) A catalyst lowers the activation energy of a reaction, making it easier for it to go to completion such as that defined by line 3.

95. (A) When a reaction strays from its optimal pH, it is slowed because of changes in the substrate and catalyst. This in turn leads to an increased activation energy for the reaction, as depicted in line 1.

APPENDIX

THE MENDELIAN EYE EXAM:
TY Ty tY ty
Ty TTYY TTYy TtYY TtYy
Ty TTYy Ttyy TtYy Ttyy
tY TtYY TtYy ttYY ttYy
ty TtYy Ttyy ttYy ttyy

GLOSSARY

abiotic nonliving, as in the physical environment

absorption the process by which water and dissolved substances pass through a membrane

acetylcholine a transmitter substance released from the axons of nerve cells at the synapse

active immunity protective immunity to a disease in which the individual produces antibodies as a result of previous exposure to the antigen

adaptation a behavioral or biological change that enables an organism to adjust to its environment

adaptative radiation the production of a number of different species from a single ancestral species

adenosine phosphate adenosine diphosphate (ADP) and adenosine triphoshate (ATP), which are energy storage molecules

ADH (vasopressin) a hormone that regulates water reabsorption

adipose fatty tissue, fat-storing tissue, or fat within cells

adrenal cortex the outer part of the adrenal gland that secretes many hormones, including cortisone and aldosterone

adrenal medulla the inner part of the adrenal gland that secretes adrenalin

adrenaline (epinephrine) an "emergency" hormone stimulated by anger or fear; increases blood pressure and heart rate in order to supply the emergency needs of the muscles

adrenocorticotrophic hormone usually referred to as ACTH and secreted by the anterior lobe of the pituitary gland; stimulates the adrenal cortex to produce its charcteristic hormones

aerobe an organism that requires oxygen for respiration and can live only in the presence of oxygen

aerobic requiring free oxygen from the atmosphere for normal activity and respiration

Agnatha jawless fish; a class of vertebrates that includes the lamprey and the hagfish

aldosterone hormone active in osmoregulation; a mineral corticoid produced by the adrenal cortex; stimulates reabsorption of Na^+ and secretion of K^+

alimentary canal an organ centrally involved in the human digestive system

allantois the extraembryonic membrane of birds, reptiles, and mammals that serves as an area of gaseous exchange and as a site for the storage of noxious excretion products

allele one of two or more types of genes, each representing a particular trait; many alleles exist for a specific gene locus

alternation of generations the description of a plant life cycle that consists of a diploid, asexual, sporophyte generation and a haploid, sexual, gametophyte generation

alveolus an air sac in the lung; the site of respiratory exchange, involving diffusion of oxygen and carbon dioxide between the air in the alveolus and the blood in the capillaries (plural = *alveoli*)

ameboid movement movement involving the flowing of cytoplasm into pseudopods, as in amoeba

amnion the extraembryonic membrane in birds, reptiles, and mammals that surrounds the embryo, forming an amniotic sac

anaerobe an organism that does not require free oxygen in order to respire

anaerobic living or active in the absence of free oxygen; pertaining to respiration that is independent of oxygen

analogous describes structures that have similar function but different evolutionary origins; e.g., a bird's wing and a moth's wing

anaphase the stage in mitosis that is characterized by the migration of chromatids to opposite ends of the cell; the stage in meiosis during which homologous pairs migrate (Anaphase I), and the stage in meoisis during which chromatids migrate to different ends of the cell (Anaphase II)

androgen a male sex hormone (e.g., testosterone)

angiosperm a flowering plant; a plant of the class Angiospermae that produces seeds enclosed in an ovary and is characterized by the possession of fruits and flowers

animal pole the end of the egg cell that contains the least amount of cytoplasm

Annelida the phylum to which segmented worms belong

anther the part of the male reproductive organ (the stamen) that produces and stores pollen

antheridium the organ of plants that produces sperm

antibiotic an antipathogenic substance (e.g., penicillin)

antibody globular proteins produced by tissues that destroy or inactivate antigens

antigen a foreign protein that stimulates the production of antibodies when introduced into the body of an organism

aorta the largest artery; carries blood from the left ventricle

aortic arch blood vessels located between ascending and descending aortas that deliver blood to most of upper body

appendage a structure that extends from the trunk of an organism and is capable of active movements

aqueous humor fluid in the eye, found between the cornea and the lens

Arachnida a class of arthropods that includes scorpions, spiders, mites, and ticks

archegonium the organ of plants that produces eggs

archenteron the central cavity of the gastrula, which is lined by endoderm and gives rise to the adult digestive cavity

artery a blood vessel that carries blood away from the heart

Arthropoda the phylum to which jointed-legged invertebrates belong, including insects, arachnids, and crustaceans

asexual reproduction the production of daughter cells by means other than the sexual union of gametes (as in budding and binary fission)

assimilation the conversion of digested foods and other materials into forms usable by the body (i.e., the conversion of amino acids into proteins)

assortative mating the type of mating that occurs when an organism selects a mating partner that resembles itself

atrium the thin-walled anterior chamber of the heart (also called the auricle)

autolysis self-digestion occurring in plant and animal tissues, particularly after they have ceased to function properly

autonomic nervous system the part of the nervous system that regulates the involuntary muscles, such as the walls of the alimentary canal; includes the parasympathetic and sympathetic nervous systems

autosome any chromosome that is not a sex chromosome

autotroph an organism that is capable of organizing organic molecules from inorganic molecules

auxin a plant growth hormone

axon a nerve fiber

bacillus bacteria that are rod shaped

bacteriophage a type of virus that can destroy bacteria by infecting, parasitizing, and eventually killing them

bile an emulsifying agent secreted by the liver

bile salts compounds in bile that aid in emulsification

binary fission asexual reproduction; in this process, the parent organism splits into two equal daughter cells

binomial nomenclature the system of naming an organism by its genus and species names

biome a habitat zone, such as desert, grassland, or tundra

biotic living, as in living organisms in the environment

blastopore the outer opening of the archenteron that develops during gastrulation, becoming the anus in deuterostomia and the mouth in protostomia

blastula a stage of embryonic development in which the embryo consists of a hollow ball of cells

Bowman's capsule part of the nephron in the kidney; involved in excretion

bryophyte a plant phylum that incorporates mosses and liverworts

bud in plants, an area of undifferentiated tissue covered by embryonic leaves

budding a process of asexual reproduction in which the offspring develop from an outgrowth of the plant or animal

buffer a substance that prevents appreciable changes in pH in solutions to which small quantities of acids or bases are added

calorie a unit of heat; the amount of heat required to raise the temperature of one gram of water by one degree centigrade (Note: a large Calorie (food calorie) = 1000 calories)

Calvin cycle cycle in photosynthesis that reduces fixed carbon to carbohydrates through the addition of electrons

cambium undifferentiated tissue in the stem of a plant that aids growth in width

capillary a tube one cell thick that carries blood from artery to vein; the site of material exchange between the blood and tissues of the body

carapace a bony or chitinous case or shield covering the back or part of the back of an animal (e.g., the shell of a crab)

carbohydrate an organic compound to which hydrogen and oxygen are attached; the hydrogen and oxygen are in a 2:1 ratio; examples include sugars, starches, and cellulose

carbon cycle the recycling of carbon from decaying organisms for use in future generations

carnivore a flesh-eating animal; a holotrophic animal that subsists on other animals or parts of animals

carotene an orange plant pigment that is the precursor of Vitamin A

cation an ion with a positive charge, or an ion that migrates towards the cathode (negative electrode) in an electric field

cell wall a wall composed of cellulose that is external to the cell membrane in plants; it is primarily involved in support and in the maintenance of proper internal pressure

cell wall plate in mitosis of higher plants, the structure that forms between the divided nuclei of the two daughter cells and eventually becomes the cell wall

central nervous system (CNS) encompasses the brain and the spinal cord

centriole the small granular body within the centrosome to which the spindle fibers attach

centromere the place of attachment of the mitotic fiber to the chromosome

centrosome a structure in animal cells containing centrioles from which the spindle fibers develop

cephalic pertaining to the head

Cephalochordata a subphylum of chordates consisting of lancelets and amphioxus (nonvertebrates)

cephalothorax the fused head and thorax of arachnids and higher crustaceans

cerebellum the hindbrain region that controls equilibrium and muscular coordination

cerebral cortex the outer layer of cerebral hemispheres in the forebrain, consisting of gray matter

cerebral hemisphere one of the paired lateral divisions of the forebrain

cerebrum the largest portion of the human brain; it is believed to be the center of intelligence, conscious thought, and sensation

chemosynthesis the process by which carbohydrates are formed through chemical energy; found in bacteria

chemotropism the orientation of cells or organisms in relation to chemical stimuli; the growth or movement response of organisms to chemical stimuli

chitin a white or colorless, amorphous, horny substance that forms part of the outer integument of insects, crustaceans, and some other invertebrates; it also occurs in certain fungi

chloroplast a plastid containing chlorophyll

chlorophyll a green pigment that performs essential functions as an electron donor and light "entrapper" in photosynthesis

chlorophyte a green alga; a member of the phylum Chlorophyta

Chondrichthyes fish that have a cartilage skeleton; a class of vertebrates that includes sharks, skates, rays, and related types

Chordata an animal phylum in which all members have a notochord, dorsal nerve cord, and pha-ryngeal gill slits at some embryonic stage; includes the Cephalochordata and the Vertebrates

chorion the outermost, extra-embryonic membrane of reptiles and birds

chromatid one of the two strands that constitute a chromosome; chromatids are held together by the centromere

chromatin a nuclear protein of chromosomes that stains readily

chromosome a short, stubby rod consisting of chromatin that is found in the nucleus of cells; contains the genetic or hereditary component of cells (in the form of genes)

chromosome map the distribution of genes on a chromosome, derived from crossover frequency experiments

chyme partially digested food in the stomach

circadian rhythms daily cycles of behavior

cleavage the division in animal cell cytoplasm caused by the pinching in of the cell membrane

climax community the stable, biotic part of the ecosystem in which populations exist in balance with each other and with the environment

clotting the coagulation of blood caused by the rupture of platelets and the interaction of fibrin, fibrinogen, thrombin, prothrombin, and calcium ions

cloaca the chamber in the alimentary canal of certain vertebrates located below the large intestine, into which the ureter and reproductive organs empty (as in frogs)

cochlea the sensory organ of the inner ear of mammals; it is coiled and contains the organ of Corti

Coelenterata an invertebrate animal phylum in which animals possess a single alimentary opening and tentacles with stinging cells; examples are jellyfish, corals, sea anemones, and hydra

coelom the space between the mesodermal layers that forms the body cavity of some animal phyla

coenzyme an organic cofactor required for enzyme activity

colon the large intestine

commensal describes an organism that lives symbiotically with a host; this host neither benefits nor suffers from the association

cone a cell in the retina that is sensitive to colors and is responsible for color vision

consumer organism that consumes food from outside itself instead of producing it (primary, secondary, and tertiary)

contractile vacuole a specialized structure that controls osmotic pressure by removing water from the cell; exists in protozoans

cornea the outer, transparent layer of the eye

corpus callosum a tract of nerve fibers connecting the two cerebral hemispheres

corpus luteum a remnant of follicle after ovulation that secretes the hormone progesterone

cortex in plants, the tissue between the epidermis and the vascular cylinder in the roots and stems of plants; in animals, the outer tissue of some organs

cortisone a hormonal secretion of the adrenal cortex

cotyledon a "seed leaf;" responsible for food digestion and storage in a plant embryo

cretinism a thyroid deficiency that results in stunted growth and feeblemindedness

crossing over the exchange of parts of homologous chromosomes during meiosis

cross-pollination the pollination of the pistil of one flower with pollen from the stamen of a different flower of the same species

Crustacea crustaceans; a large class of arthropods, including crabs and lobsters

cuticle a waxy protective layer secreted by the outer surface of plants, insects, etcetera

cytochrome a hydrogen carrier containing iron that functions in many cellular processes, including respiration

cytokinesis a process by which the cytoplasm and the organelles of the cell divide; the final stage of mitosis

cyton the cell body of a neuron

cytoplasm the living matter of a cell, located between the cell membrane and the nucleus

cytoskeleton the organelle that provides mechanical support and carries out motility functions for the cell

cytosine a nitrogen base that is present in nucleotides and nucleic acids; it is paired with guanine

deamination the removal of an amino group from an organism, particularly from an amino acid

deletion the loss of all or part of a chromosome

deme a small, local population

dendrite the part of the neuron that transmits impulses to the cell body

deoxyribose a five carbon sugar that has one oxygen atom less than ribose; a component of DNA (deoxyribose nucleic acid)

diastole the passive, rhythmical expansion or dilation of the cavities of the heart (atria or ventricles) that allows these organs to fill with blood; preceded and followed by systole (contraction)

dicotyledon a plant that has two seed leaves or cotyledons

diencephalon the hind portion of the forebrain of vertebrates

differentiation a progressive change from which a permanently more mature or advanced state results; for example, a relatively unspecialized cell's development into a more specialized one

diffusion the movement of particles from one place to another as a result of their random motion

digestion the process of breaking down large organic molecules into smaller ones

dihybrid an organism that is heterozygous for two different traits

dimorphism the instance of polymorphism in which there is a difference of form between two members of a species, as between males and females

diploid describes cells that have a double set of chromosomes in homologous pairs ($2n$)

disaccharide a sugar composed of two combined monosaccharides (e.g., sucrose, lactose)

DNA deoxyribonucleic acid; found in the cell nucleus; its basic unit is the nucleotide; contains coded genetic information; can replicate on the basis of heredity

dominance an allele that suppresses the expression of the other member of an allele pair when both members are present; a dominant gene exerts its full effect regardless of the effect of its allelic partner

dorsal root the sensory branch of each spinal nerve

ductus arteriosus an artery present in the fetus and embryo of mammals that conducts blood from the pulmonary artery to the aorta; it shrivels at birth, when the lungs become functional

duodenum the most anterior portion of the small intestine of vertebrates, adjacent to the stomach; the continuation of the stomach into which the bile duct and pancreatic duct empty

dura mater the outermost of the three membranes covering the brain and spinal cord

Echinodermata the phylum of spiny-skinned animals that includes starfish and sea urchins

ecological succession the orderly process by which one biotic community replaces another until a climax community is established

ecology the study of organisms in relation to their environment

ectoderm the outermost embryonic germ layer that gives rise to the epidermis and the nervous system

egg (ovum) the female gamete; it is nonmotile, large in comparison to male gametes, and stores nutrients

electron transport chain a complex carrier mechanism located on the inside of the inner mitochondrial membrane of the cell; releases energy, and is used to for ATP

embolus a blood clot that is formed within a blood vessel

emulsion a colloidal system involving the dispersion of a liquid within a liquid

endemic pertaining to a restricted locality; ecologically, occurring only in one particular region

endocrine gland a ductless gland that secretes hormones directly into the bloodstream

endocytosis a process by which the cell membrane is invaginated to form a vesicle which contains extracellular medium

endoderm the innermost embryonic germ layer that gives rise to the lining of the alimentary canal and to the digestive and respiratory organs

endoplasmic reticulum a network of membrane-enclosed spaces connected with the nuclear membrane; transports materials through the cell

enzyme an organic catalyst and protein

endoplasm the inner portion of the cytoplasm of a cell or the portion that surrounds the nucleus

endosperm the triploid tissue in some seeds that contains stored food and is formed by the union of one sperm nucleus with two nuclei of the female's gametophyte

epidermis the outermost surface of an organism

epididymis the coiled part of the sperm duct, adjacent to the testes in mammals

epiglottis in mammals, a flap of tissue above the glottis; it folds back over the glottis in swallowing to close the air passages of the lungs; contains elastic cartilage

epicotyl the portion of seed plant embryo above the cotyledon

epinephrine see *adrenalin*

epithelium the cellular layer that covers external and internal surfaces

epiphyte a plant that lives on another plant commensalistically

erythrocyte an anucleate red blood cell that contains hemoglobin

esophagus the portion of alimentary canal connecting the pharynx and the stomach

estrogen a female sex hormone secreted by the follicle

ethanol fermentation a form of anaerobic respiration found in yeast and bacteria

ethylene a hormone that ripens fruit and induces aging

eukaryotes multicellular organisms

Eustachian tube an air duct from the middle ear to the throat that equalizes external and internal air pressure

excretion the elimination of metabolic waste matter

exocrine pertaining to a type of gland that releases its secretion though a duct; e.g., the salivary gland, the liver

exocytosis a process by which the vesicle in the cell fuses with the cell membrane and releases its contents to the outside

exoskeleton describes arthropods and other animals whose skeletal or supporting structures are outside the skin

eye a sensory organ capable of detecting light

F_1 the first filial generation (first offspring)

F_2 the second filial generation; offspring resulting from the crossing of individuals of the F_1 generation

fallopian tube the mammalian oviduct that leads from the ovaries to the uterus

feedback mechanism the process by which a certain function is regulated by the amount of the substance it produces

femur the thigh bone of vertebrates

fermentation anaerobic respiration that yields 2 molecules of ATP, lactic acid, ethyl alcohol and carbon dioxide, or some similar compound via the glycolytic pathway

fertilization the fusion of sperm and the egg to produce a zygote

fibrin protein threads that form in the blood during clotting

fibrinogen blood protein that is transformed to fibrin upon clotting

flagellate an organism that possesses one or more whiplike appendages called flagella

flagellum a microscopic, whiplike filament that serves as a locomotor structure in flagellate cells

follicle the sac in the ovary in which the egg develops

food vacuole a vacuole in the cytoplasm in which digestion takes place (in protozoans)

frame shift mutation a mutation involving the addition or loss of nucleotides

fruit a mature ovary

FSH an anterior pituitary hormone that stimulates the follicles in females and the function of the seminiferous tubules in males

functional groups chemical groups attached to carbon skeletons that give compounds their functionality

gall bladder an organ that stores bile

gamete a sex or reproductive cell that must fuse with another of the opposite type to form a zygote, which subsequently develops into a new organism

gametophyte the haploid, sexual stage in the life cycle of plants (alternation of generations)

ganglion a grouping of neuron cell bodies that acts as a coordinating center

gastrula a stage of embryonic development characterized by the differentiation of the cells into the ectoderm and endoderm germ layers and by the formation of the archenteron

gene the portion of a DNA molecule that serves as a unit of heredity; found on the chromosome

gene frequency a decimal fraction that represent the presence of an allele for all members of a population that have a particular gene locus

genetic code a four-letter code made up of the DNA nitrogen bases A, T, G, and C; each chromosome is made up of thousands of these bases

genetic drift random evolutionary changes in the genetic makeup of a (usually small) population

genotype the genetic makeup of an organism without regard to its physical appearance; a homozygous dominant and a heterozygous organism may have the same appearance but different genotypes

genus in taxonomy, a classification between species and family; a group of very closely related species, e.g., homo, felis

geographical barrier any physical feature that prevents the ecological niches of different organisms (not necessarily different species) from overlapping

geotropism any movement or growth of a living organism in response to the force of gravity

germ cell a reproductive cell

germ layer one of the primary tissues of the embryo

gibberellin a hormone that stimulates plant stem elongation

gill slit a perforation leading from the pharynx to the outside environment that is a characteristic of chordates at one stage of their development

glomerulus a network of capillaries in the Bowman's capsules of the kidney

glottis in mammals, the slitlike opening formed by the vocal folds in the larynx

glycogen a starch form in animals; glucose is converted to glycogen in the liver

glycolysis the anaerobic respiration of carbohydrates

goiter (simple) an enlargement of the thyroid gland due to lack of iodine

Golgi apparatus membranous organelles involved in the storage and modification of secretory products

gonads the reproductive organ that produces sex cells (e.g., ovary, testes)

Graffian follicle the cavity in the mammalian ovary in which the egg ripens

granum the smallest particle that is capable of carrying out photosynthesis; the functional unit of a chloroplast

gray matter a portion of CNS consisting of cytons (cell bodies), their dendrites, and synaptic connections

guanine a purine (nitrogenous base) component of nucleotides and nucleic acids; it links up with cytosine in DNA

guard cell one of a pair of kidney-shaped cells that surround a stomate and regulate the size of the stomate in a leaf

gymnosperm a plant that belongs to the class of seed plants in which the seeds are not enclosed in an ovary; includes the conifers

haploid describes cells (gametes) that have half the chromosome number typical of the species (*n* chromosome number)

hemoglobin a protein compound containing iron that is found in red blood cells; hemoglobin pigment combines with oxygen and gives the red blood cells their respiratory function

hepatic portal system the veins that carry blood from the digestive organs to the liver

herbivore a plant-eating animal

hermaphrodite an organism that possesses both the male and the female reproductive organs

heterotroph an organism that must get its inorganic and organic raw materials from the environment; a consumer

heterozygous describes an individual that possesses two contrasting alleles for a given trait (*Tt*)

homeotherm an animal with a constant body temperature

homologous describes two or more structures that have similar forms, positions, and origins despite the differences between their current functions; examples are the arm of a human, the flipper of a dolphin, and the foreleg of a horse

homozygous describes an individual that has the same gene for the same trait on each homologous chromosome (*TT* or *tt*)

hormone a chemical messenger that is secreted by one part of the body and carried by the blood to affect another part of the body, usually a muscle or gland

host any organism that is the victim of a parasite

humerus a bone of the upper arm

hybrid an offspring that is heterozygous for one or more gene pairs

hydrostatic skeleton fluid skeleton of annelids

hyperthyroidism an oversecretion of thyroid that leads to high metabolism and exophthalmia goiter

hypertonic describes a fluid that has a higher osmotic pressure than another fluid it is compared to; it exerts greater osmotic pull than the fluid on the other side of a semipermeable membrane; hence, it possesses a greater concentration of particles, and acquires water during osmosis

hypocotyl the portion of the embryonic seed plant below the point of attachment of the cotyledon; forms the root

hypothalamus a section of the posterior forebrain associated with the pituitary gland

hypotonic describes a fluid that has a lower osmotic pressure than a fluid it is compared to; it exerts lesser osmotic pull than the fluid on the other side of a semipermeable membrane; hence, it possesses a lesser concentration of particles, and loses water during osmosis

ilium the dorsal part of the hip girdle

immunity a resistance to disease developed through immune system

imprinting the process by which environmental patterns or objects presented to a developing organism during a "critical period" of its growth is accepted as a permanent element of its behavior

incomplete dominance genetic blending; each allele exerts some influence on the phenotype (for example, red and white parents may yield pink offspring)

independent assortment the law by which genes on different chromosomes are inherited independently of each other

ingestion the intake of food from the environment into the alimentary canal

inner ear a fluid-filled sensory apparatus that aids in balance and hearing

insulin a hormone produced by the Islets of Langerhans in the pancreas; regulates blood sugar concentration by converting glucose to glycogen (in the process lowering glucose level)

integument refers to protective covering, such as the covering of an ovule, that develops into the seed coat, or an animal's skin

interphase a metabolic stage between mitoses in which genetic material is reproduced

interstitial cells cells which in the female are located between the ovarian follicles, and in the male are located between the seminiferous tubules of the testes; in both cases, these cells produce male sex hormones

inversion occurs when a segment of genetic material on a chromosome becomes reversed

iris the colored part of the eye that is capable of contracting and regulating the size of the pupils

irritability the ability to respond to a stimulus

isolation the separation of some members of a population from the rest of their species; prevents interbreeding and may lead to the development of a new species

isomer one of a group of compounds that is identical in atomic composition, but different in structure or arrangement

isotonic describes a fluid that has the same osmotic pressure as a fluid it is compared to; it exerts the same osmotic pull as the fluid on the other side of a semipermeable membrane; hence it neither gains nor loses net water during osmosis, and possesses the same concentration of particles before and after osmosis occurs

Krebs cycle process of aerobic respiration that fully harvests the energy of glucose; also known as the citrid acid cycle

lactase the enzyme that acts upon lactose

lacteal a lymph tubule located in the villus that absorbs fatty acids

lactid acid fermentation a type of anaerobic respiration found in fungi, bacteria, and human muscle cells

larva a period in the development of animals between the embryo and adult stages; starts at hatching and ends at metamorphosis

legume a flowering plant with simple dry fruit, characterized by nodes on their roots that contain nitrogen-fixing bacteria (e.g., beans, clover)

lens a structure of the eye that focuses images on the retina by changing its convexity

levels of structure different relationships that are formed in proteins between the original sequence of amino acids and more complex three-dimensional compounds

lichen an association between an algae and a fungus that is symbiotic and mutualistic in nature

linkage occurs when different traits are inherited together more often than they would have been by chance alone; it is assumed that these traits are linked on the same chromosome

lipase a fat-digesting hormone

lipid a fat or oil

littoral zone a marine biome; a region on the continental shelf that contains an ocean area with depths of up to 600 ft

Loop of Henle the thin, bent part of the renal tubule that is the site of the counter-current flow and the sodium gradient

luteinizing hormone (LH) secreted by the anterior pituitary gland, this hormone stimulates the conversion of a follicle into the corpus luteum and the secretion of progesterone by the corpus luteum; it also stimulates the secretion of sex hormones by the testes

lymph a body fluid that flows in its own circulatory fluid in lymphatic vessels separate from blood circulation

lymph capillary one of many tubules that absorb tissue fluid and return it to the bloodstream via the lymphatics

lymphocyte a kind of white blood cell in vertebrates that is characterized by a rounded nucleus; involved in the immune response

lysosome an organelle that contains enzymes that aid in intracellular digestion

macula a sensory hair structure in the utriculus and the sacculus of the inner ear; orients the head with respect to gravity

malleus the outermost bone of the middle ear (hammer)

malpighian tubules tubules that excrete metabolic wastes into the hindgut in arthropods

maltase an enzyme that acts upon maltose and converts it into glucose

maltose a 12-carbon sugar that is formed by the union of two glucose units (a disaccharide)

marsupial a pouched mammal, such as the kangaroo or opossum

medulla the inner layer of an organ surrounded by the cortex

medulla oblongata the posterior part of the brain that controls the rate of breathing and other autonomic functions

medusa a jellyfish; the bell-shaped, free-swimming stage in the life cycle of coelenterates

meiosis a process of cell division whereby each daughter cell receives only one set of chromosomes; the formation of gametes

Mendelian laws laws of classical genetics established through Mendel's experiments with peas

meninges three membranes that envelop the brain and spinal cord (pia mater, dura mater, and arachnoid)

meristem an undifferentiated, growing region of a plant that is constantly undergoing cell division and differentiation

mesoderm the primary germ layer, developed from the lip of the blastopore, that gives rise to the skeleton, the circulatory system, and many organs and tissues between the epidermis and the epithelium

metabolism a group of life-maintaining processes that includes nutrition, respiration (the production of usable energy), and the synthesis and degradation of biochemical substances

metamorphosis the transformation of an immature animal into an adult; a change in the form of an organ or structure

metaphase a stage of mitosis; chromosomes line up at the equator of the cell

microbodies organelles that serve as specialized containers for metabolic reactions

micron (micrometer) one-thousandth of a millimeter; a unit of microscopic length

mitochondria cytoplasmic organelles that serve as sites of respiration; a rod-shaped body in the cytoplasm known to be the center of cellular respiration

mitosis a type of nuclear division that is characterized by complex chromosomal movement and the exact duplication of chromosomes; occurs in somatic cells

monocotyledon a plant that has a single cotyledon or seed-leaf

monohybrid an individual that is heterozygous for only one trait

monosaccharide a simple sugar; a 5- or 6-carbon sugar (e.g., ribose or glucose)

morphology the study of form and structure

morula the solid ball of cells that results from cleavage of an egg; a solid blastula that precedes the blastula stage

mucosa a mucus-secreting membrane, such as the inner intestinal lining

mutagenic agent agent that induces mutations; typically carcinogenic

mutation changes in genes that are inherited

mutualism a symbiotic relationship from which both organisms involved derive some benefit

myelin sheath a fatty sheath surrounding the axon of a neuron that aids in stimulus transmission; it is secreted by the Schwann cells

NAD an abbreviation of nicotinamide-adenine-dinucleotide, also called DPN; a respiratory oxidation-reduction molecule

NADP an abbreviation of nicotinamide-adenine-dinucleotide-phosphate, also called TPN; an organic compound that serves as an oxidation-reduction molecule

nephron functional urinary tubules responsible for excretion in the kidney of vertebrates

nerve a bundle of nerve axons

nerve cord a compact linear organization of nerve tissues with ganglia in the CNS

nerve net a multidirectional sensory system of lower animals such as the hydra, consisting of nerve fibers spread throughout the ectoderm

neural tube an embryonic structure that gives rise to the central nervous system

neuron a nerve cell

niche the functional role and position of an organism in an ecosystem; embodies every aspect of the organism's existence

nictitating membrane a thin, transparent, eyelid-like membrane that opens and closes laterally across the cornea of many vertebrates (the third eyelid)

nitrogen cycle the recycling of nitrogen from decaying organisms for use in future generations

nondisjunction the failure of some homologous pairs of chromosomes to separate following meiotic synapsis

notochord a flexible, supportive rod running longitudinally through the dorsum ventral to the nerve cord; found in lower chordates and in the embryos of vertebrates

nuclear membrane a membrane that envelops the nucleus and separates it from the cytoplasm; present in eukaryotes

nucleolus a dark-staining small body within the nucleus; composed of RNA

nucleotide an organic molecule consisting of joined phosphate, 5-carbon sugar (deoxyribose or ribose), and a purine or a pyrimidine (adenine, guanine, uracil, thymine, or cytosine)

nucleus an organelle that regulates cell functions and contains the genetic material of the cell

olfactory related to the sense of smell

oogenesis a process of formation of ova

organelle a specialized structure that carries out particular functions for eukaryotic cells; examples include the plasma membrane, the nucleus, and ribosomes

osmoregulation the ways in which organisms regulate their supply of water

osmosis the diffusion of water through a semipermeable membrane, from an area of greater concentration to an area of lesser concentration

ovary the female gonad in animals; the base of the pistil in plants

oviduct a tube connecting the ovaries and the uterus

oxidation the removal of hydrogen or electrons from a compound or addition of oxygen; half of a redox (oxidation or reduction) process

pairing (synapsis) an association of homologous chromosomes during the first meiotic division

parasitism a relationship in which one organism benefits at the expense of another

parasympathetic pertaining to a subdivision of the autonomic nervous system of vertebrates

parathyroid an endocrine gland of vertebrates, usually paired, and located near or within thethyroid; it secretes parathormone, which controls the metabolism of calcium

parenchyma plant tissue consisting of large thin-walled cells for storage

passive immunity a resistance to disease produced through the injection of antibodies

parthenogenesis a form of asexual reproduction in which the egg develops in the absence of sperm

pathogen a disease-causing organism (pathogenic = disease inducing)

pedigree a family tree depicting the inheritance of a particular genetic trait over several generations

pelagic zone a marine biome typical of the open seas

pepsin a stomach enzyme that partially digests proteins

peptide the kind of bond formed when two amino acid units are jointed end to end; a double unit is called a dipeptide; the joining of many amino acid units into a chain results in a polypeptide that is the structural unit of a protein molecule

peripheral nervous system comprises somatic and autonomic nervous systems; consists of cranial nerves and spinal nerves

peristalsis waves of contraction and relaxation passing along a tubular structure, such as the digestive tube

permeability degree of penetrability, as in membranes that allow given substances to pass through; the ability to penetrate

pH a symbol that denotes the relative concentration of hydrogen ions in a solution: the lower the pH, the more acidic a solution; the higher the pH, the more basic is a solution; pH is equal to $-\log(H^+)$

phagocyte any cell capable of ingesting another cell

pharynx the part of the alimentary canal between the mouth and the esophagus

phenotype the physical appearance or makeup of an individual, as opposed to its genetic makeup

pheromone substances secreted by organisms that influence the behavior of other members of the same species

phloem the vascular tissue of a plant that transports organic materials (photosynthetic products) from the leaves to other parts of the plant

photolysis a process of photosynthesis in which water is split into H^+ and OH^-; the hydrogen ion is then joined to NADP

photoperiodism a response by an organism to the duration and timing of light and dark conditions

photosynthesis the process by which light energy and chlorophyll are used to manufacture carbohydrates out of carbon dioxide and water; an autotrophic process using light energy

phototropism plant growth stimulated by light (stem: +, towards light; root: –, away from light)

phylogeny the study of the evolutionary descent and interrelations of groups of organisms

phylum a category of taxonomic classification that is ranked above class; kingdoms are divided into phyla

physiology the study of all living processes, activities, and functions

pineal body a structure found between the cerebral hemispheres of vertebrates; secretes melatonin, which may help regulate the pituitary by regulating hypothalamic releasing factors

pinocytosis the intake of fluid droplets into a cell

pistil the part of the flower that bears the female gametophyte

pith the central tissue of a stem, used for food storage

pituitary a gland composed of two parts, anterior and posterior, each with its own secretions; called the "master gland" because its hormones stimulate secretion by other glands

placenta a structure formed by the wall of uterus and the chorion of embryo; serves as the area in which the embryo obtains nutrition from the parent

planarian a member of the class of free-living flatworms

plankton passively floating or drifting flora and fauna of a body of water; consists mainly of microscopic organisms

plasma the liquid part of blood

plasma membrane the cell membrane

plasmodium a motile, multinucleate mass of protoplasm resulting from fusion of uninuclear amoeboid cells; an organism consisting of such a structure, e.g., a slime mold

plastid cytoplasmic bodies within a plant cell that are often pigmented (e.g., chloroplasts)

platelet small disc-shaped bodies in the blood that play a chief role in coagulation

pleural cavity the cavity between the lungs and the wall of the chest

plexus a network, particularly of nerve or blood vessels

point mutation a mutation in which a single nucleotide base is substituted for another nucleotide base

polar body nonfunctional haploid cells created during meiosis in females; they have very little cytoplasm—most has gone into the functional egg cell

pollen the microspore of a seed plant

pollination the transfer of pollen to the micropyle or to a receptive surface that is associated with an ovule (such as a stigma)

polymer a large molecule that is composed of many similar molecular units (e.g., starch)

polymorphism the individual differences of form among the members of a species

polyp a typical coelenterate individual with a hollow tubular body whose outer ectoderm is separated from its inner endoderm by mesoglea

polyploidy a condition in which an organism may have a multiple of the normal number of chromosomes ($4n$, $6n$, etcetera)

polysaccharide a carbohydrate that is composed of many monosaccharide units joined together, such as glycogen, starch, and cellulose

pons the part of the hindbrain located in the brain stem

population all the members of a given species inhabiting a certain locale

Porifera the phylum of sponges

primary oocyte a cell that divides to form the polar body and the secondary oocyte

primary spermatocyte a cell that divides to form two secondary spermatocytes

producer organism that produces its own food; first stage in the food chain

progesterone the hormone secreted by the corpus luteum of vertebrates and the placenta of mammals; its function is to maintain the endometrium

prokaryote unicellular organism with simple cell structure

prophase a mitotic or meiotic stage in which the chromosomes become visible and during which the spindle fibers form; synapsis takes place during the first meiotic prophase

protein one of a class of organic compounds that is composed of many amino acids; contains C, H, O, and N

prothrombin a constituent of the plasma of the blood of vertebrates; it is converted to thrombin by thrombokinase in the presence of calcium ions, thus contributing to the clotting of blood

Protista a kingdom of unicellular living organisms that are neither animals nor plants; includes some groups of algae, slime molds, and protozoa

ptyalin a digestive enzyme of the saliva that turns starch into maltose (salivary amylase)

pulmonary relating to the lung

pupil an opening in the eye whose size is regulated by the iris

purine a nitrogenous base such as adenine or guanine; when joined with sugar and phosphate, a component of nucleotides and nucleic acids

pyrimidine a nitrogen base such as cytosine, thymine, and uracil; when joined with sugar and phosphate, a component of nucleotides and nucleic acids

pyloric valve a muscular valve regulating the flow of food from the stomach to the small intestine

recessive pertains to a gene or characteristic that is masked when a dominant allele is present

recombinant DNA technology technology that allows for manipulation of genetic material

reduction a change from a diploid nucleus to a haploid nucleus, as in meiosis

regeneration the ability of certain animals to regrow missing body parts

respiration a chemical action that releases energy from glucose to form ATP

respiratory center the area of medulla that regulates the rate of breathing

reticulum a network or mesh of fibrils, fibers, or filaments, as in the endoplasmic reticulum

retina the innermost tissue layer of the eyeball that contains light-sensitive receptor cells

Rh factor an antigen in blood; can cause erythroblastosis fetalis when the mother is Rh^- and the fetus is Rh^+

rhizome an underground stem

ribosome an organelle in the cytoplasm that contains RNA; serves as the site of protein synthesis

rhodopsin the pigment in rod cells that causes light sensitivity

rickettsia a kind of microorganism that is between a virus and a bacterium; parasitic within the cells of insects and ticks

RNA an abbreviation of ribonucleic acid, a nucleic acid in which the sugar is ribose; a product of DNA transcription that serves to control certain cell activities; acts as a template for protein translation; types include mRNA, tRNA, and rRNA

rod a cell in the retina that is sensitive to weak light

root hair outgrowths of a root's epidermal cells that allow for greater surface area for absorption of nutrients and water

saprophyte an organism that obtains its nutrients from dead organisms

secondary tissue tissue formed by the differentiation of cambium that causes a growth in width of a plant stem

selective breeding the creation of certain strains of specific traits through control of breeding

self-pollination the transfer of pollen from the stamen to the pistil of the same flower

semicircular canals fluid-filled structures in the inner ear that are associated with the sense of balance

seminal vesicle an organ containing seminal fluid

seminiferous tubules structures in the testes that produce sperm and seminal fluid

sensory neuron a neuron that picks up impulses from receptors and transmits them to the spinal cord

serum the fluid that remains after fibrinogen is removed from the blood plasma of vertebrates

sex chromosome there are two kinds of sex chromosomes, *X* and *Y*; *XX* signifies a female and *XY* signifies a male; there are fewer genes on the *Y* chromosome than on the *X* chromosome

sex linkage occurs when certain traits are determined by genes on the sex chromosomes

sinus a space in the body (e.g., blood sinus or maxillary sinus)

small intestine the site of most digestion of nutrients and absorption of digested nutrients (e.g. the wall of the alimentary canal)

smooth muscle involuntary muscle (e.g., the wall of the alimentary canal)

somatic cell any cell that is not a reproductive cell

species a group of populations that can interbreed

spermatogenesis the process of forming the sperm cells from primary spermatocytes

spindle a structure that arises during mitosis and helps separate the chromosomes; composed of tubulin

spiracle the external opening of the trachea in insects, opening into respiratory system

sphincter a ring-shaped muscle that is capable of closing a tubular opening by constriction; one example is the orbicularis oris muscle around the mouth

spore a reproductive cell that is capable of developing directly into an adult

sporophyte an organism that produces spores; a phase in the diploid-haploid life cycle that alternates with a gametophyte phase

stamen the part of the flower that produces pollen

steroid one of a class of organic compounds that contains a molecular skeleton of four fused rings of carbon; includes cholesterol, sex hormones, adrenocortical hormones, and Vitamin D

stigma the uppermost portion of pistil upon which pollen grains alight

stoma (stomate) a microscopic opening located in the epidermis of a leaf and formed by a pair of guard cells; the guard cells interact physically and regulate the passage of gas between the internal cells and the external environment

stomach the portion of alimentary canal in which some protein digestion occurs; its muscular walls of stomach churn food so that it is more easily digested; its low pH environment activates certain protein-digesting enzymes

stroma a dense fluid within the chloroplast; the site at which CO_2 is converted into sugars in photosynthesis

style a stalk-like or elongated body part, usually pointed at one end; part of the pistil of the flower

substrate a substance that is acted upon by an enzyme

sucrase an enzyme that acts upon sucrose

symbiosis the living together of two organisms in an intimate relationship; includes commensalism, mutualism, and parasitism

sympathetic pertaining to a subdivision of the autonomic nervous system

synapse the gap between the axon terminal of one neuron and the dendrites of another neuron

synergistic describes organisms that are cooperative in action, such as hormones or other growth factors that reinforce each other's activity

synaptic terminal the swelling at the end of an axon

synapsis the pairing of homologous chromosomes during meiosis

systole the contraction of the atria or ventricles of the heart

taiga a terrestrial habitat zone that is characterized by large tracts of coniferous forests, long and cold winters, and short summers; bounded by tundra in the north and found particularly in Canada, northern Europe, and Siberia

taxonomy the science of classification of living things

telophase a mitotic stage in which nuclei reform and nuclear membrane reappears

test cross the breeding of an organism with a homozygous recessive in order to determine whether an organism is homozygous dominant or heterozygous dominant for a given trait

testes the male gonads that produce sperm and male hormones

tetrad a pair of chromosome pairs present during the first metaphase of meiosis

thalamus a lateral region of the forebrain

thermoregulation the ways in which organisms regulate their internal heat

thoracic duct a major lymphatic that empties lymph into a vein in the neck

thorax the part of the body of an animal that is between the neck or head and the abdomen

thrombin a substance that participates in the clotting of blood in vertebrates; formed from prothrombin, it converts fibrinogen into fibrin

thrombokinase the enzyme released from the blood platelets in vertebrates during clotting; transforms prothrombin into thrombin in the presence of calcium ions; also known as thromboplastin

thymine a pyrimidine component of nucleic acids and nucleotides; pairs with adenine in DNA

thymus a ductless gland in upper chest region concerned with immunity and the maturation of lymphocytes

thyroid an endocrine gland located in the neck that produces thyroxin

thryoxin a hormone of the thyroid that regulates basal metabolism

tissue a mass of cells that have similar structures and perform similar functions

trachea an air-conducting tube, e.g., the windpipe of mammals or the respiratory tubes of insects

transcription the first stage of protein synthesis, in which the information coded in the DNA base is transcribed onto a strand of mRNA

translation the final stages of protein synthesis in which the genetic code of nucleotide sequences is translated into a sequence of amino acids

translocation the transfer of a piece of chromosome to another chromosome

transpiration the evaporation of water from leaves or other exposed surfaces of plants

trilobite a marine arthropod, now extinct, that lived during the Paleozoic era

trypsin an enzyme from the pancreas that digests proteins in the small intestine

tundra the biome located between the polar region and the taiga; characterized by a short growing season, no trees, and frozen ground

turgor pressure the pressure exerted by the contents of a cell against the cell membrane or cell wall

umbilicus the navel; the former site of connection between the embryo and the umbilical cord

ungulate a hoofed animal

uracil a pyrimidine found in RNA (but not in DNA); pairs with DNA adenine

urea an excretory product of protein metabolism

ureter a duct that carries urine from the kidneys to the bladder

urethra a duct through which the urine passes from the bladder to the outside

urinary bladder an organ that stores urine temporarily before it is excreted

urine fluid excreted by the kidney containing urea, water, salts, etcetera

uterus the womb in which the fetus develops

vacuole a space in the cytoplasm of a cell that contains fluid

vagus nerve the tenth cranial nerve that innervates digestive organs, heart, and other areas

vegetal pole the end of the egg cell containing the most yolk; undergoes less division than the animal pole

vein a blood vessel that carries blood back to the heart from the capillaries

ventral root the basal branch of each spinal nerve; carries motor neurons

ventricle the more muscular chamber(s) of the heart that pump blood to the lungs and to the rest of the body

vestigial organ an organ that is not functional in an organism, but was functional at some period in its evolution

villus a small projection in the walls of the small intestine that increases the surface area available for absorption *(pl.: villi)*

vitamin an organic nutrient required by organisms in small amounts to aid in proper metabolic processes; may be used as an enzymatic cofactor; since it is not synthesized, it must be obtained prefabricated in the diet

white matter an accumulation of axons within the CNS that is white because of its fatty, myelin sheath

wood xylem that is no longer being used; gives structural support to the plant

xylem vascular tissue of the plant that aids in support and carries water

yolk sac a specialized structure that leads to the digestive tract of a developing organism and provides it with food during early development

zygote a cell resulting from the fusion of gametes

Index

The following index provides page number references for all the headings in the five biology review chapters of this book.